Gas Turbine Handbook: Principles and Practice

5th Edition

Gas Turbine Handbook: Principles and Practice

5th Edition

by

Tony Giampaolo, MSME, PE

Library of Congress Cataloging-in-Publication Data

Giampaolo, Tony, 1939-
Gas turbine handbook : principles and practice / by Tony Giampaolo, MSME, PE. -- 5th edition.
pages cm
Includes bibliographical references and index.
ISBN-10: 0-88173-712-7 (alk. paper)
ISBN-13: 978-1-4822-2888-5 (Taylor & Francis distribution : alk. paper)
ISBN-10: 0-88173-713-5 (electronic)
1. Gas-turbines--Handbooks, manuals, etc. I. Title.

TJ778.G48 2014
621.43'3--dc23 2013034004

Published by The Fairmont Press, Inc.
700 Indian Trail
Lilburn, GA 30047
tel: 770-925-9388; fax: 770-381-9865
http://www.fairmontpress.com

Distributed by Taylor & Francis Ltd.
6000 Broken Sound Parkway NW, Suite 300
Boca Raton, FL 33487, USA
E-mail: orders@crcpress.com

Distributed by Taylor & Francis Ltd.
23-25 Blades Court
Deodar Road
London SW15 2NU, UK
E-mail: uk.tandf@thomsonpublishingservices.co.uk

Printed in the United States of America
10 9 8 7 6 5 4 3 2 1

ISBN-10: 0-88173-712-7 (The Fairmont Press, Inc.)
ISBN-13: 978-1-4822-2888-5 (Taylor & Francis Ltd.)

Dedication

This 5th edition is dedicated to my wife, Gay. We are now some years past our 50th Wedding Anniversary, and you are still the light of my life.

Contents

Preface to the Fifth Edition xiii

Acknowledgments xv

Chapter 1

Evolution of the Gas Turbine **1**

Technical Improvements 7

Chapter 2

Applications **11**

Jet Engines 11

Mechanical Drive 15

Chapter 3

Hardware **23**

Similarities And Differences 23

Compressors 27

Turbines 30

Combustors 37

Bearing Design 39

Inspection And Replacement 41

Operation 42

Chapter 4

Gas Turbine Systems Theory **45**

Gas Turbine Operating Cycle 45

Component Efficiencies 49

Compressor 52

Burner Section 72

Turbine 73

Chapter 5

Gas Turbine Controls **79**

Chapter 6

Accessories **93**

Starting System 93

Ignition System 98

Lubrication Systems 102

Characteristics of Lube Oils 108

Chapter 7
Parameter Characteristics 113
Vibration 113
Vibration Measurement 115
Exhaust Gas Temperature (EGT) 117
Rotor Speed 117
Oil Pressure And Temperature. 119

Chapter 8
Gas Turbine Inlet Treatment 123
The Environment 123
Inlet Air Filters 125
Filter & Gas Turbine Match 130
Inlet Air Cooling 130
Wet Compression 136
Chillers 139

Chapter 9
Gas Turbine Exhaust Treatment. 141
Water or Steam Injection. 146
Water Injection 147
Steam Injection 148
Selective Catalytic Reduction 149
Effects of Operating Parameters 152

Chapter 10
Combustion Turbine Acoustics and Noise Control 155
Combustion Turbines—Lots of Power in a Little Space. 155
Noise Sources 157
Some Fundamentals of Acoustics 158
Noise Criterion 164
Noise Control 166
Summary 176

Chapter 11
Microturbines. 179
Applications 180
Hardware 182

Chapter 12
Waste Heat Recovery 193
Introduction 193
Definitions. 194

Climatic Influences 197
Advantages of WHR 201
Review of Basic Thermodynamics 205
Thermodynamics of Waste Heat Recovery. 207
Heat Recovery Boilers 216
Natural Circulation Boiler 217
Forced Circulation 218
Forced Recirculation Boiler. 222
Supplementary Fired Boiler 223
Boiler Hardware 224
Condensers 226
Boiler Tubing 226
Steam Turbines 228
Off Design Considerations 229

Chapter 13
Detectable Problems. 231

Chapter 14
Boroscope Inspection 253
Objectives and Expectations 253
A Backup to Confirming Suspected Problems 256
Assessing Damage to Internal Engine Components 257
Some Safety Advice 276

Chapter 15
Case Study #1 277
Abstract 277
Introduction 278
Plant Configuration and Performance 282
Gas Turbine Design Details 286
Summary 294
Doe Disclaimer 295

Chapter 16
Case Study #2 297
Abstract 297
Introduction 297
Refinery Processes 300
Conclusions 308

Chapter 17
Case Study #3 309
Plant Configuration 310
Overview of Operating Plan 312
Operating Reliability 313
Maintenance Policy 314
Commitment To Partnership 315

Chapter 18
Case Study #4 317
Abstract 317
Introduction 318
Project Design 321
Summary 324

Chapter 19
The Gas Turbine's Future 327

Appendix A-1
Gas Turbine Manufacturers 335

Appendix A-2
Manufacturers 360

Appendix B
Accessory Manufacturers 361
Acoustic Enclosures 362
Bearings 363
Boroscopes 363
Castings 364
Coatings 364
Component Parts 365
Compressor Wash 366
Control Instruments 367
Control Systems 367
Couplings 368
Expansion Joints 370
Filters (Fuel, Oil) 371
Fuel Nozzles 371
Fuel Treatment 372
Gearboxes 372
Inlet Air Cooling 373

Inlet Air Filters 373
Lubrication Systems 375
Pumps 375
Selective Catalytic
Reduction (SCR) 377
Starters 377
Switchgear 378
Vibration Equipment 378

Appendix C-1
Characteristics of Particles and Particle Dispersoids 381

Appendix C-2
Material Chart 384

Appendix C-3
Interrelationship of Gas Turbine Engine Parameters 402

Appendix C-4
Classification of Hazardous Atmospheres 410

Appendix C-5
Guidelines for Selecting. 411
Gas Turbine Inlet Air Filtration 411

Appendix C-6
Air/Oil Cooler Specifications Check List 414

Appendix C-7
Gaseous Fuel Properties 419
Fuel Gas Properties. 419

Appendix C-8
Liquid Fuel Properties 426

Appendix C-9
List of Symbols 428

Appendix C-10
Conversion Factors 431

Appendix C-11
Guide for Estimating
Inlet Air Cooling Affects of Water Fogging 436

Appendix C-12
Overall A-weighted Sound Level Calculation 439

Appendix C-13
Occupational Noise Exposure 441

Appendix C-14
Resume: Elden F. Ray Jr., P.E.. 443

Appendix C-15
Microturbine Manufacturers 445

Appendix C-16
Estimate WHR Boiler Size and Cost 446

Appendix D-1
Technical Organizations 457

Appendix D-2
Technical Articles 461
Materials 461
Combustion 465
Fuels. 467
Controls 469
Diagnostics 470
Rotors, Bearings, And Lubrication. 474
Filtration And Cleaning 475
Design Techniques 476
Inlet And Turbine Cooling 480
Emissions 484
Performance. 486
Repair Techniques 488
Instrumentation 489
Fire Protection 490
Additional Articles 490

Glossary 493

Index 495

Preface to the Fifth Edition

The supply of nonrenewable fuels such as oil and natural gas are basically fixed since they are formed at a negligible rate in relation to drilling activity. The future predicts increasing fuel cost and fuel shortages, and the requirement and incentives for heat and energy recovery will therefore be that much more important. As a result every effort must be made to improve utilization of our accessible fuel supply.

The topic of waste heat recovery and combined thermodynamic cycles was originally intended for a new book. However, since this subject is so closely associated with gas turbines is seemed appropriate to include an abbreviated version in this book. The principles of combining thermodynamic cycles may be applied to the reciprocating engine exhaust, incinerator exhaust, fluidized bed exhaust, and low efficiency boilers.

This effort is being provided to assist engineers involved in the design, specification, operation, maintenance, and installation of cogeneration equipment and systems. Chapter 12 will provide the basis for understanding the interrelationship of thermodynamics, heat transfer and power generation in the design, operation & maintenance of a combined cycle/cogeneration plant.

Because of its high exhaust temperature and high mass flow the most obvious source for waste heat energy recovery is the gas turbine. These two parameters (temperature and mass flow), which are released to the atmosphere and wasted in the simple cycle engine, combine to provide a significant source of potential, recoverable, power. Considering that 30% to 40% of the power produced by the gas turbine can be created just by recovering its exhaust heat without consuming additional fuel or generating any additional emissions is a benefit that should not be ignored. A gas turbine producing 100 megawatts in its simple cycle configuration can product from 130 to 140 megawatts in a combined cycle configuration. In so doing the overall plant efficiency would be increased from 35% to 45%. The actual economic advantage of any form of waste heat recovery system depends on the cost of the fuel being considered regardless of whether it is a fossil fuel or a plant based fuel. To be effective the project should pay for itself in a reasonable amount of time.

The heat generated by these units is often referred to as the topping units or topping cycle. The sensible heat in the waste exhaust may be recovered in a bottoming cycle. The bottoming cycle may be utilized directly as process heat or in a steam generator.

An example of a waste heat recovery estimate is included in Appendix C-16. Charts, tables and curves are used whenever possible to avoid complex computations and to improve the explanation of the WHR system.

Tony Giampaolo
Spring, Texas

Acknowledgments

I wish to thank the following people for their assistance in proofreading this book and for their suggestions and ideas: Robin Kellogg, Editor of Western Energy Magazine; Carl Gipson, Jr., MSME, my sister Loretta Giampaolo, BAN, RN; and my son Vince Giampaolo, MSM&PO, MANS&SS, Lt. Comdr., USN. I must also thank Paul Crossman for his support in arranging the first Gas Turbine Seminar, upon which this book is based.

I am also grateful for the support I have received from my family: my wife, my children and their spouses.

I would like to acknowledge the contribution of Mr. Eldon F. Ray, Jr., P.E., for providing the chapter on acoustics (Chapter 10).

Finally I would like to thank the following companies for their confidence and support by providing many of the photographs and charts that are in this book:

ABB STAL AB
American Air Filter International
ASEA Brown Boveri AG
Baltimore Aircoil Company
Capstone Turbine Corp.
Catalytica Combustion Systems
Cooper Cameron Corporation
DEMAG DELAVAL Turbomachinery Corp.
Diesel & Gas Turbine Worldwide Catalog
Elliott Energy System, Inc.
Freeport-McMoran Oil & Gas LLC
Farr Filter Company
General Electric Company
Harbor Cogeneration Company
Ingersoll-Rand Energy Systems
Olympus America Inc.
Plains Exploration and Production Company
Rolls Royce Industrial & Marine Gas Turbines Limited
Solar Turbines Incorporated
SRI International
United Technologies Corporation Hamilton Standard Division
United Technologies Corporation Pratt & Whitney Aircraft
United Technologies Corporation Pratt & Whitney Canada

Chapter 1

Evolution of the Gas Turbine

Through the design experience developed for steam turbines and available to gas turbines, it is not surprising that gas generator compressors, turbines, and power-extraction turbines bear a striking resemblance to each other and to the steam turbine. Nor should it be surprising that the axial flow compressors of today's gas turbines resemble the reaction steam turbine with the flow direction reversed. While many people today recognize the similarities between steam and gas turbine components, most do not fully appreciate the common history these two products share. History tells us that the idea for the gas turbine and the steam turbine were conceived simultaneously. As early as 1791, John Barber's patent for the steam turbine described other fluids or gases as potential energy sources. "John Barber invented what may be considered a gas turbine in which gas was produced from heated coal, mixed with air, compressed and then burnt. This produced a high speed jet that impinged on radial blades on a turbine wheel rim."[1] John Barber's ideas, as well as those before him (Giovanni Branca's impulse steam turbine—1629, Leonardo da Vinci's "smoke mill"—1550, and Hero of Alexandria's reaction steam turbine—130 BC)[2] were just ideas. Even though the gas turbines described by these early visionaries would today be more accurately termed 'turboexpanders' (the source of compressed air or gas being a by-product of a separate process), there is no evidence that any of these ideas were ever turned into working hardware until the late 19th Century.

For the next 90 years ideas abounded, but all attempts to produce working hardware were unsuccessful. As Norman Davy stated in 1914, "The theory of the gas turbine was as fully grasped by Barber at the end of the eighteenth century, and by Bresson in the beginning of the nineteenth century, as it is by experts today. The success of the gas turbine as a heat engine rest solely upon practical limitations."[3]

However, even in this period of unsuccessful attempts at producing a working prototype, progress was still being made.

- In 1808 John Dumball envisioned a multi-stage turbine. Unfortunately his idea consisted only of moving blades without stationary airfoils to turn the flow into each succeeding stage.[1,3,4] Had he realized the need for a stationary stage between each rotating stage he would have originated the concept of an axial flow turbine.

- In Paris in 1837, Bresson's idea was to use a fan to drive pressurized air into a combustion chamber. Here, the air was mixed with fuel gas and burnt. These combustion products were cooled by the addition of more air, and this final product was used to drive turbine blades.[1,2]

- In 1850, in England, Fernimough suggested a mixed steam and gas turbine, in which air was blown through a coal grate while water was sprayed into the hot gases. The gas and steam mixture then acted to drive a two-bladed rotor.[1]

- Not until 1872 did Dr. Franz Stolze combine the ideas of Barber and Dumball to develop the first axial compressor driven by an axial turbine.[1] Due to a lack of funds, he did not build his machine until 1900. Dr. Stolze's design consisted of a multi-stage axial flow compressor, a single combustion chamber, a multi-stage axial turbine, and a regenerator utilizing exhaust gases to heat the compressor discharge air. This unit was tested between 1900 and 1904, but never ran successfully.

As will be discussed in Chapter IX, Bresson's ideas are the basis of air cooling (to extend hot gas path part life), Fernimough's ideas are the basis for water injection (for power augmentation and later NO_x control), and Stolze's ideas led the way for application of both the present day trends in gas turbine design and the regenerator for improved efficiency.

It was not until 1884 with Sir Charles Parsons' patent for a reaction steam turbine and gas turbine, and 1888, with Charles de Laval's application of Giovanni Branca's idea for an impulse steam turbine

(Figure 1-1) did workable hardware emerge.[1,3] In the 1895/1896 time frame variations in the impulse turbine designs were developed by August C. Rateau, Charles Curtis, and Dr. Zoelly. The experience gained in the development of hardware for steam turbines was directly transferable to gas turbines. At the end of the 19th Century the ideas of the previous centuries were finally being transformed into working hardware.

In 1903, Rene Armengaud and Charles Lemale built and successfully tested a gas turbine using a Rateau rotary compressor and a Curtis velocity compounded steam turbine. Armengaud and Lemale went on to build and test several experimental gas turbines. Originally they used a 25 HP de Laval steam turbine driven by compressed

Figure 1-1. Courtesy of DEMAG DELAVAL Turbomachinery Corp. This graphic, the first trademark of De Laval Steam Turbine Company, illustrates the impulse-type turbine wheel with expanding nozzles. This single-wheel, high-speed, impulse turbine operated at the then incredible speed of 30,000 rpm.

gases from a combustion chamber, which was fed from a compressor. The turbine and compressor ran at 4,000 rpm and, in another early example of steam injection, temperatures were kept down by injecting steam upstream of the turbine nozzle.[1] Even at the turn of the century turbine blade cooling was being integrated into the turbine design as documented in 1914 by N. Davy who wrote, "In the experimental turbine of Armengaud and Lemale the turbine wheel was a double wheel of the Curtis type, water cooled throughout, even the blades themselves being constructed with channels for the passage of the water."[3] Out of necessity (they did not possess the metallurgy to withstand high temperatures) these early pioneers used steam and water injection, and internal air and water cooling to reduce the temperature effects on the combustor, turbine nozzles, and turbine blades.

Later Brown Boveri and Co. went on to build a 500 horsepower gas turbine with a three-stage centrifugal compressor, each stage having 25 impellers in series. This centrifugal compressor, specifically built for a gas turbine application, was modeled from a A.C. Rateau design.[4] It is sometimes difficult to separate, in retrospect, whether these pioneers were augmenting their steam turbines with hot gas, or their gas turbines with steam. But one thing is evident—their ideas are still an important part of today's gas turbine operation.

Throughout most of the first half of the 20th century the development of the gas turbine continued slowly. Advances were hampered primarily by manufacturing capability and the availability of high strength, high temperature resistant materials for use in compressor, turbine, and combustor components. As a result of these limitations compressor pressure ratios, turbine temperatures, and efficiencies were low. To overcome the turbine temperature limits, the injection of steam and water to cool the combustor and turbine materials was used extensively. As N. Davy noted in 1914, "From the purely theoretical valuation of the cycle, the efficiency is lowered by any addition of steam to the gaseous fluid, but in actual practice there is a considerable gain in economy by so doing. Limits of temperature, pressure, and peripheral speed, together with the inefficiencies inherent in pump and turbine, reduce the efficiency of the machine to a degree such that the addition of steam (under the conditions of superheat, inaugurated by its injection into the products of combustion) is of considerable economic value. It is also a great utility in reducing the temperature of the hot gases on the turbine wheel to a limit compatible with the material of which the blades are made."[3] These techniques continue to be used even with today's technology.

In 1905, the first gas turbine and compressor unit built by Brown Boveri was installed in the Marcus Hook Refinery of the Sun Oil Company near Philadelphia, PA,. It provided 5,300 kilowatts (4,400 kilowatts for hot pressurized gas and 900 kilowatts for electricity)[1]. Brown Boveri also constructed the first electricity generating turbine for a power station at Neuchatel in Switzerland[1]. This 4,000-kilowatt turbine (Figure 1-2), which was exhibited in 1938, consisted of an axial flow compressor delivering excess air at around 50 pounds per square inch (3.5 kg/cm^2) to a single combustion chamber and driving a multi-stage reaction turbine. The excess air was used to cool the exterior of the combustor and to heat that air for use in the turbine.

Figure 1-2. Courtesy of ASEA Brown Boveri AG. The world's first industrial gas turbine-generator, a 4,000-kilowatt unit, was presented at the Swiss National Exhibition in Zurich in 1939. Afterwards it was installed at Neuchatel, Switzerland.

Another early gas turbine industrial installation was a central power plant in the U.S. at the Huey Station of the Oklahoma Gas & Electric Co., Oklahoma City. This 3,500-kilowatt unit, commissioned in July 1949, was a simple-cycle gas turbine consisting of a fifteen stage axial compressor, six straight flow-through combustors placed circumferentially around the unit, and a two stage turbine.[5]

The First World War demonstrated the potential of the airplane as an effective military weapon. But in this time frame (1918-1920) the reciprocating gasoline engine was being developed as the power plant for the small, light aircraft of the time. The gas turbine was too big and bulky, with too large a weight-to-horsepower output ratio to be considered for an aircraft power plant. However, the turbo-charger became a highly developed addition to the aero-piston-engine. Following the work of A.C. Rateau, Stanford Moss developed the exhaust-driven turbo-charger (1921), which led to the use of turbo-charged piston engine aircraft in the Second World War.[6,7]

As the stationary gas turbines started to achieve recognition (largely due to the ability to improve performance and efficiency in stationary applications with complex arrangements of heat exchangers and water or steam injection) development of the gas turbine aero-engine was beset with one obstacle after another. In 1919, Britain's Air Ministry went to Dr. W.J. Stern, Director of the South Kensington Laboratory, to look into the possibility of the gas turbine for aircraft propulsion. Dr. Stern declared the idea unworkable. His review indicated a totally unacceptable weight-to-power ratio of 10:1, and a fuel consumption of 1.5 pounds (oil) per brake horsepower-hour (an efficiency of approximately 9%).[8] In 1929, Dr. A.A. Griffith (previously transferred to the South Kensington Laboratory after proposing a contrarotating contraflow engine) was asked to review Frank Whittle's jet engine designs. Frank Whittle's designs were again put aside as unworkable. Frank Whittle persisted and filed patent number 347,206, "Improvements in Aircraft Propulsion," on January 16, 1930. This patent was for a compound axial-centrifugal compressor and a single axial stage turbine. But it was not until 1937, with funding from Power Jets Ltd., that the Whittle engine (designated WU for Whittle Unit) built by the British Thomson-Houston Co. of Rugby was successfully tested. Whittle's WU engine consisted of a double entry centrifugal compressor and a single stage axial turbine.[8]

Some six years behind Whittle, Germany's Hans Pabst von Ohain put forth his ideas for a turbojet engine in 1935. It consisted of a compound axial-centrifugal compressor similar to Whittle's patent design and a radial turbine. H.P. von Ohain's designs were built by aircraft manufacturer Ernst Heinkel. August 24, 1939 marked the first flight of a turbojet aircraft, the He 178, powered by the HeS 3B engine.

Throughout the war years various changes were made in the design of these engines: radial and axial turbines, straight through and reverse flow combustion chambers, and most notably the axial compressor. The compressor pressure ratio, which started at 2.5:1 in 1900, went to 5:1 in 1940, 15:1 in 1960, and is currently approaching 40:1. Since the Second World War, improvements made in the aero gas turbine-jet engine industry have been transferred to the stationary gas turbine. Following the Korean War, the Pratt & Whitney Aircraft designed JT3 (the military designation was the J57) provided the cross-over from the aero gas turbine to the stationary gas turbine. The JT3 became the FT3 aeroderivative. In 1959, Cooper Bessemer installed the world's first base-load aeroderivative industrial gas turbine, the FT3, in a compressor drive application for Columbia Gulf Transmission Co. at their Clementsville, Ky., mainline compressor station (Figure 1-3). The unit, designated the RT-246, generated 10,500 brake horsepower (BHP) driving a Cooper-Bessemer RF2B-30 pipeline compressor.[9] In 1981, that unit had accumulated over 100,000 operating hours and is still active.

Table 1-1 is a chronology of key events in the development of the gas turbine as it evolved in conjunction with the steam turbine. Absent from this list are an unknown number of inventors such as John Dumball. Their contribution was not in demonstrating to the engineering community what worked, but what did not work.

TECHNICAL IMPROVEMENTS

The growth of the gas turbine in recent years has been brought about most significantly by three factors:

- metallurgical advances that have made possible the employment of high temperatures in the combustor and turbine components,

Figure 1-3. Courtesy of Cooper Cameron Corporation (formerly Cooper Bessemer Company). Partial view of the Cooper-Bessemer 10,500 horsepower, Model RT-248 gas turbine-compressor at Clementsville, Kentucky. Cooper-Bessemer teamed up with Pratt & Whitney Aircraft to employ a modified J57 jet engine (later identified as a GG3) as the hot gas generator for the RT-248 power turbine. J57 turbojet engines were used extensively on the Boeing 707 jet aircraft.

- the cumulative background of aerodynamic and thermodynamic knowledge, and
- the utilization of computer technology in the design and simulation of turbine airfoils and combustor and turbine blade cooling configurations.

Combining the above has led directly to improvements in compressor design (increases in pressure ratio), combustor design (regenerators, low NO_x), turbine design (single crystal blades, cooling), and

Table 1-1. Chronology of the Gas Turbine Development

Date	Name	Invention
130BC	Hero of Alexandria	Reaction Steam Turbine
1550	Leonardo da Vinci, Italy	Smoke Mill
1629	Giovanni Branca, Italy	Impulse Steam Turbine
1791	John Barber, England	Steam Turbine and Gas Turbine
1831	William Avery, USA	Steam Turbine
1837	M. Bresson	Steam Turbine
1850	Fernimough, England	Gas Turbine
1872	Dr. Stolze, Germany	Gas Turbine
1884	Charles A. Parsons	Reaction Steam Turbine & Gas Turbine
1888	Charles G.P. de Laval	Impulse Steam Turbine Branca type
1894	Armengaud+Lemale, France	Gas Turbine
1895	George Westinghouse	Steam Turbine Rights
1896	A.C. Rateau, France	Multi Impulse Steam Turbine
1896	Charles Curtis	Velocity Compound Steam Turbine/Gas Turbine
1895	Dr. Zoelly, Switzerland	Multi Impulse Steam Turbine
1900	F. Stolze, Germany	Axial Compressor & Turbine Gas Turbine
1901	Charles Lemale	Gas Turbine
1902	Stanford A. Moss, USA	Turbo-Charger/Gas Turbine
1903	A. Elling	Gas Turbine
1903	Armengaud+Lemale	Gas Turbine
1905	Brown Boveri	Gas Turbine
1908	Karavodine	Gas Turbine with deLaval Steam Turbine
1908	Holzwarth	Gas Turbine with Curtis + Rateau Compressor
1930	Frank Whittle, England	Aero Gas Turbine (Jet Engine)
1938	Brown Boveri—Neuchatel, Switzerland	1st Commercial Axial Compressor & Turbine

overall package performance. Gas turbines, which have always been tolerant of a wide range of fuels—from liquids to gases, to high and low Btu heating values—are now functioning satisfactorily on gasified coal and wood. This is significant considering that coal is the largest source of energy, at least in the USA. Another contributing factor to the success of the gas turbine is the ability to simplify the control of this highly responsive machine through the use of computer control technology. Computers not only start, stop, and govern the minute-to-minute operation of the gas turbine (and its driven equipment) but can also report on the unit's health (diagnostics), and predict future failures (prognostics).

The various gas turbine manufacturers and packagers available in the market today for electric generation, cogeneration, and mechanical drive applications are listed in Appendix A.

References

1. "Engines—The Search For Power," John Day, 1980.
2. "A History Of Mechanical Inventions," Abbott Payson Usher, 1988.
3. "The Gas Turbine," Norman Davy, 1914.
4. "Modern Gas Turbines," Author W. Judge, 1950.
5. "Gas Turbine Analysis And Practice," B.H. Jennings & W.L. Rogers, 1953.
6. Encyclopedia America.
7. General Electric Power Systems: geps@www.ge.com.
8. "The Development Of Jet And Turbine Aero Engines," Bill Gunston, 1995.
9. Cooper Industries 1833-1983, David N. Keller, 1983.

Chapter 2

Applications

Gas turbines are used in diversified services from jet engines and simple mechanical drives (on land, sea and air) to sophisticated gas lasers and supersonic wind tunnels. For simplicity the gas turbine will be considered for airborne applications and surface (land and sea) based applications. In the airborne applications these units are referred to as jets, turbojets, turbofans, and turboprops. In land and sea based applications these units are referred to as mechanical drive gas turbines. Each category will be discussed in detail.

JET ENGINES

In the strict sense all gas turbines are gas generators. Their hot gases are expanded either through a turbine to generate shaft power or through a nozzle to create thrust. Some gas generators expand their hot gases only through a nozzle to produce thrust—these units are easily identified as jet engines (or turbojets). Other gas turbines expand some of the hot gas through a nozzle to create thrust and the rest of the gas is expanded through a turbine to drive a fan —these units are called turbofans. When a unit expands virtually all of its hot gases through the turbine driving the compressor and the attached propeller and no thrust is created from the gas exiting the exhaust nozzle—it is called a turboprop. All of the above describe flight engines. However, turboprops have much in common with land and sea based gas turbines. This should not be surprising as in many cases the basic gas turbine is identical for both applications. The engines used in aircraft applications may be either turbojets, turbofans, or turboprops, but they are all commonly referred to as "jet engines."

The turbojet is the simplest form of gas turbine in that the hot gases generated in the combustion process escape through an exhaust nozzle to produce thrust. While jet propulsion is the most common usage for the turbojet, it has been adapted to direct drying applications, to power a supersonic wind tunnel, and as the energy source in a gas laser. The turbofan (Figure 2-1) combines the thrust provided by expanding the hot gases through a nozzle (as in the turbojet) with the thrust provided by the fan. In this application the fan acts as a ducted propeller. In recent turbofan designs the turbofan approaches the turboprop in that all the gas energy is converted to shaft power to drive the ducted fan (Figure 2-2). Turboprops (Figure 2-3) utilize the gas turbine to generate the shaft power to drive the propeller (there is virtually no thrust from the exhaust). Therefore, the turboprop is not, strictly speaking, a jet engine.

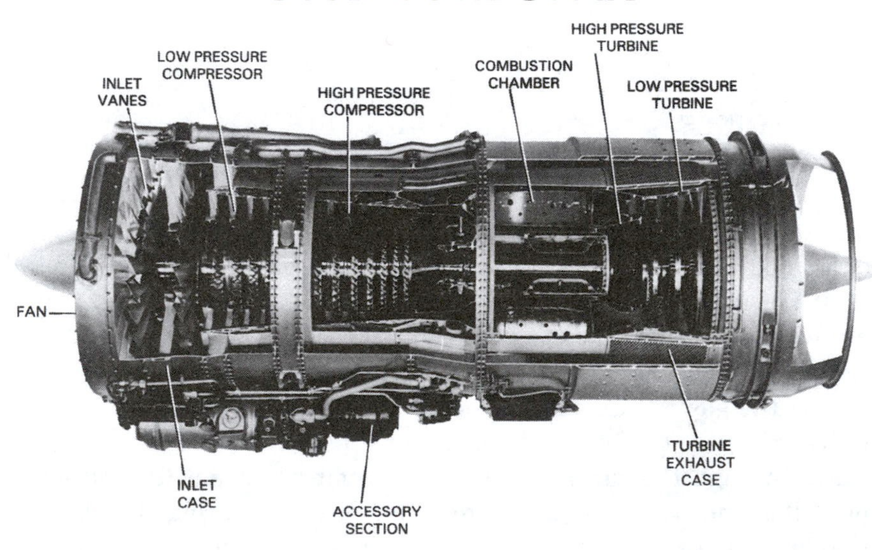

Figure 2-1. Courtesy of United Technologies Corporation, Pratt & Whitney Aircraft. The JT8D turbofan engine was one of the early "bypass" engines (BPR 1.7:1). The JT8D-200 series produces over 20,000 pounds "take-off" thrust and powers the McDonnell Douglas MD-80. The FT8 is the industrial, aero-derivative, version of this engine.

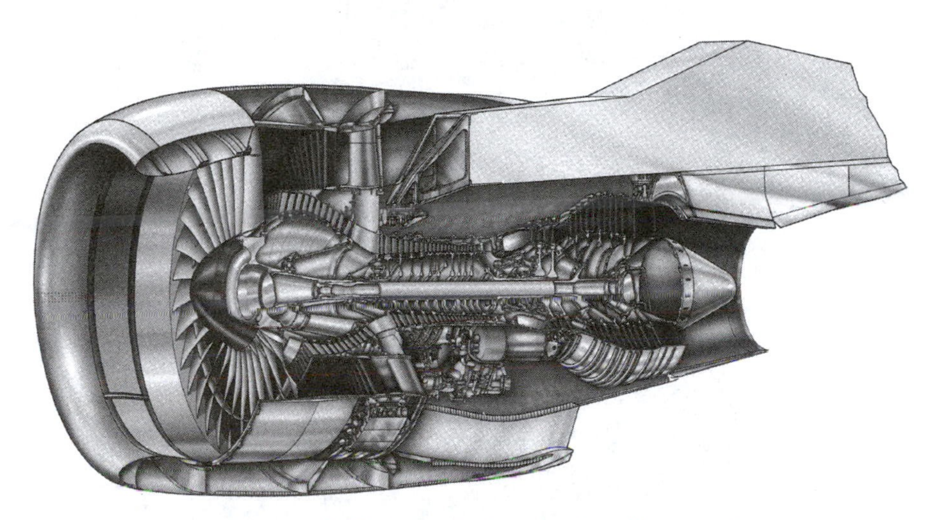

Figure 2-2. Courtesy of United Technologies Corporation, Pratt & Whitney Aircraft. The PW4000 turbofan engine is a high "bypass ratio" engine (BPR 5.1:1). This engine produces 68,000 pounds "take-off" thrust and currently powers the Airbus A330 wide body twinjet aircraft.

Some turboprop engines have made the transition from flight engines to land based applications (Pratt & Whitney Aircraft Canada PT6/ST6). Indianapolis 500 Race fans may recall the introduction of the Pratt & Whitney type ST6B-62 to that race in 1967. The car, owned by Andy Granatelli and driven by Pernelli Jones, led the race for 171 laps, only to fail a gearbox bearing in the 197th lap[1]. That car had an air inlet area of 21.9 square inches. Later, the Indianapolis 500 Race Officials modified the rules by restricting the air inlet area to 15.999 square inches or less. A year later race officials further restricted the air inlet area to 12.99 square inches. This effectively eliminated gas turbines from ever racing again. The aero-engines that have been most successful in making the transition from flight applications to land based applications have been the turbojets (Pratt & Whitney Aircraft J75/FT4, General Electric J79/LM1500, and Rolls Royce Avon) and the turbofans (General Electric CF6/LM2500, CF6/

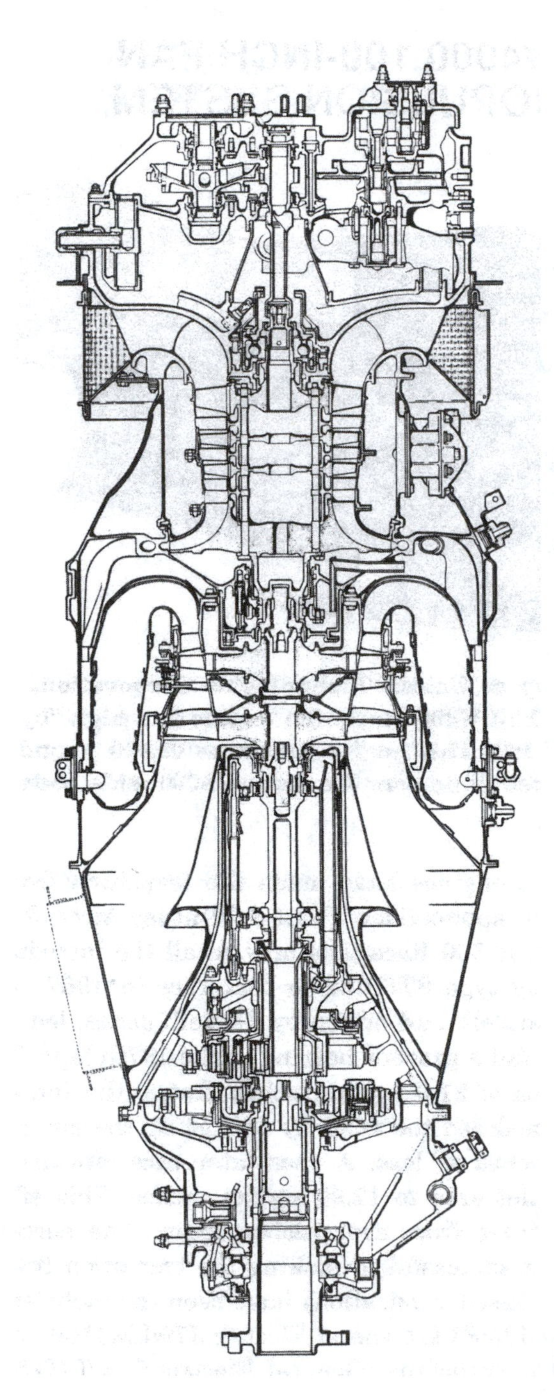

Figure 2-3. Courtesy of United Technologies Corporation, Pratt & Whitney Canada. The PT6/ST6 turboprop engine is a 1,500 shaft horsepower unit used primarily in aircraft, helicopter, and marine applications.

LM5000, CF6/LM6000, Rolls Royce RB211, and Pratt & Whitney JT8/FT8). These engines are commonly referred to as aero-derivatives. Of these aero-derivative engines, the LM2500 (shown in Figure 2-4) has been the most commercially successful. However, not all land based gas turbines were derived from aero engines, the majority of them were derived from the steam turbine as discussed in Chapter 1. Like the steam turbines, these gas turbines have large, heavy, horizontally split cases (hence the designation "heavy industrial gas turbine") and operate at lower speeds and higher mass flows than the aero-derivatives (at equivalent horsepower). A number of hybrid gas turbines in the small and intermediate size horsepower range have been developed to incorporate features of the aero-derivative and the heavy industrial gas turbines. The heavy industrial and the hybrid industrial gas turbines will be addressed in the next section.

MECHANICAL DRIVE

Mechanical drive gas turbines include the steam turbine derived heavy industrial gas turbines, aero-derivative gas turbines, and the hybrid industrial gas turbines. These turbines are listed in Appendix

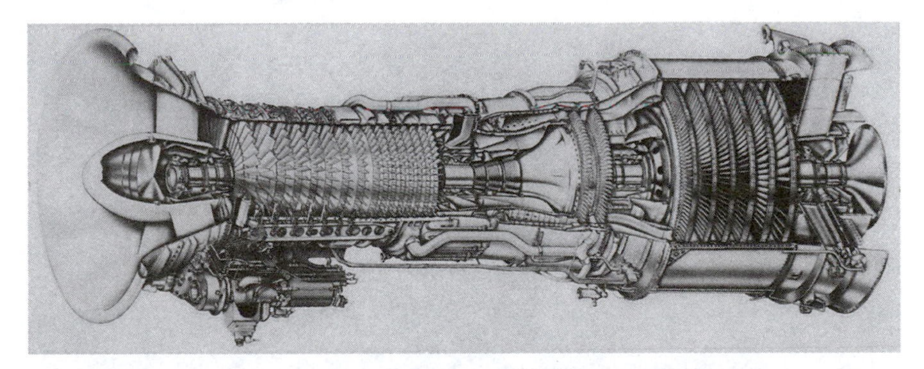

Figure 2-4. Courtesy of General Electric Company. The LM2500 stationary gas turbine was derived from the CF6 "high bypass ratio" flight engine that powered the C5A, the largest military cargo aircraft built in the USA. The LM2500 is ISO base load rated at 23 megawatts. This cross section of the LM2500 includes the six stage power turbine. This power turbine was derived from the six stage fan-turbine used in a CF6 flight engine.

A by application (mechanical drive, electric generation, combined cycle, and marine propulsion). Typical of the steam turbine derived heavy industrial gas turbine is the General Electric Frame 7001 and the Solar MARS hybrid industrial gas turbine. The mechanical drive gas turbines (Figure 2-5 and Figure 2-6) are available in three configurations: single spool-integral output shaft, single spool-split output shaft, and dual spool-split output shaft. In a single spool-integral output shaft unit the output shaft is an extension of the main shaft, which connects the compressor and turbine components. The output shaft may be an extension of the turbine shaft (as shown in Figure 2-7) or it may be an extension of the compressor shaft (as shown in Figure 2-8). When the output drive shaft is an extension of the turbine component shaft it is referred to as a "hot end drive." Likewise, when the output drive shaft is an extension of the compressor component shaft it is referred to as a "cold end drive." There are disadvantages to each configuration.

Hot End Drive

In this configuration the output shaft extension is at the turbine end where exhaust gas temperatures can reach 800°F to 1,000°F (427°C to 538°C). These temperatures affect bearing operation and

Figure 2-5. Courtesy of General Electric Company. The General Electric Model Series (MS)7000 heavy frame industrial gas turbine is currently in service throughout the world. The various models of this machine are ISO base load rated from 80 megawatts to 120 megawatts.

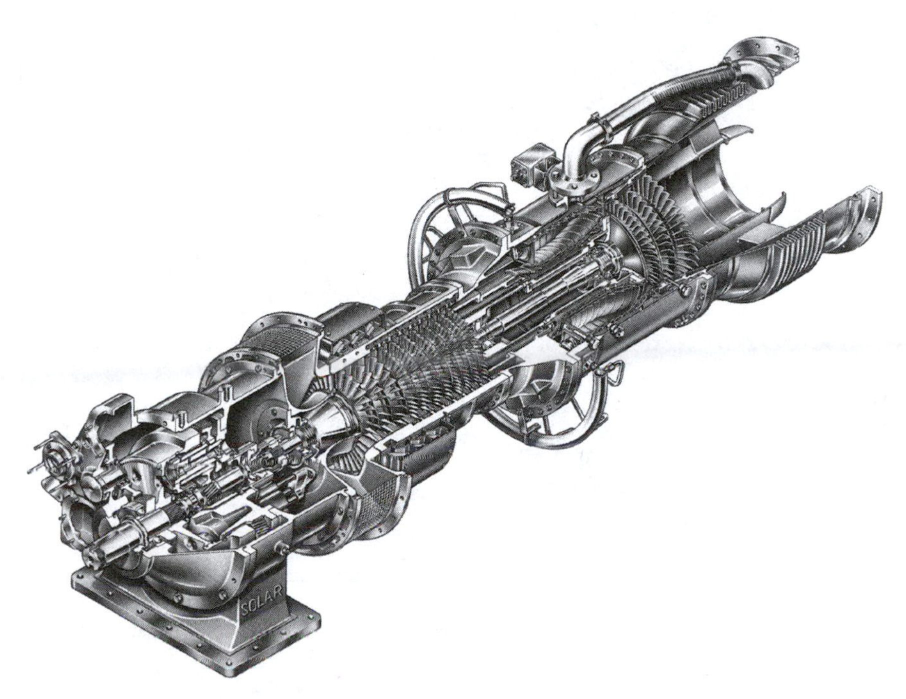

Figure 2-6. Courtesy of Solar Turbines Incorporated. The Solar Centaur 50 is an industrial hybrid gas turbine with an ISO base load rating of 4.35 megawatts.

life. Also, this configuration is difficult to service as the assembly must be fitted through the exhaust duct. The designer is faced with a number of constraints such as: output shaft length, high temperatures (800°F to 1000°F), exhaust duct turbulence, pressure drop, and maintenance accessibility. Insufficient attention to any of these details, in the design process, often results in power loss, vibration, shaft or coupling failures, and increased down-time for maintenance.

Cold End Drive

In the cold end drive configuration the output shaft extends out the front of the compressor. Here the driven equipment is accessible, relatively easy to service, and exposed to ambient temperatures only. The single drawback to this configuration is that the compressor inlet must be configured to accommodate the output shaft and the driven equipment (electric generator, pump, compressor, or speed increaser/ decreaser as necessary). This inlet duct must be turbulent free and

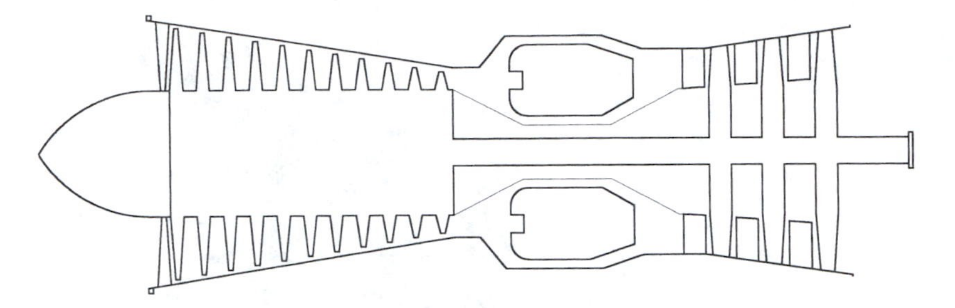

Figure 2-7. Sketch of a single spool gas turbine hot end drive.

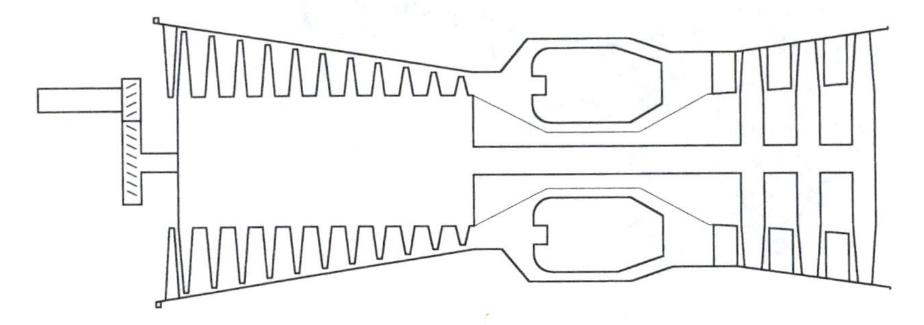

Figure 2-8. Sketch of a single spool gas turbine cold end drive.

provide uniform, vortex free, flow throughout the operating speed range. Inlet duct turbulence is the major concern facing the designer in this configuration. The problems resulting from a poor design can be catastrophic. For example, inlet turbulence can induce surge in the gas turbine compressor resulting in complete destruction of the unit. Inlet duct turbulence is often eliminated at the expense of pressure drop (ΔP). As inlet ΔP increases, power output decreases. This is discussed in more detail in Chapters 8 and 9.

In either case the output shaft speed is the same as the compressor/turbine speed. The output shaft speed may be geared up or down depending on the rated speed of the driven equipment. Single spool-integral output shaft gas turbines, both 'hot end drive' and 'cold end drive,' are used primarily to drive electric generators. While there have been applications where 'single spool-integral output shaft gas turbines' have driven pumps and compressors, they are uncommon. The high torque required to start pumps and compressors under full pressure results in high turbine temperatures during the start cycle (when cooling air flow is low or non-existent). Some publications identify this configuration as 'generator-drives' and list them separate from mechanical drive units. The single-spool gas turbine with the output shaft as an extension of the main shaft is conceptually identical to the turboprop used in fixed wing aircraft applications.

A single spool-split output shaft gas turbine (sometimes referred to as a split-shaft mechanical drive gas turbine) is a single-spool gas turbine driving a free power turbine as shown in Figure 2-9. The compressor/turbine component shaft is not physically connected to the power output (power turbine) shaft, but is coupled aerodynamically. This aerodynamic coupling (also referred to as a liquid coupling) is advantageous in that starts are easier (cooler) on the turbine components. This is due to the fact that the gas turbine attains self-sustaining operation before it "picks-up" the load of the driven equipment. In fact, the gas turbine can operate at this low idle speed without the driven equipment even rotating. In this configuration the power turbine output shaft runs at speeds that can be very different from the gas generator speed. This configuration is most often seen in process compressor and pump drives. However, it is also used in electric generator drive application.

One of the advantages of this arrangement is that the power turbine can be designed to operate at the same speed as the driven equipment. Therefore, for generator drive applications the power turbines operate at either 3,000 or 3,600 rpm (in order to match 50 cycle or 60 cycle generators). For centrifugal compressor and pump applications, speeds in the 4,000 to 6,000 rpm range are common. Matching the speeds of the driver and driven equipment eliminates the need for a gearbox. Gearbox losses are typically in the 2% to 4% range of power output, therefore, eliminating the gearbox constitutes

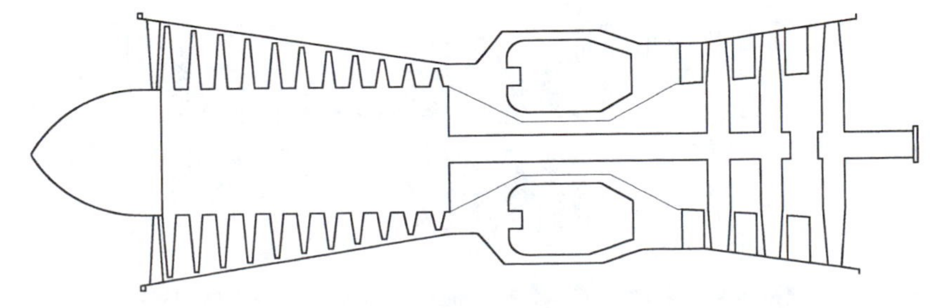

Figure 2-9. Sketch of a single spool gas turbine with a hot end drive power turbine.

a significant cost savings over the life of the plant. When the driver-driven equipment speeds cannot be matched exactly, the variation in power turbine operating speeds results in a lower gear ratio. Gearboxes with low gear ratios are less expensive to manufacture and maintain, and provide lower efficiency losses. By its very nature this design is limited to the hot end drive configuration. As such it shares all the problems discussed earlier in this chapter.

Gas turbines originally designed for jet aircraft applications as turbojets (or as the core engine of turbofans) have been successfully adapted to ground based applications using the split output shaft configuration. In helicopters, the turbojet is used to drive a free power turbine, which drives the helicopter rotor via a speed reducing gearbox.

The dual spool-split output shaft gas turbine is similar to the single spool-split output shaft type except that independent low and high pressure compressors and turbines generate the hot gases that drive the free turbine (Figure 2-10). The free power turbine runs at different and variable speed compared to the high pressure compressor-turbine rotor and the low pressure compressor-turbine rotor. Therefore, in this gas turbine there are three shafts, each operating at different speeds. The dual spool units are utilized in similar applications as the single spool units (compressor, pump, and generator drives) but are generally higher horsepower applications.

Gas turbines are being used throughout the world for power generation in stationary, land based power plants. As of 1993 approximately 25 billion kilowatts of electric power were generated by gas turbines[2]. The Harbor Cogeneration Power Plant shown in Figure 2-11 is

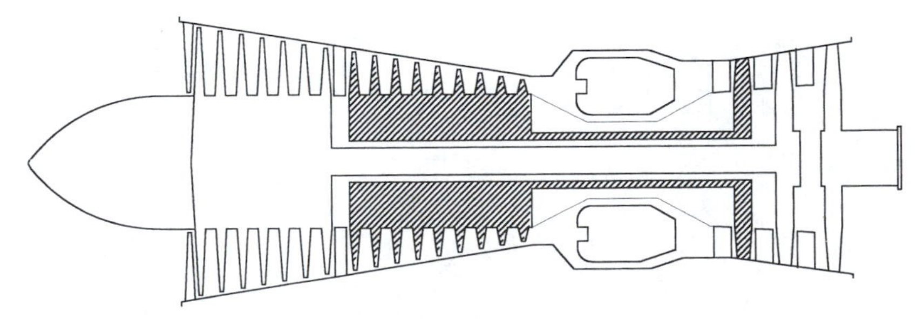

Figure 2-10. Sketch of a dual spool gas turbine and free power turbine.

a typical application. In California, a very high percentage of the power produced is generated by gas turbine power plants. In addition, many of these facilities also utilize cogeneration to recover the waste heat from the gas turbine exhaust. Gas turbines have also found a niche in pipeline pumping and compression applications in some of the harshest environments in the world. The Alyeska Pipeline pumps approximately 2 million barrels of crude oil per day from the oil fields at Prudhoe Bay to the shipping port in Valdez, Alaska. This pipeline, installed in the 1970s, extends 800 miles over some of the coldest landscape in the world. The pipeline utilizes aero-derivative single spool-split output shaft gas turbines driving centrifugal pumps. Another application of aero-derivative gas turbines is the Saudi Arabian East-West Pipeline. This pipeline extends across Saudi Arabia, east-to-west approximately 900 miles, over some of the hottest landscape in the world. The Saudi Arabian East-West Pipeline transports natural gas liquids from Abquaiq to the port facility in Yanbu.

Figure 2-11. Courtesy of Harbor Cogeneration Company. Cogeneration facility utilizing a General Electric MS 7001 gas turbine, and a 400,000 pound per hour Heat Recovery Steam Generator.

In the mid 1960s, the U.S. Navy implemented a program to determine the effectiveness of gas turbines as a ship's propulsion power plant. The first actual combat ship authorized for construction by the U.S. Navy was the *USS Achville* (PG-84). This was a Patrol Gunboat commissioned in 1964[3]. The U.S. Navy has outfitted larger class ships, including the Arleigh Burke Class Destroyer (DDG-51), and the Ticonderoga Class (CG-47). The Arleigh Class Destroyer utilizes four LM2500-30 aero-derivative gas turbines as the main propulsion units (100,000 total shaft horsepower). By the end of 1991, the U.S. Navy had 142 gas turbine propelled combat ships.[3] As of 1996, 27 navies of the world are powering 338 ships with 873 LM2500 gas turbines.[4] Some of the worst effects of turbine hot section corrosion is experienced in these ocean going applications. The prevention or reduction of sea-salt-instigated sulfidation corrosion is addressed in the design of the inlet air filter system and the selection of turbine materials and material coatings.

Gas turbines have also been used to power automobiles, trains, and tanks. The M1A1 Abrams tank, equipped with the AGT-1500 gas turbine engine, won acclaim for its service in Desert Storm. This 63 ton, battle-loaded, tank can travel up to 41.5 miles per hour on level ground.[5] A new power source coming into the market place is the 20 kilowatt to 60 kilowatt, regenerated, gas turbine power package. This package, in combination with a battery pack, promises to deliver low emission power for use in automobiles.

References

1. "Floyd Clymer's Indianapolis Official Yearbook"—1967.
2. 1995 Statistical Abstract of the United States published by the Department of Commerce. In 1993 total generation was 2,883 billion kilowatts of which 25 billion kilowatts was produced by gas turbines and IC engines.
3. "Sawyer's Turbomachinery Maintenance Handbook," First Edition, Vol. I pp. 4-1.
4. Naval Institute Proceedings, May 1996
5. "Tanks," Thomas Fleming, American Heritage of Invention & Technology, Winter 1995.

Chapter 3

Hardware

Aero-derivative and industrial gas turbines have demonstrated their suitability for heavy duty, continuous, base load operation in power generation, pump and compressor applications. While they share many similarities, there are times when their differences make them uniquely more suitable for a specific application. These differences are not always adequately considered during the equipment selection phase. As a result operations and maintenance personnel must deal with them throughout the plants useful life. As shown in the previous chapter, schematically there are little differences between the various types of gas turbines. However, considering the actual hardware, primarily in the 20,000-horsepower and above range, the differences are very significant.

SIMILARITIES AND DIFFERENCES[1]

In spite of their common background, there are variations between the aero-derivative and heavy industrial gas turbines. These are weight, combustor design, turbine design, and bearing design (including the lube-oil system). Grouping units in the same or similar horsepower output range, the most obvious difference is in the physical size of the heavy industrial compared to the aero-derivative gas turbines (Figure 3-1, 3-2, and 3-2a). The heavy industrial General Electric MS 5000 (26 MW) takes up several times more laydown space than either the General Electric LM 6000 (42 MW) or the Rolls Royce 211-H63 (59,000BHP). This physical size difference leads to the comparisons in Table 3-1.

The differences between the aero-derivatives and the hybrid industrial gas turbines are less significant (Figure 3-3). Size, ro-

Table 3.1. Industrial Gas Turbines Compared To Aero-Derivative Gas Turbines

Observation	Industrial Compared To Aero-Derivative
Shaft speed	slower
Air flow	higher
Maintenance time	longer
Maintenance lay-down space	larger

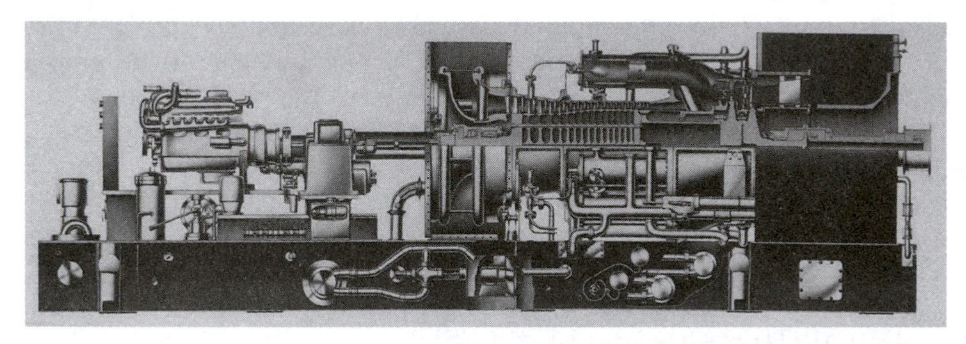

Figure 3-1. Courtesy of General Electric Company. The General Electric Model Series (MS)5000 heavy frame industrial gas turbine is currently in service throughout the world. Although General Electric is no longer producing this machine, it is available through its partners and licensees. This machine is ISO base load rated at 26 megawatts.

tating speed, air flow, and maintenance requirements are similar for both type machines. The primary distinction is in the bearing selection where the hybrids use hydrodynamic bearings (similar to the heavy industrial) and the aero-derivatives use anti-friction bearings.

The technology that resulted in putting a man on the moon had its beginnings in the aero-engine industry. That technology developed super alloys, achieved light weight without sacrificing

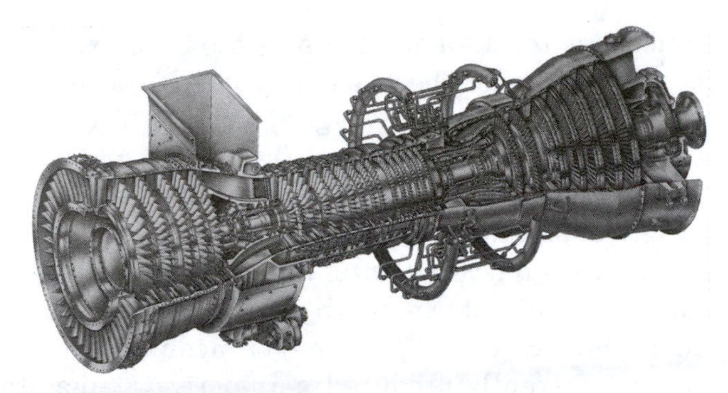

Figure 3-2. Courtesy of General Electric Company. The LM6000 is ISO base load rated at 42 megawatts. This engine was derived from the CF6-80C2 flight engine. The flight engine has over 6 million flight hours in aircraft such as the 747, A300, and MD-11.

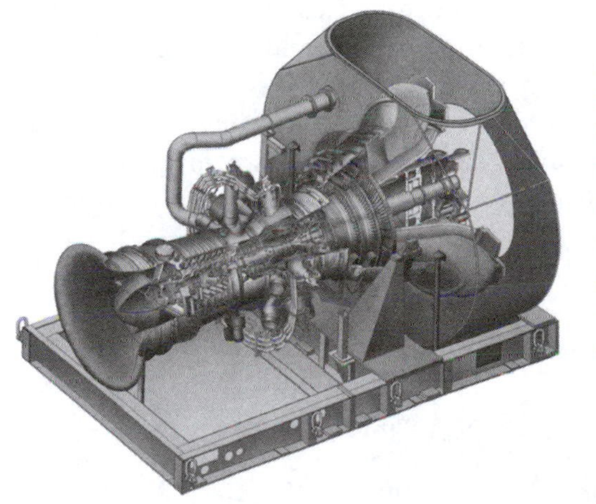

Figure 3-2a. Courtesy of Rolls Royce Industrial & Marine Gas Turbines Limited. The RB211-H63 WLE is ISO Rated for Continuous Duty at 59,005 brake horsepower. This engine was derived from the RB-211 flight engine

Figure 3-3. Courtesy of Solar Turbines Incorporated. The Solar MARS 100 is an industrial hybrid gas turbine with an ISO base load rating of 11 megawatts.

strength, and computer designed components for maximum performance. The technology has been used to provide a maintainable, flexible, industrial aero-derivative gas turbine. The key to this maintainability is the modular concept (Figure 3-4 and 3-4a), which provides for the removal of a component (compressor, combustor, or turbine) and replacement with a like kind without removing the engine from its support mounts. Some variations exist among the aero-derivative units in the method of removing and replacing a particular module. Still, the principle of modular replacement is common in most currently produced aero-derivative gas turbines. As a result, the effort to remove and replace a compressor module is not significantly different from the effort to remove and replace a turbine module.

The heavy industrial units, by contrast, require the least amount of effort to remove and replace the combustor parts, more effort to inspect or repair the turbine section, and the most effort to inspect or repair the compressor section.

The flexibility of the aero-derivative unit is "weight" and "size" related. For example, an application convenient to a large source of experienced manpower (with a well-defined baseload requirement,

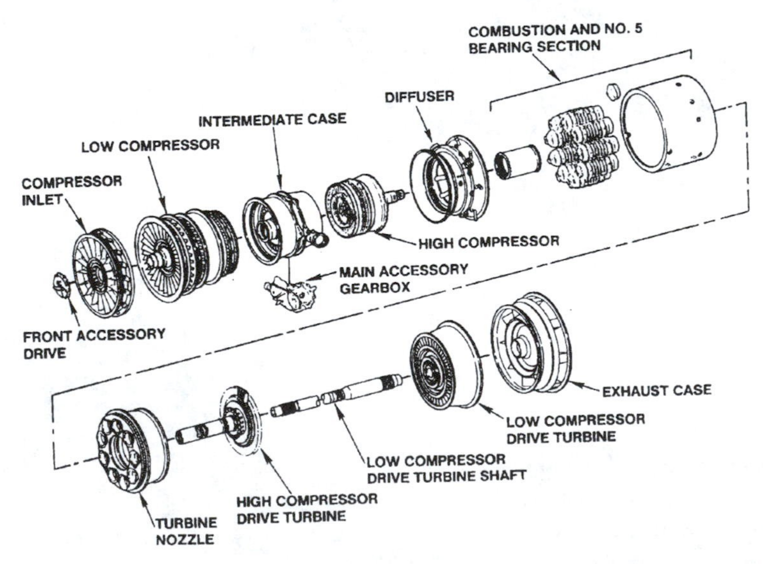

Figure 3-4. Courtesy of United Technologies Corporation, Pratt & Whitney Canada. Exploded view of the modules that make up the FT8 aero-derivative gas turbine.

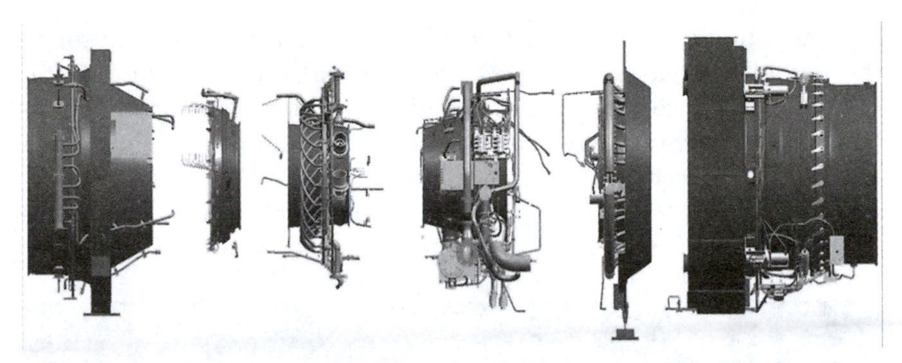

Figure 3-4a. Courtesy of Rolls Royce Industrial & Marine Gas Turbines Limited. This is a modular view of the Trent 60 engine. This engine is ISO Rated at 79,120 brake horsepower.

and good quality fuel) is considerably different from an application in a remote environment (away from skilled labor, good roads, and subject to varying qualities of fuel and loading conditions). The user must weigh his needs and requirements against the variety of machines offered. The preference has been to place the aero-derivative units in remotely located applications and to place the heavy industrial unit in easily accessible base-load applications. This is changing with the aero-derivative and the heavy frame industrial gas turbines competing on the same economic level.

COMPRESSORS

Compressors are either the axial design (with up to 19 stages) or the centrifugal design (with one or two impellers). In the axial compressor designs, beam and cantilever style stator vanes are utilized. Cantilever style stator vanes are used in compressors where stage loading is relatively light. Compressor pressure ratios have increased significantly over the past forty years with the aero-derivative consistently leading the way to higher levels. Pressure ratios, which were 5:1 at the start of World War II have increased to 12:1 for the newer industrial gas turbines. Through the use of increased stage loading (variable geometry and dual-spool techniques), compressor pressure ratios of most recent aero-derivatives have been increased to greater than 30:1 (Figure 3-5).

This advancement in the state of the art is a prime contributor in the overall increase in simple-cycle thermal efficiency to 35% for aero-derivative gas turbines. To achieve similar efficiencies the industrial gas turbines have had to use regenerators and other forms of waste heat recovery.

Typical materials used in the compressor are listed in Table 3-2.

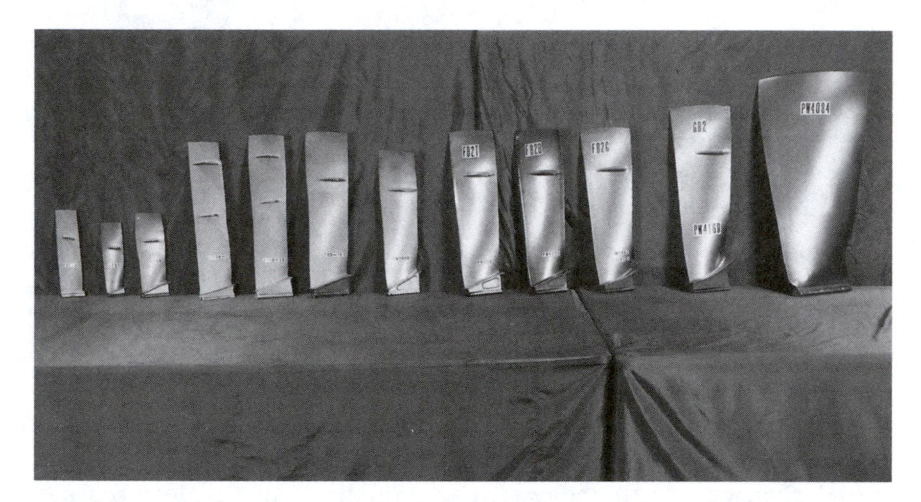

Figure 3-5. Courtesy of United Technologies Corporation, Pratt & Whitney Aircraft. A pictorial summary portraying the history of compressor blades from the early JT3 turbojet compressor blade on the left through to the most recent PW4084 blade on the right. This photograph represents a three-fold increase in compressor pressure ratios.

Table 3-2.

COMPONENT	MATERIAL	TRADE NAMES
Air Inlet Housing	Aluminum	
Forward Bearing Support	Aluminum	RR350, L51
	Iron	Nodular
	Stainless Steel	Jethete M.152, 17-4 Ph, 410
Housing	Aluminum	RR350, RR390, L51
	Titanium	6A1-4V
	Iron	MSRR6078, FV 448, FV 507

Table 3-2. (*Continued*)

COMPONENT	MATERIAL	TRADE NAMES
	Stainless Steel	Jethete M.152, Chromally
	Precipitation Hardening Super Alloy	Inco 718
Exit Housing Diffuser	Aluminum	
Rear Bearing Support	Aluminum	RR350, L51
	Iron	Nodular
	Stainless Steel	310, 321, FV 448, Chromally 410, Jethete M.152, MSRR 6078
	Precipitation Hardening Super Alloy	Inco 718
Stator Vanes	Aluminum	RR 58
	Titanium	6A1-4V
	Stainless Steel	A286, Chromally, Jethete M. 152, Greek Ascoloy, FV 535, FV500, 18/8,
	Precipitation Hardening Super Alloy	Nimonic 75, Nimonic 105
Rotor Blades	Aluminum	RR 58
	Titanium	6A1-4V, TBB
	Stainless Steel	A286, Greek Ascoloy, FV 535, FV520, 17-4 Ph, 403
	Precipitation Hardening Super Alloy	Inco 718, Nimonic 901
Discs, Spool, Drum	Titanium	6A1-4V, TBA (IMI 679), IMI 381
	Steel	4340, FV 448, B5-F5, 9310
	Stainless Steel	410, 17-4 Ph, Jethete M. 152, Chromally (FV 535)
	Precipitation Hardening Super Alloy	Incoloy 901, Inco 718, Nimonic 901
Shafts, Hubs	Steel	Hykoro, 4340, 9310, B5-F5
	Precipitation Hardening Super Alloy	Inco 718

Additional information on material specifications and material compositions is listed in Appendix C-2

TURBINES

Turbine nozzle and blade design was, and still is, a function of the performance match between the turbine and the compressor, and the strength and temperature resistance of available materials. Present production gas turbines (aero-derivative, heavy industrial, and hybrids) use an impulse-reaction turbine design. Turbine blade designs in the aero-derivative unit use high aspect ratio (long, thin) blades incorporating tip shrouds to dampen vibration and improve blade tip sealing characteristics (Figure 3-6). The heavy industrial machine incorporates a low aspect ratio (short, thick) blade with no shroud. Where long thin airfoils have been used, lacing wire has been employed to dampen vibration as shown in Figure 3-7. Improvements in metallurgy and casting techniques have allowed designers to eliminate mid-span shrouds and lacing wires. On many units, aero-derivative, hybrid, and heavy frame industrial, the nozzle guide vanes are manufactured in cast segments of two airfoils per segment up to

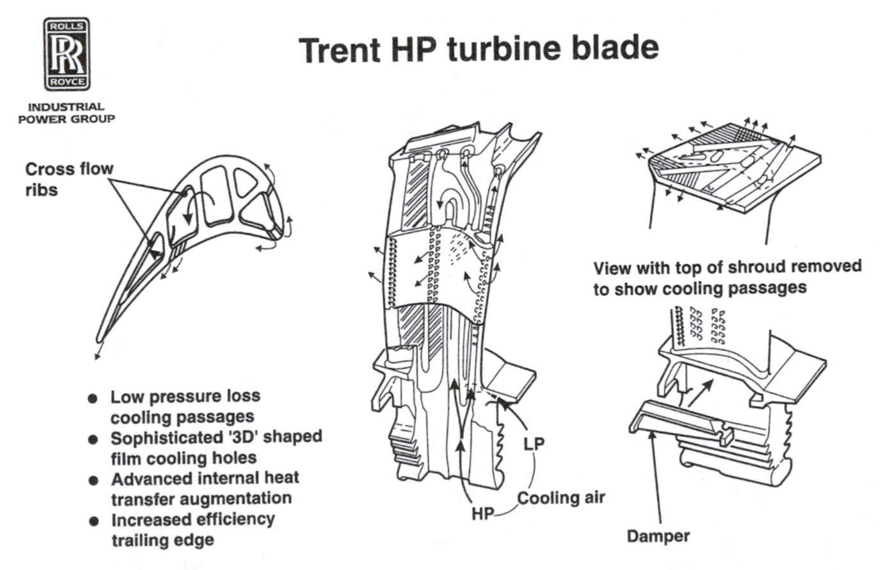

Figure 3-6. Courtesy of Rolls-Royce Industrial & Marine Gas Turbines Limited. Composite view of the Trent high pressure turbine blade showing its internal cooling scheme. The Trent is a growth version of the RB211 aero and industrial engine. The industrial Trent gas turbine is a 70,000 shaft horsepower engine.

Figure 3-7. Courtesy of Power & Compression Systems. This is a 1970 vintage gas turbine designed with lacing wire to dampen the second stage turbine blades. Note the hollow 1st stage turbine blade indicating blade cooling and the corrosion on both the 1st and the 2nd-stage blades.

half the nozzles per stage (Figure 3-8). The larger number of airfoils per segment do not facilitate airfoil coating and results in very high fabricating and coating cost. The large cross-sectional area of blades and vanes in heavy industrial turbines does not resist sulfidation corrosion attack, but can tolerate more corrosion than the thin, high aspect ratio, turbine blades of the aero-derivative machine. The heavy industrial machine of comparable horsepower pumps more fuel and approximately 50% more air than the aero-derivative unit. As a result, the turbine is exposed to a greater quantity of the elements that cause sulfidation corrosion (that is, it is exposed to more airborne salts and more fuel borne sulfur because it is pumping more air and consuming more fuel). This increased exposure to the elements that cause sulfidation corrosion, to a degree, negates the advantage that might have been assumed from the large cross-sectional area of the blade and nozzle airfoils.

Turbine blades are subject to stresses resulting from high temperatures, high centrifugal forces, and thermal cycling. These stresses accelerate the growth of defects or flaws that may be present in the material. This is the basis for the demand for materials that can withstand high temperatures without losing their resistance to centrifugal forces, vibration, thermal cycling, oxidation, or corrosion. Typical super-alloy materials used in the turbine are listed in Table 3-3.

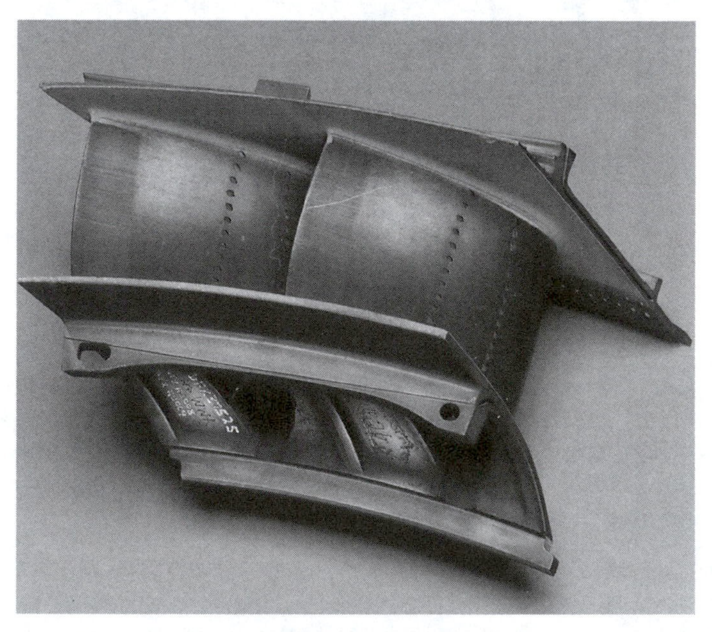

Figure 3-8. Courtesy of Rolls-Royce Industrial & Marine Gas Turbines Limited. Trent gas turbine first stage turbine nozzle vane segment. Note the cooling holes in the airfoil mid-span and in the outer platform upstream of the airfoil.

Table 3-3.

COMPONENT	MATERIAL	TRADE NAME
Casing, Housing	Iron	Ni Resis D5B
	Steel	5 Cr 1/2 Mo, 1 Cr 1/2 Mo, 410
	Stainless Steel	N 155, 310, 18/8, MSRR 6522 C263 (MSRR 7036), FV 507, 304, 321, A286, CA-6NM
	Precipitation Hardening Super Alloy	Nimonic 80A, Inco 718, Hastelloy X, Hastelloy C, Hastelloy S, Rene' 41 Inco 903,
Nozzle Vanes	Stainless Steel	304, 310, 347, 625
	Precipitation Hardening Super Alloy	HS31/X40, Inco 738, C1023 (MSRR 7046), N-155, Mar-M 509, Rene' 41, Rene' 77, Rene' 80, Hastelloy X, (X45M) FSX 414
Rotor blades	Precipitation Hardening Super Alloy	Nimonic 80A. Nimonic 90, Nimonic 105, Nimonic 108, Nimonic 115, Inconel 738, Inconel 792, Udimet 500, Udimet 700, Rene' 77, Rene' 80, Waspaloy, Mar M-246, Mar M-252, Mar M-421, A151-422, A-286, S-816, U-500
Discs, Wheels, Drums	Steel	422, FV 448, FV 535 Chromalloy, FHR Steel, Greek Ascoloy V57
	Precipitation Hardening Super Alloy	Waspaloy, Incoloy 901, Inco 706, Inco 718
Hubs, Shafts	Stainless Steel	4140, A151-4340, A286
Exhaust Collector	Stainless Steel	Nimonic 75, 310, 321, 403, 405, 409, 410

Additional information on material specifications and material compositions is listed in Appendix C-2.

Improvement in the properties of creep and rupture strength was steady from the late 1940s to the early 1970s, with the largest improvement of 390°F (200°C) in allowable operating temperature achieved in 1950 (Figure 3-9). The largest jump resulted from age-hardening or precipitation strengthening (a technique utilizing aluminum and titanium in the nickel matrix to increase strength). Since 1960, there has been increased dependence on sophisticated cooling techniques for turbine blades and nozzles. Since 1970, turbine inlet temperatures have increased as much as 500°F (260°C), some as high as 2,640°F (1,450°C). The increase in turbine inlet temperature was made possible by new air cooling schemes and the incorporation of complex ceramic core bodies used in production of hollow, cooled cast parts as shown in Figure 3-10.[2] Turbine blades and nozzles are formed by investment casting. This is not new technology. For centuries bronze statues have been cast using this

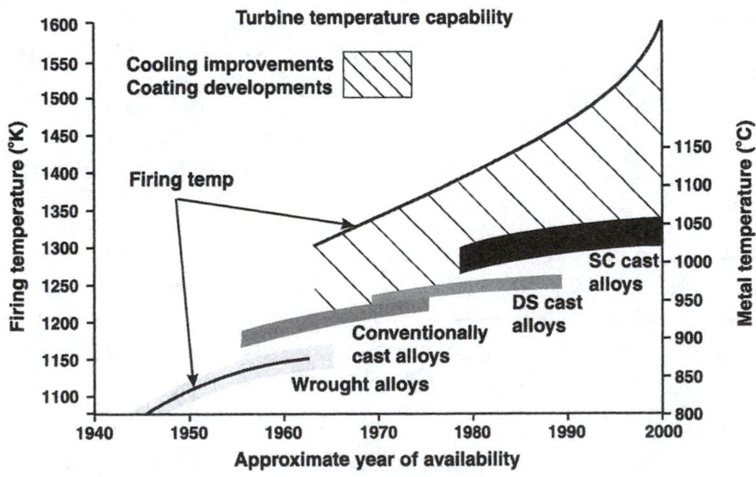

Figure 3-9. Courtesy of Rolls-Royce Industrial & Marine Gas Turbines Limited. This chart represents a pictorial history of the development of turbine materials from early wrought alloys to the latest single crystal cast alloy blades. Improvements in material strength and durability is the single most contributing factor to the advances made in gas turbines.

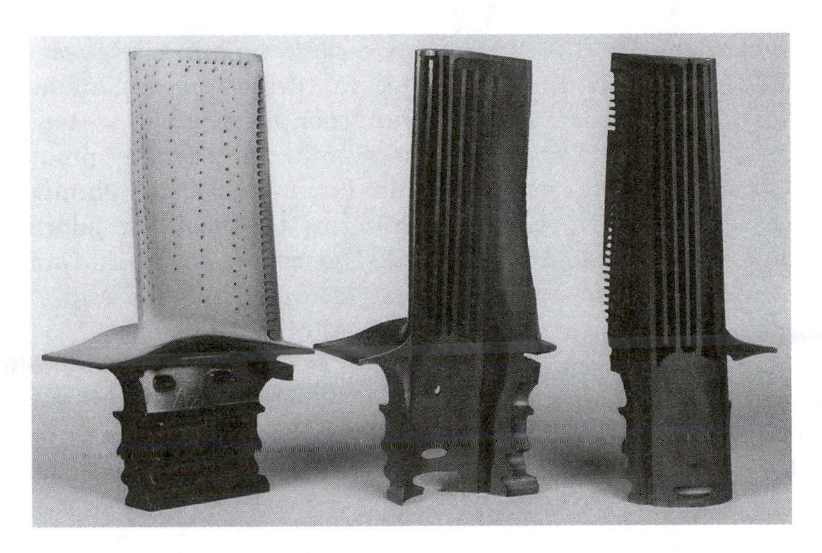

Figure 3-10. Courtesy of United Technologies Corporation, Pratt & Whitney Aircraft. Internal structure of high pressure turbine blade showing the cooling distribution throughout the core of the blade airfoil and root.

process. The critical part is the solidification of the liquid metal alloy after it is poured into the mold. It is during solidification that the alloy acquires its crystalline structure, which is a major determinant of the properties of the finished part. Many substances in nature have an atomic structure that is referred to as "crystalline," which metallurgists refer to as "grains."

Undesirable grain sizes, shapes, and transition areas were responsible for premature cracking of turbine parts. This led to the development of the equiaxed casting process. The equiaxed process assures uniformity of the grain structure along all their axes. At elevated temperatures, component failure begins within and progresses through grain boundaries. Therefore, if failures occur at grain boundaries rather than within the grains, the full strength of the crystal itself is not being utilized. Further study led to the conclusion that strength could be improved if grain boundaries were aligned in the direction normal to the applied force (most of the stress in the blade is in the direction of centrifugal force-along the length of the blade). This elongated or columinar grain formation in a preferred direction, called directional solidification or DS, was introduced by Pratt & Whitney Aircraft in 1965.[2] Because grain boundaries remain the weak link in turbine blades, numerous techniques have been used to strengthen them. Even better than strengthening grain boundaries is eliminating them by producing parts consisting of a single crystal. Furthermore, the elimination of the grain boundary, also eliminates the need for additional boundary-strengthening elements. The reduction in the amount of boundary-strengthened elements (at least in some alloys) has raised the incipient melting point by 120°F (50°C).

The evolution of equiaxed structure to directionally solidified structure to single crystal casting is shown for the same turbine blade in all three forms in Figure 3-11. Each advancement in this technology has increased the high temperature strength by approximately 85°F (30°C).[2]

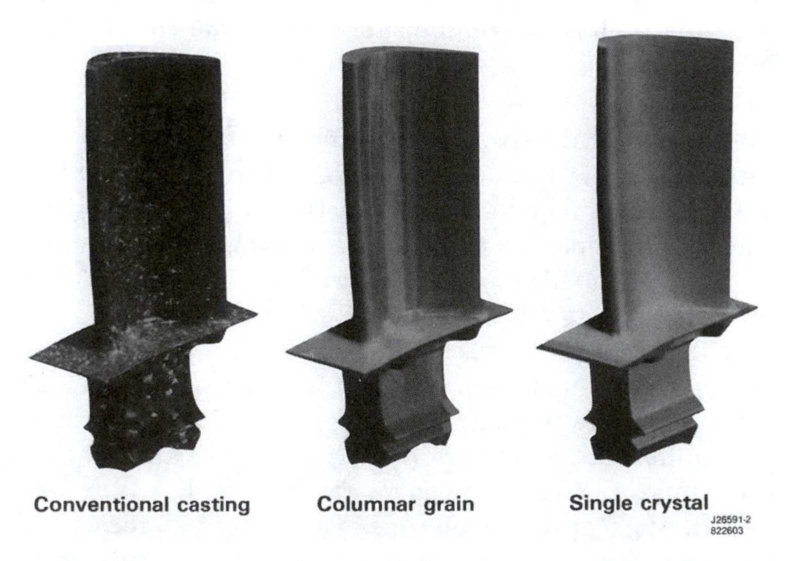

Figure 3-11. Courtesy of United Technologies Corporation, Pratt & Whitney Aircraft. Three high pressure turbine blades, conventional (equiaxed), columinar grain (directional solidified), and single crystal. This concept of growing the grain structure continuously along the blade's long, most highly stressed axis is called directional solidification. The single crystal blade has no grain boundaries because the entire part is grown as a single crystal.

COMBUSTORS

Combustor design is a complex task, often referred to as a "black art." The combustor design took two distinct configurations fairly early in the evolution of the gas turbine. These are the can-annular combustor and the annular (including the single combustor) combustor sections. There are generally two types of can-annular combustors: one is the more efficient straight flow-through; the other is the reverse flow combustor (Figure 3-12). The advantage of the reverse flow combustor, as used in the heavy industrial gas turbine, is that this design facilitates the use of a regenerator, which improves overall thermal efficiency. A further distinctive design approach within the can-annular concept is a single fuel nozzle and multi-fuel nozzles per combustion chamber. In theory, a large

Figure 3-12. Courtesy of United Technologies Corporation, Pratt & Whitney Aircraft. A combustor outlet view of one of the nine combustors in the, 25.5 megawatt, FT8 aero-derivative gas turbine. Flame temperature at the center of the combustor approaches 3,000°F. The small cooling holes enable a flow of compressor air to continuously cool the combustor liner walls.

number of fuel nozzles provide better distribution of the fuel gas (or greater atomization of the liquid fuel particles) and more rapid and uniform burning and heat release (Figure 3-13). But the problems of equally distributing fuel to each fuel nozzle significantly limit the number of fuel nozzles employed.

The second combustor concept is the annular or single combustor. The single combustor is a stand-alone combustor, generally outside the envelope of the compressor and turbine. The annular combustor is also a single combustor but within the envelope of the compressor and turbine (Figure 3-14). Two main distinctions exist between these designs. The single combustor is usually configured with a single fuel nozzle, while the annular combustor must be configured with multiple fuel nozzles. Secondly, the annular combustor incorporates an inner wall or liner and heat shield, which is necessary to protect the rotor. The heavy industrials use a longer combustion chamber, which makes it more suitable for burning poor quality fuel. However, the longer combustor also sacrifices combustor efficiency due to reverse flow pressure losses, a greater surface area to be cooled, and greater heat loss through convection and radiation. Typical materials used in the combustor are listed in Table 3-4.

Figure 3-13. Courtesy of Solar Turbines Incorporated. Comparison of the low NO_x and standard fuel injectors for the Centaur 50 gas turbine. SoLoNO_x™ is Solar's designation for it's low NO_x emission engine configuration.

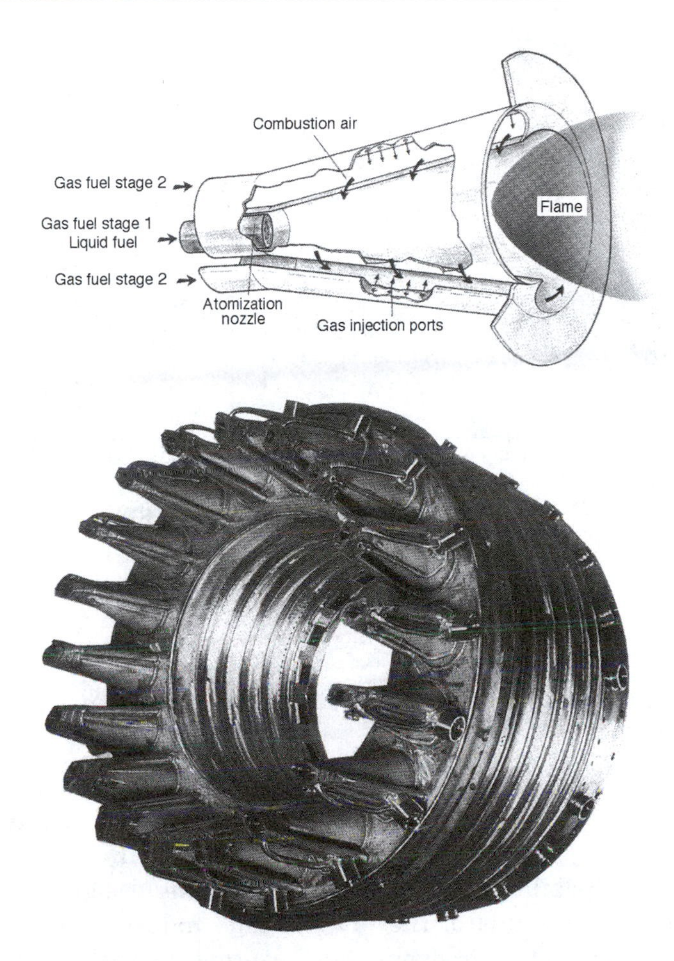

Figure 3-14. Courtesy of ABB STAL AB. The low NO_x, double-coned, EV burner design is used in both the GT10 (25 megawatt) and the GT35 (17 megawatt) gas turbines. The EV burner reduces NO_x to 25 ppm - dry.

BEARING DESIGN

The last major hardware difference between the aero-derivative and heavy and hybrid industrial gas turbines is in bearing design. The aero-derivatives use anti-friction roller and ball bearings while the heavy and hybrid industrials use hydrodynamic bearings. The use of anti-friction bearings in the aero-derivative gas turbines is a

Table 3-4.

COMPONENT	MATERIAL	TRADE NAMES
Housing	Steel	MSRR 6078, FV 448
	Stainless Steel	Jethete M.152, 410
	Precipitation Hardening Super Alloy	Inco 718
Liner, Basket, Flame Tube	Stainless Steel	321, C242 (MSRR 7143), C263 (MSRR 7036)
	Precipitation Hardening Super Alloy	Hastelloy X, N-155, HS-188, Nimonic 75,

Additional information on material specifications and material compositions is listed in Appendix C-2

carryover from the flight engine application. In flight engines, where weight is a major concern, rotor weights are small and well within the loading capabilities of ball and roller bearing design.

The rotor weights of the heavy frame industrial gas turbines dictates the use of hydrodynamic sleeve-type journal bearings and plain or Kingsbury-type thrust bearings. Hydrodynamic bearings incorporate tilting pad bearing designs to aid in rotor stability and alignment. The bearing designs used in the heavy industrial gas turbines are similar to the designs used in pumps and compressors. Therefore, in pump or compressor drive applications, the lubrication media (oil) can be, and usually is, the same. The heavy industrial gas turbine packages will use 1,500 to 2,500 gallons of turbine light mineral oil for the gas generator, power turbine, and driven equipment. This is four times that required of the aero-derivative package of similar output (note also that aero-derivatives use a synthetic oil). For example, the flight version of the aero-derivative engine requires as little as 25 gallons of synthetic oil for airborne operation, while the land based aero-derivative version is normally provided with a 120-gallon synthetic oil reservoir. An aero-derivative gas turbine package requires 400 gallons of turbine light oil for the power turbine and the driven equipment (pump, compressor, or generator). Even though synthetic oils have better heat transfer capabilities and are fire resistant, they are more expensive. Lubrication systems are addressed in detail in Chapter 6.

INSPECTION AND REPLACEMENT

One major asset of the aero-derivative industrial engine is the boroscope capability built into the design. Boroscoping facilitates inspection of critical internal parts without disassembling the engine. Complete boroscoping of the aero-derivative engine can be accomplished in an hour. There are no provisions for boroscoping the heavy industrial gas turbine. The heavy-industrial units must be disassembled for all inspections. The inspection and replacement of the fuel nozzles, combustor, and transition duct of a heavy industrial gas turbine with 10 to 12 can-annular combustors takes about 24 hours (96 man-hours not including cool-down time). Some very resourceful field technicians have developed techniques to boroscope the combustor, first stage turbine nozzles, and last stage compressor stators of the heavy industrial gas turbine. However, their efforts have not been embraced by the equipment manufacturers. Boroscoping techniques are discussed in greater detail in Chapter 13.

In contrast, inspection of the aero-derivative combustors and nozzles can be accomplished in 4 1/2 hours (9 man-hours). Removal and replacement of the combustors on the aero-derivative gas turbine can take 4 to 30 hours (8 to 75 man-hours) depending on the combustor configuration. Also, removal and replacement of the turbine section can be accomplished in 18 to 30 hours (up to 75 man-hours). On a heavy industrial, a turbine inspection alone takes 80 hours (480 man-hours). If blades are removed and replaced, an additional 8 hours is required. If more than several new or replacement blades are installed, the unit must be field balanced, which is time consuming. If field balancing can not be satisfactorily completed, the rotor must be removed, transported to a repair facility, balanced, and returned to the site.

The aero-derivative gas turbine stators or diaphragms are constructed in one, two, or three vane segments and not the large diaphragm segments used on most heavy industrial units. This facilitates part replacement and keeps cost down. The skill levels for on-site maintenance of the heavy industrials is not different from the skill level required to remove and replace the aero-derivative engine or to remove and replace modules from the aero-derivative engine. Replacing components within an aero-derivative module requires a "shop-level" skill. However, the modules can easily be transported to a maintenance base where the skill level is available. This same skill level is required at maintenance bases where detailed repair and replacement of heavy industrial gas turbines is presently being carried out.

Inspection of the compressor on the heavy industrial units requires removal of the upper cover and the compressor diaphragms, a 72 hour (720 man-hours) operation. If more than two or three rows of blades need to be replaced, the rotor must be removed, balanced, and then replaced. With some heavy industrial units, in order to remove the compressor cover, the combustor and turbine covers must also be removed. This makes the total operation a time consuming 125 hours (1,200 man-hours). Obviously, heavy lift equipment is also required!

In contrast, inspection and replacement of the aero-derivative compressor will take from 6 to 13 hours (13 to 30 man-hours). An alternative possible only with the aero-derivative gas turbine is complete removal and replacement of the gas generator. This can be accomplished in 6 hours (15 man-hours).

OPERATION

Starting and accelerating the aero-derivative unit is fast, and loading time is limited only by the driven equipment. Shutdown is also fast, about 3-5 minutes. Cool-down time approaches 2-3 hours, but can be shortened by "motoring" the gas generator after shutdown.

Starting and accelerating the heavy industrial unit is slow (15 to 20 minutes for starting and accelerating to a minimum load). Also, load-time is slow (about 20°F per minute) in order to keep thermal stresses to a minimum. Shutdown time approaches 30 minutes. The heavy-industrial unit requires an elaborate emergency lube system to protect the bearings through this long shutdown period. Furthermore, cool-down time approaches 48 hours during which time the unit must be slow-rolled to achieve uniform shaft cooling and to avoid bowing of the shaft. Shaft bowing is extremely critical on these heavy industrial machines. It is not practical to "motor" the heavy machines at cranking speeds (in an attempt to shorten the cool-down time) because the starting motors are not designed for prolonged running.

The aero-derivative units can be started with 50-150 horsepower starting motors. All the aero-derivative gas generators can be factory tested to maximum turbine inlet temperature and maximum gas horsepower using an exhaust "jet" nozzle, if necessary. Some manufacturers have the capability of testing to maximum shaft horsepower. In very few instances are the heavy industrial units tested to either maximum temperature or maximum horsepower. At most, manufacturers provide testing to rated speed only.

American Petroleum Institute (API) Standard 616 was written and published specifically for the heavy industrial gas turbine, although it is also used for the aero-derivative industrial turbine.

References

1. "How Lightweight And Heavy Gas Turbines Compare," Anthony J. Giampaolo, *Oil & Gas Journal*, January 1980.
2. "The Crystallography of Cast Turbine Airfoils," by D.H. Maxwell and T.A. Kolakowski, TRW/DSSG/Quest, 1980.

Chapter 4

Gas Turbine Systems Theory

GAS TURBINE OPERATING CYCLE

The gas turbine cycle is best depicted by the Brayton Cycle. The characteristics of the operating cycle are shown on the pressure-temperature map, the pressure-specific volume map, and the temperature-entropy map (Figure 4-1a to 4-1c).

The gas turbine, as a continuous flow machine, is best described by the first law of thermodynamics.

$$W\left({}_1Q_2\right) \equiv W\left(h_2 - h_1 + KE_2 - KE_1 + PE_2 - PE_1 + \frac{{}_1W_2}{J}\right) \qquad (4\text{-}1)$$

where

$\boldsymbol{W}$* = Mass flow rate, lb_m/sec
${}_1Q_2$ = Heat transferred to or from the system, Btu/lb_m
h_2 = enthalpy of the fluid leaving, Btu/lb_m
h_1 = enthalpy of the fluid entering, Btu/lb_m
KE_2 = kinetic energy of the fluid leaving, Btu/lb_m
KE_1 = kinetic energy of the fluid entering, Btu/lb_m
PE_2 = potential energy of the fluid leaving, Btu/lb_m
PE_1 = potential energy of the fluid entering, Btu/lb_m
${}_1W_2$ = Work per unit mass on or by the system, $ft\text{-}lb_f/lb_m$
J = Ratio of work unit to heat unit, 778.2 $ft\ lb_f/Btu$
${}_1W_2/J$ = Work, Btu/lb_m

*Here mass flow is designated by $\boldsymbol{W}$ to distinguish it from ${}_1W_2$, which is used to designate work. Later in this text W_a and W_f (which are industry standards) are used to designate air flow and fuel flow.

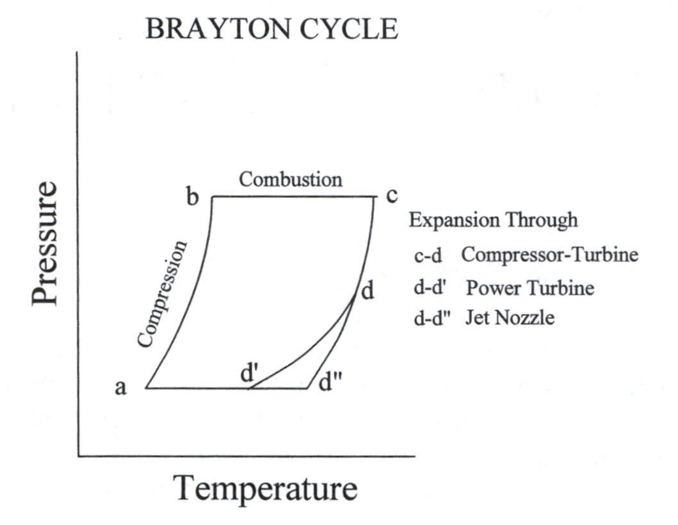

Figure 4-1a.

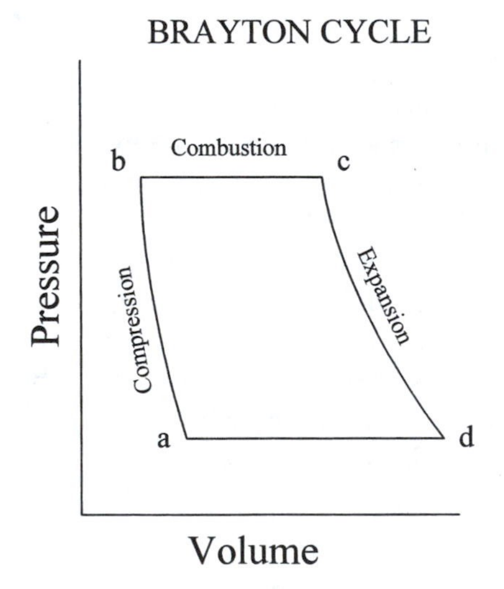

Figure 4-1b.

Figure 4-1a, b, c. Brayton Cycle pressure-temperature, pressure volume, and temperature-entropy plots.

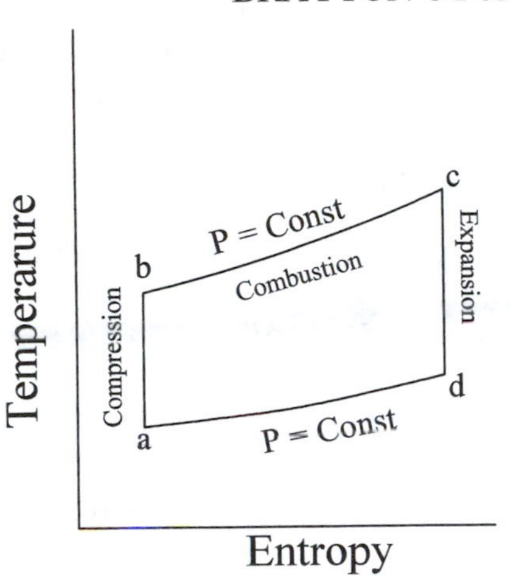

Figure 4-1c.

In most gas turbine applications, the numerical magnitude of the difference in potential energy is so small, relative to the other values in the equation, that it is customary to disregard it. This first law equation is rewritten as follows:

$$\mathbf{W}\,({}_1Q_2) = \mathbf{W}\left(h_2 - h_1 + KE_2 - KE_1 + \frac{{}_1W_2}{J}\right) \tag{4-2}$$

For adiabatic processes (no heat transfer)

$${}_1Q_2 = 0 \tag{4-3}$$

$$\mathbf{W}\left(\Delta h + \Delta KE + \frac{{}_1W_2}{J}\right) = 0 \tag{4-4}$$

$$\mathbf{W}\,\frac{{}_1W_2}{J} = \mathbf{W}\,(h_1 - h_2 + KE_1 - KE_2) = Power \tag{4-5}$$

$$HP = \left(\frac{\mathbf{W}}{0.707}\right)(h_1 - h_2 + KE_1 - KE_2) \quad \text{and} \tag{4-6}$$

$$KE = \frac{C^2}{2g_cJ} \tag{4-7}$$

where HP is horsepower, C is the velocity of the air entering the compressor or air and combustion products leaving the turbine, and g_c is the gravitational constant 32.17 $ft\ lb_m/lb_f sec^2$.

The engineering thermodynamics can be summarized with the following relationships:

Gas Turbine Horsepower Output

$$HP = \frac{\mathbf{W}}{0.707}\left(h_1 - h_2 + KE_1 - KE_2\right) \tag{4-8}$$

where 0.707 (more exactly 0.7068) converts Btu/sec to horsepower.

Gas Turbine Efficiency

Thermal efficiency (η_t) of a gas turbine, considering the compression and expansion processes as being irreversible, is defined as the work output divided by the fuel energy input. The work output is the total turbine work minus the work on the compressor (note compressor work is negative). Therefore,

$$\eta_t = \frac{\Sigma W}{{}_1Q_2} \quad \text{or} \tag{4-9}$$

$$\eta_t = \frac{2540}{Btu/HP-Hr} = \frac{3600}{KJ/kW-Hr} = \frac{3414}{Btu/kW-Hr} \tag{4-10}$$

where KJ is kilojoules, kW is kilowatts, and Hr is hours.

This expression is the most used tool in comparing one gas turbine with another. As used here, this expression represents the simple cycle gas turbine efficiency. This equation is also used to demonstrate that a particular engine is (or is not) deteriorating with use. If the overall engine simple cycle efficiency is deteriorating then an examination of each component is necessary to determine the cause of the problem.

COMPONENT EFFICIENCIES

The efficiency of each module contributes to the overall efficiency of the gas turbine. The ability to examine the efficiency of each module or component is a necessary tool in isolating engine problems.

Compressor Efficiency

Compressor efficiency (η_c) is directly proportional to the compressor pressure ratio and inversely proportional to the compressor discharge temperature. The following equation more exactly defines compressor efficiency:

$$\eta_c = \frac{R_c^{\sigma} - 1}{\frac{T_o}{T_i} - 1} \tag{4-11}$$

where

$$\sigma = \frac{k-1}{k} \quad \text{and} \tag{4-12}$$

R_c = Compressor pressure ratio, P_o/P_i
P_o = Compressor total discharge pressure, psia
P_i = Compressor total inlet pressure, psia
k = Ratio of specific heats, c_p/c_v
c_p = Specific heat at constant pressure, *Btu / lb °F*
c_v = Specific heat at constant volume, *Btu / lb °F*
T_o = Compressor total discharge temperature, °R
T_i = Compressor total inlet temperature, °R

Compressor Horsepower (Required)

Compressor horsepower is the power that the compressor consumes in compressing the air and moving it into the combustor.

$$HP_c = \frac{J}{550} W_a c_p \frac{T_i}{\eta_c} \left[\left(\frac{P_o}{P_i} \right)^{\sigma} - 1 \right] = \frac{J}{550} W_a c_p (\Delta T) \tag{4-13}$$

where W_a is the air flow entering the compressor in lb/sec.

Turbine Efficiency

Tracking or trending turbine efficiency would be an excellent method to monitor the health of a unit. However, as turbine inlet temperatures (TIT) have climbed higher and higher, they have become virtually impossible to measure on a long term basis. In fact, many manufacturers measure an intermediate turbine temperature for gas turbine control. Where this is the case the turbine inlet temperatures are calculated.

$$\eta_t = \frac{1 - \frac{T_{EXH}}{TIT}}{1 - R_t^{\sigma}} \qquad (4\text{-}14)$$

where

T_{EXH} = Turbine total exhaust temperature, °R
TIT = Turbine total inlet temperature, °R
R_t = Turbine total inlet pressure/Turbine total exhaust pressure

Turbine efficiency can be closely approximated by substituting

$$\frac{1}{R_c} = R_t$$

$$\eta_t = \frac{1 - \frac{T_{EXH}}{TIT}}{1 - \left(\frac{1}{R_c}\right)^{\sigma}} \qquad (4\text{-}15)$$

Turbine Horsepower (Produced)

This is the total horsepower produced by the turbine. It includes the horsepower to drive the compressor and, for single shaft machines, the power used by the driven load. For units with separate power turbines, this horsepower should equal the power absorbed by the compressor plus losses.

$$HP_t = \frac{J}{550} W_g c_p \eta_t T_i \left[1 - \left(\frac{P_o}{P_i}\right)^{\sigma}\right] = \frac{J}{550} W_g c_p T_i \left[1 - \frac{T_o}{T_i}\right] \tag{4-16}$$

where

W_g = Turbine inlet gas flow, lbs/sec
c_p = Specific heat at constant pressure, Btu/lb °F
η_t = Gas generator turbine adiabatic efficiency
P_o = Gas generator discharge total pressure, psia
P_i = Turbine inlet total pressure, psia
T_o = Gas generator discharge total temperature, °R
T_i = Turbine inlet total temperature, °R

Power Turbine Efficiency

Decreases in power turbine efficiency are primarily the result of loss of material due to erosion, corrosion, or foreign object damage.

$$\eta_{PT} = \frac{1 - \frac{T_o}{T_i}}{1 - \left(\frac{P_{AMB}}{P_i}\right)^{\sigma}} \tag{4-17}$$

where

T_i = Power turbine inlet total temperature, °R
T_o = Power turbine discharge total temperature, °R
P_{AMB} = Atmospheric total pressure*, psia

Power Turbine Horsepower

On free power turbines units, this is the horsepower generated to drive the driven load.

$$HP_{PT} = \frac{J}{550} W_g c_p T_i \eta_{PT} \left[1 - \left(\frac{P_{AMB}}{P_i}\right)^{\sigma}\right] \tag{4-18}$$

*Simple Cycle Gas Turbine Application. For combined cycle applications, or any application resulting in significant exhaust duct backpressure, it is more accurate to use P_{out} instead of P_{AMB}.

The internal pressure, temperature, and velocity variations within the gas turbine are shown in Figures 4-1 and 4-2. Note that compressor work is shown as the pressure rise from **a** to **b**. From **b** to **c**, the energy addition due to combustion is shown as a temperature rise at near constant pressure. In actuality there is some pressure drop through the combustor. The hot gases are then expanded through the turbine from point **c** to **d**, as evidenced by a drop in pressure and temperature. In a jet-type gas turbine the temperature and pressure decreases from **d** to **d"** as the gas expands through a jet exhaust nozzle and creates thrust. In mechanical drive type applications, this energy is expanded from point **d** to **d'** in the form of shaft horsepower. Finally, the Brayton Cycle can be considered a closed cycle for the gas turbine if we consider the surrounding atmosphere as the heat sink (as depicted by the constant pressure from point **d'** or **d"** to **a**).

COMPRESSOR

Description

The compressor provides the high pressure, high volume air which, when heated and expended through the turbine section, pro-

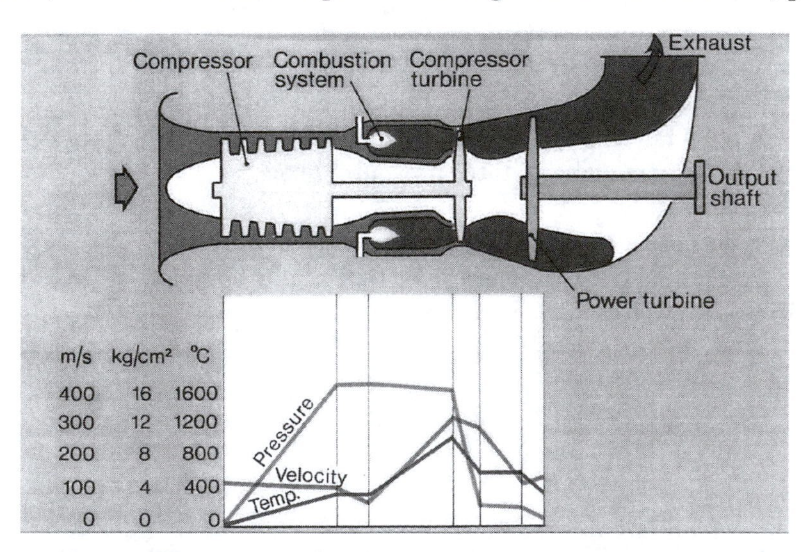

Figure 4-2. Courtesy of Rolls-Royce Industrial & Marine Gas Turbines Limited. Typical pressure, temperature, and velocity profiles relative to gas turbine engine position.

vides the power output required by the process (mechanical drive, generator drive, etc.). Compressor performance is generally shown as pressure ratio plotted against airflow. (Note: it is more accurate to use head instead of pressure ratio, as head takes into account the compressibility, molecular weight, temperature, and the ratio of specific heat of air—and corrected airflow—all at constant speed.) This is discussed in more detail later in this chapter.

Two types of compressors are in use today—they are the axial compressor and the centrifugal compressor. The axial compressor is used primarily at medium and high horsepower applications, while the centrifugal compressor is utilized in low horsepower applications.

Both the axial and centrifugal compressors are limited in their range of operation by what is commonly called *stall* (or *surge*) and *stone wall*. The stall phenomena occurs at certain conditions of airflow, pressure ratio, and speed (rpm), which result in the individual compressor airfoils going into stall similar to that experienced by an airplane wing at a high angle of attack. The stall margin is the area between the steady state operating line and the compressor stall line. Surge or stall will be discussed in detail later in this chapter. Stone wall occurs at high flows and low pressure. While it is difficult to detect it is manifested by increasing air (gas) temperature.

Considering the Axial Compressor

Air flowing over the moving airfoil exerts lift and drag forces approximately perpendicular and parallel to the surface of the airfoil (Figure 4-3). The resultant of these forces can be resolved into two components:

1) The component parallel to the axis of the compressor represents an equal and opposite rearward force on the air—causing an increase in pressure.

2) A component in the plane of rotation represents the torque required to drive the compressor.

From the aerodynamic point of view there are two limiting factors to the successful operation of the compressor. They are the angle of attack of the airfoil and the speed of the airfoil relative to the approaching air (Figure 4-4). If the angle of attack is too steep, the airflow

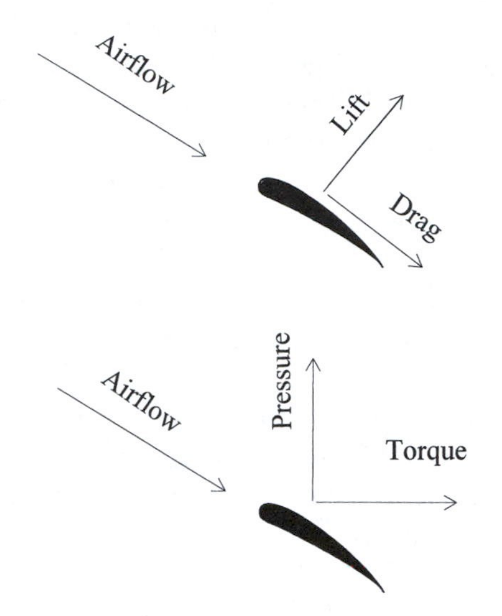

Figure 4-3. Forces acting on blades.

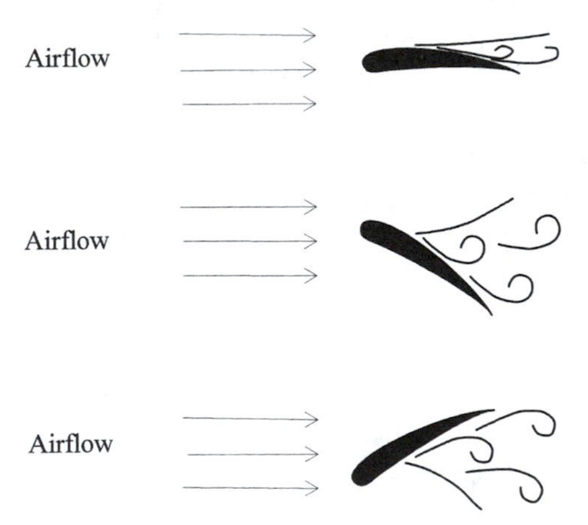

Figure 4-4. Airfoil angle of attack relative to approaching air.

will not follow the convex surface of the airfoil. This will reduce lift and increase drag. If the angle of attack is too shallow, the airflow will separate from the concave surface of the airfoil. This also results in increased drag.

If the speed of the airfoil relative to the air is too high, a shockwave will develop as the air exceeds the speed of sound trying to accelerate as it passes around the airfoil. This shockwave will cause turbulent flow and result in an increase in drag. Depending on the length of the airfoil, this excessive speed could apply only to the tip of the compressor blade. Manufacturers have overcome this, in part, by decreasing the length of the airfoil and increasing the width (or chord).

For single stage operation, the angle of attack depends on the relation of airflow to speed. It can be shown that the velocity relative to the blade is composed of two components: the axial component depends on the flow velocity of the air through the compressor, and the tangential component depends on the speed of rotation of the compressor (Figure 4-5). Therefore, if the flow for a given speed of rotation (rpm) is reduced, the direction of the air approaching each blade is changed so as to increase the angle of attack. This results in more lift and pressure rise until the stall angle of attack is reached.

This effect can be seen on the compressor characteristic curve. The characteristic curve plots pressure against airflow (Figure 4-6). The points on the curve mark the intersection of system resistance, pressure, and airflow. (Note that opening the bleed valve reduces system resistance and moves the compressor operating point away from surge.) The top of each constant speed curve forms the loci for the compressor stall (surge) line.

Therefore, the overall performance of the compressor is depicted on the compressor performance map, which includes a family of constant speed (rpm) lines (Figure 4-7). The efficiency islands are included to show the effects of operating on and off the design point. At the design speed and airflow, the angle of attack relative to the blades is opti-

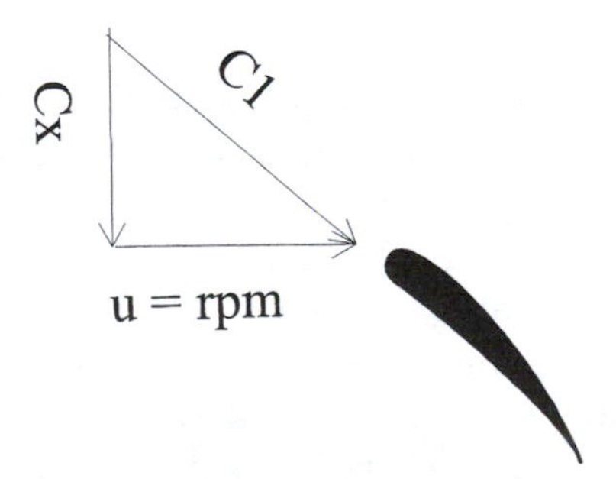

Figure 4-5. Velocity components relative to airfoil.

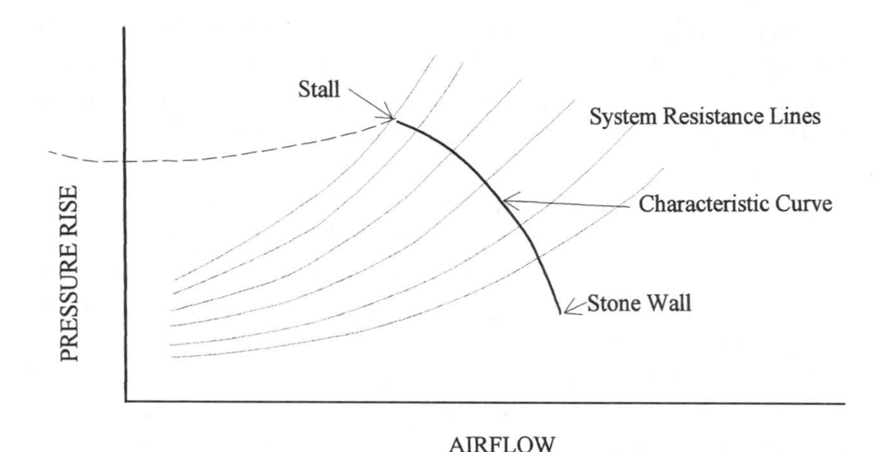

Figure 4-6. Compressor system curve.

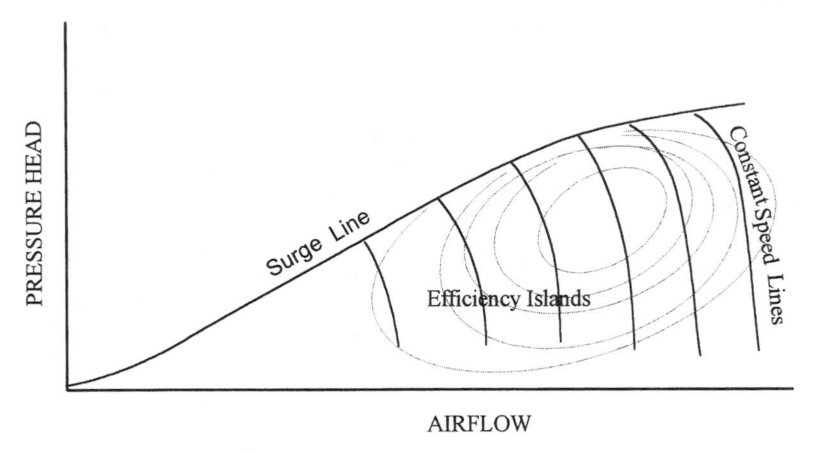

Figure 4-7. Compressor performance curve.

mum and the compressor operates at peek efficiency. If flow is reduced at a constant speed, the angle of attack increases until the compressor airfoil goes into stall.

As flow is increased at a constant speed the compressor characteristic curve approaches an area referred to as "stone wall." Stone wall does not have the dynamic impact that is prevalent with stall, but it is a very inefficient region. Furthermore, operation at or near stone wall will result in over-temperature conditions in the turbine section.

From the mechanical point of view, blade stresses and blade vibration are limiting factors. The airfoil must be designed to handle the varying loads due to centrifugal forces, and the load of compressing air to higher and higher pressure ratios. These are conflicting requirements. Thin, light blade designs result in low centrifugal forces, but are limited in their compression-load carrying ability, while thick, heavy designs have high compression-load carrying capability, but are limited in the centrifugal forces they can withstand. Blade vibration is just as complex. There are three categories of blade vibration: resonance, flutter, and rotating stall. They are explained here.

- *Resonance*—As a cantilever beam, an airfoil has a natural frequency of vibration. A fluctuation in loading on the airfoil at a frequency that coincides with the natural frequency will result in fatigue failure of the airfoil.

- *Flutter*—A self-excited vibration usually initiated by the airfoil approaching stall.

- *Rotating Stall*—As each blade row approaches its stall limit, it does not stall instantly or completely, but rather stalled cells are formed (Figure 4-8). These tend to rotate around the flow annulus at about half the rotor speed. Operation in this region is relatively short and usually proceeds into complete stall. Rotating stall can excite the natural frequency of the blades.

The best way to illustrate airflow through a compressor stage is by constructing velocity triangles (Figure 4-9). Air leaves the stator vanes at an absolute velocity of C_1 and direction θ_1. The velocity of this air relative to the rotating blade is W_1 at the direction β_1. Air leaves the rotating stage with an absolute velocity C_2 and direction θ_2, and a relative velocity W_2 and direction β_2. Air leaving the second stator stage has the same velocity triangle as the air leaving the first stator stage. The projection of the velocities in the axial direction are identified as Cx, and the tangential components are Cu. The flow velocity is represented by the length of the vector. Velocity triangles will differ at the blade hub, mid-span, and tip just as the tangential velocities differ.

Pressure rise across each stage is a function of the air density, ρ, and the change in velocity. From the velocity triangles the pressure rise per stage is determined.

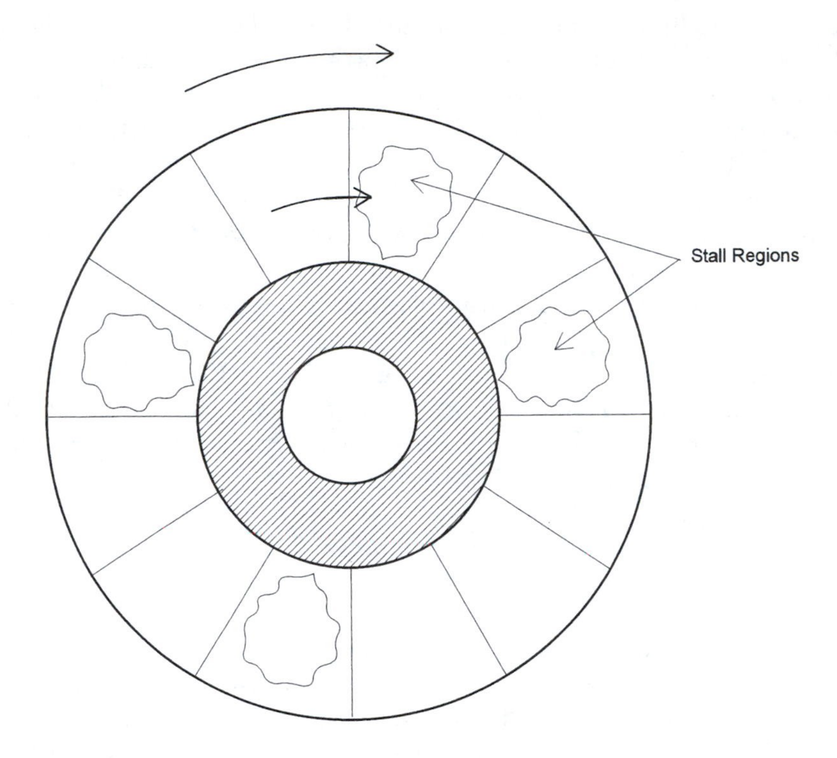

Figure 4-8. Rotating stall.

$$\Delta P = \frac{\rho}{2g_c}\left(W_1^2 - W_2^2\right) + \frac{\rho}{2g_c}\left(C_2^2 - C_3^2\right) \tag{4-19}$$

This expression can be further simplified by combining the differential pressure and density, and referring to feet of head.

$$Head = \frac{\mu}{g_c}\,\Delta C = \frac{\Delta P}{\rho} \tag{4-20}$$

where $\frac{\Delta P}{\rho}$ is the pressure rise across the stage and head is the pressure rise of the stage measured in feet head of the flowing fluid. The standard equation for compressor head is given below.

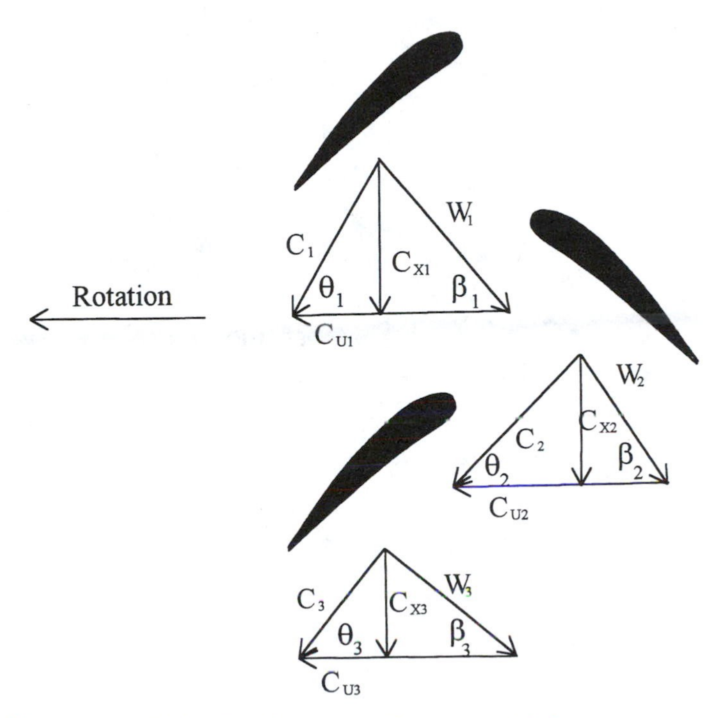

Figure 4-9. Velocity diagrams for an axial flow compressor.

$$Head = \frac{Z_{ave}T_s}{MW}\left(\frac{R_c^{\sigma} - 1}{\sigma}\right) \tag{4-21}$$

where Z_{ave} is the average compressibility factor of air, and MW is the mole weight.

Before proceeding further, we will define the elements of an airfoil (Figure 4-10).

Fixed Blade Row
Moving Blade Row
Camber Line
Camber or Blade Angle $\equiv \beta_2 - \beta_1$
Inlet Blade Angle $\equiv \beta_1$
Exit Blade Angle $\equiv \beta_2$
Inlet Flow Angle $\equiv \beta_1'$
Exit flow angle $\equiv \beta_2'$

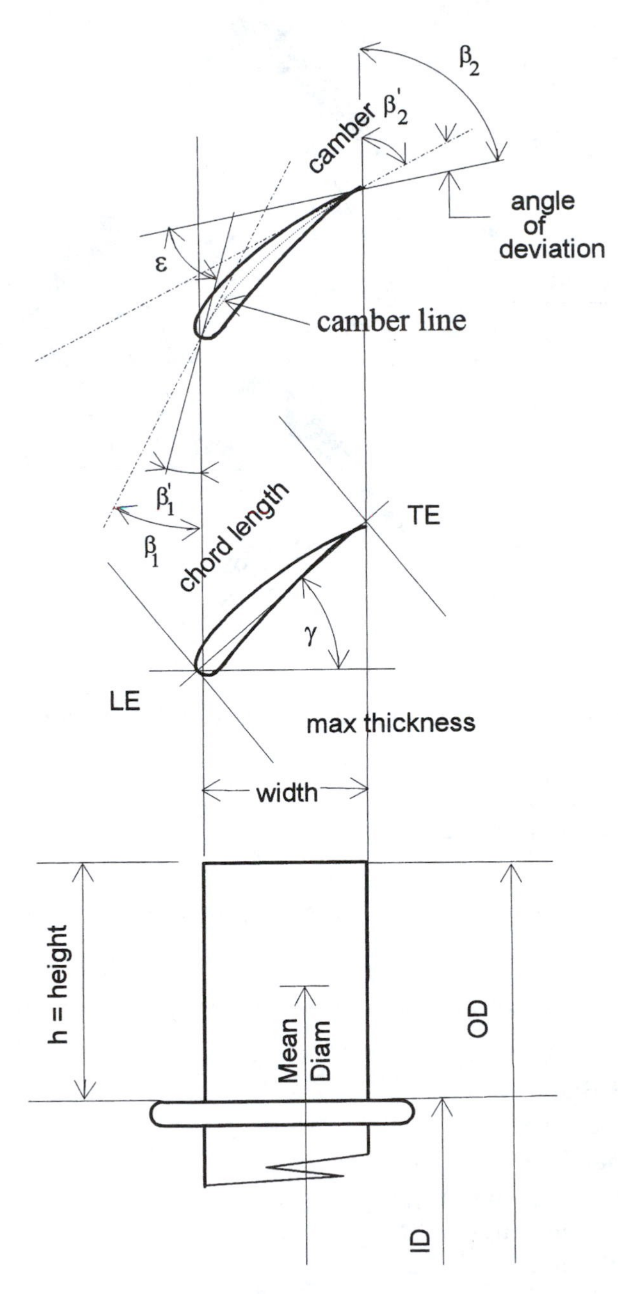

Figure 4-10. Elements of an airfoil.

Angle of Deviation ≡ $\beta_2 - \beta_2'$
Stagger Angle ≡ γ
Pitch ≡ s
Turning Angle ≡ $\varepsilon = \beta_2' - \beta_1'$
Chord Length
Leading Edge
Trailing Edge
Maximum Thickness ≡ t
Width ≡ w
Height ≡ h
Aspect Ratio ≡ ratio of blade height to blade chord
Angle of Attack or angle of incidence ≡ $i = \beta_1 - \beta_1'$

Considering the Centrifugal Compressor

The centrifugal compressor, like the axial compressor, is a dynamic machine that achieves compression by applying inertial forces to the air (acceleration, deceleration, turning) by means of rotating impellers.

The centrifugal compressor is made up of one or more stages, each stage consisting of an impeller and a diffuser. The impeller is the rotating element and the diffuser is the stationary element. The impeller consist of a backing plate or disc with radial vanes attached to the disc from the hub to the outer rim. Impellers may be either open, semi-enclosed, or enclosed design. In the open impeller the radial vanes attach directly to the hub. In this type of design the vanes and hub may be machined from one solid forging, or the vanes can be machined separately and welded to the hub. In the enclosed design, the vanes are sandwiched between two discs. Obviously, the open design has to deal with air leakage between the moving vanes and the non-moving diaphragm, whereas the enclosed design does not have this problem. However, the enclosed design is more difficult and costly to manufacture.

Generally air enters the compressor perpendicular to the axis and turns in the impeller inlet (eye) to flow through the impeller. The flow through the impeller than takes place in one or more planes perpendicular to the axis or shaft of the machine. This is easier to understand when viewing the velocity diagrams for a centrifugal compressor stage. Although the information presented is the same, Figure 4-11 demonstrates two methods of preparing velocity diagrams.

Centrifugal force, applied in this way, is significant in the devel-

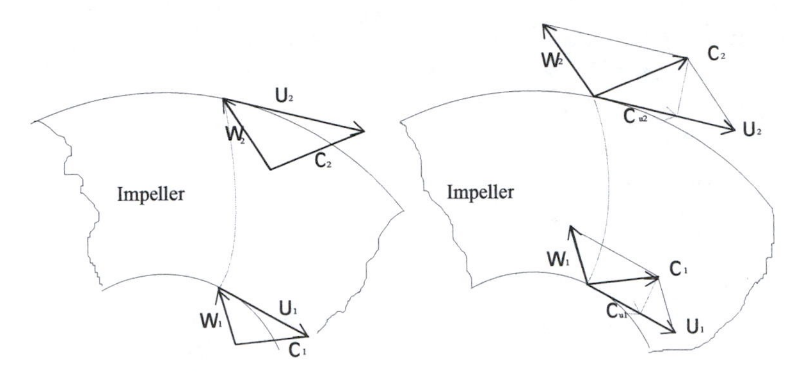

Figure 4-11. Velocity diagrams for a centrifugal compressor.

opment of pressure. Upon exiting the impeller, the air moves into the diffuser (flow decelerator). The deceleration of flow or "diffuser action" causes pressure build-up in the centrifugal compressor. The impeller is the only means of adding energy to the air and all the work on the air is done by these elements. The stationary components, such as guide vanes and diffusers, can only convert velocity energy into pressure energy (and incur losses). Pressure from the impeller eye to the impeller outlet is represented by the following:

$$P_m = \frac{G}{2g_c}\left[\left(C_2^2 - C_1^2\right) + \left(U_2^2 - U_1^2\right) + \left(W_1^2 - W_2^2\right)\right] = \frac{G}{2g_c}\left(C_{2u}U_2 - C_{1u}U_1\right) \tag{4-22}$$

The term $\left(C_2^2 - C_1^2\right)/2g_c$ represents the increase in kinetic energy contributed to the air by the impeller. The absolute velocity C_1 (entering the impeller) increases in magnitude to C_2 (leaving the impeller). The increase in kinetic energy of the air stream in the impeller does not contribute to the pressure increase in the impeller. However, the kinetic energy does convert to a pressure increase in the diffuser section. Depending on impeller design, pressure rise can occur in the impeller in relation to the terms $\left(U_2^2 - U_1^2\right)/2g_c$ and $\left(W_1^2 - W_2^2\right)/2g_c$. The term $\left(U_2^2 - U_1^2\right)/2g_c$ measures the pressure rise associated with the radial/centrifugal field, and the term $\left(W_1^2 - W_2^2\right)/2g_c$ is associated with the relative velocity of the air entering and exiting the impeller. The ideal head is defined by the following relationship:

$$Head_{ideal} = P_m = \frac{1}{g_c}\left(C_{u2}U_2 - C_{u1}U_1\right) \quad (4\text{-}23)$$

$$\text{Flow Coefficient } \phi \text{ is defined as} = \frac{Q}{AU} \quad (4\text{-}24)$$

where Q = Cubic Feet per Second (CFS)
A = Ft^2
U = Feet per sec

$$\text{rewriting this equation} \quad \phi = \frac{700Q}{D^3N} \quad (4\text{-}25)$$

where D is diameter, and N is rotor speed.

The flow coefficients are used in designing and sizing compressors and in estimating head and flow changes resulting as a function of tip speed (independent of compressor size or rpm). Considering a constant geometry compressor, operating at a constant rpm, tip speed is also constant. Therefore, any changes in either coefficient will be directly related to changes in head or flow. Changes in head or flow under these conditions can result only from dirty or damaged compressor airfoils (or impellers). This is one of the diagnostic tools used in defining machine health.

The thermodynamic laws underlying the compression of gases are the same for all compressors—axial and centrifugal. However, each type exhibits different operating characteristics (Figure 4-12). Specifically, the constant speed characteristic curve for compressor pressure ratio relative to flow is flatter for centrifugal compressors than for axial compressors. Therefore, when the flow volume is decreased (from the design point) in a centrifugal compressor, a greater reduction in flow is possible before the surge line is reached. Also, the centrifugal compressor is stable over a greater flow range than the axial compressor, and compressor efficiency changes are smaller at off design points.

For the same compressor radius and rotational speed the pressure rise per stage is less in an axial compressor than in a centrifugal compressor. But, when operating within their normal design range, the efficiency of an axial compressor is greater than a centrifugal compressor.

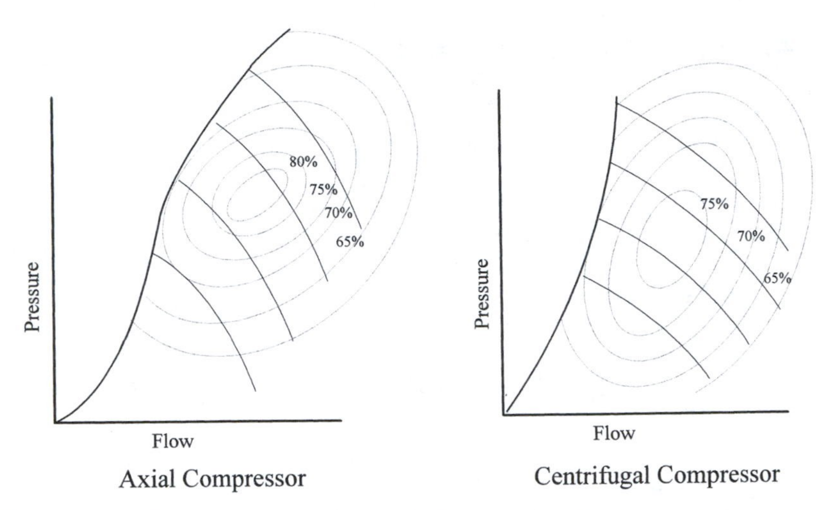

Figure 4-12. Comparing characteristic curves of axial and centrifugal compressors.

Variations in Compressor Design

Variable Guide Vane Compressor

The variable guide vanes (VGVs) are used to optimize compressor performance by varying the geometry of the compressor. This changes the compressor characteristic curve and the shape and location of the surge line. In this way the compressor envelope can cover a much wider range of pressure and flow. In centrifugal compressors the variable guide vane (that is, the variable inlet vane) changes the angle of the gas flow into the eye of the first impeller. In the axial compressor up to half of the axial compressor stages can or may incorporate variable guide vanes. In this way the angle of attack of the air leaving each rotating stage is optimized for the rotor speed and air flow.

Variable guide vane technology, as it pertains to flight (aircraft) engines & stationary (land or marine based) gas turbine, enables the designers to apply the best design features in the compressor for maximum pressure ratio & flow. By applying VGV techniques the designer can change the compressor characteristics at starting, low-to-intermediate and maximum flow conditions. Thereby maintaining surge margin throughout the operating range. Thus creating the best of all worlds. The compressor map in Figure 4-13 demonstrates how the surge line changes with changes in vane angle.

Variable guide vanes have an effect on axial and centrifugal compressors similar to speed: the characteristic VGV curves of the axial compressor being steeper than the characteristic VGV curves of the centrifugal compressor. The steeper the characteristic curve the easier it is to control against surge whereas the flatter characteristic curve the easier it is to maintain constant pressure control (or pressure ratio control).

Compressor Operation

Previously in this chapter compressor maps were depicted using pressure rise or pressure head and flow or air flow (Figures 4-6, 4-7 & 4-12). Attempts to control a compressor are often compromised by varying conditions, such as air temperature, air pressure, molecular weight and specific heat ratio.

A discharge pressure versus flow map of the compressor characteristic curve would consist of a family of curves for different suction pressures, temperatures, molecular weights and specific heat ratios. Since the molecular weight and specific heat ratios of air are fairly con-

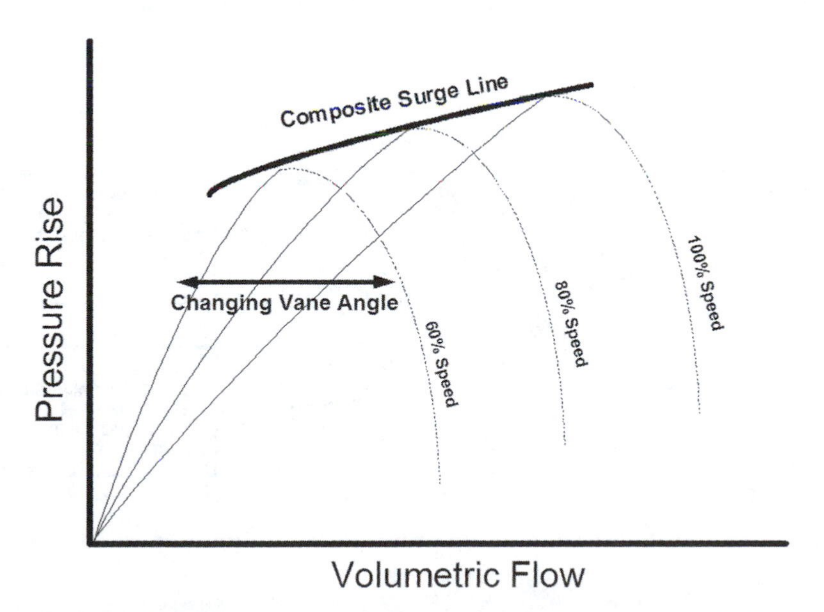

Figure 4-13. Composite surge limit line resulting from variations in vane position

stant in the normal range of atmospheric conditions only pressure and temperature need be considered. However this is still a daunting task as each of these conditions defines a different surge point. A typical sketch (Figure 4-14) demonstrates the problem. This situation is easily resolved by considering head and volumetric flow.

Therefore, for a given speed and compressor geometry, a performance map of head vs. volumetric flow more accurately represents the compressor as it has one surge limit point, regardless of all inlet conditions (that is suction pressure, suction temperature, molecular weight and specific heat ratio).

Head is defined in equation 4-21. In this approach the surge point becomes a line at the extreme left of the compressor curves made up of the loci of points of each curve. Throughput control (or capacity control) is achieved by controlling the energy input to the compressor (rotor speed and guide vane position). In this way the compressor capacity is matched to the load.

At any given condition the compressor pressure and throughput are a function of speed and, where applicable, variable guide vane position. Therefore accelerating the gas turbine from start to base load power is managed by regulating speed and the VGV position. Once the

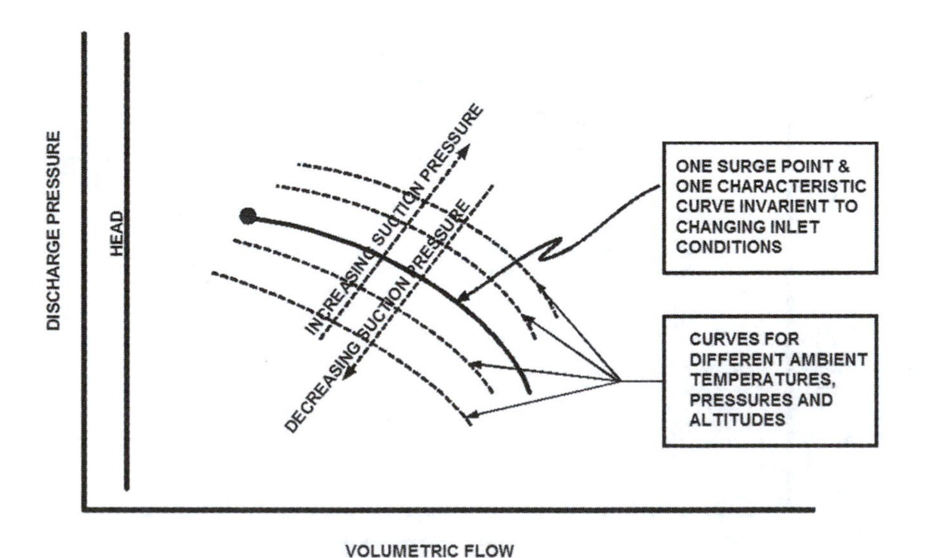

Figure 4-14. Changes in the characteristic curves with changes in ambient temperature and pressure

gas turbine is at or near base load power the VGVs are at their full open position and no longer contribute to engine control (unless compressor deterioration or contamination resulted in the surge line moving close or into the operating line field). For constant speed machines (i.e. generator drives) the VGVs further contribute to controlling emissions.

Effect of Operating Conditions

Operation of a gas turbine is achieved by regulating fuel flow and variable vane position to affect changes in rotor speed, compressor throughput, pressure ratio and emissions while biasing flow and VGV as a function of ambient pressure & temperature.

As design compression ratios are increased the surge line rotates clockwise with increases in volumetric flow. (Figures 4-15a, 4-15b, and 4-15c).

Speed affects axial and centrifugal compressors differently; the characteristic speed curves of the axial compressor being steeper than the characteristic speed curves of the centrifugal compressor. The steeper the characteristic curve the easier it is to control against surge. Whereas the flatter the characteristic curves the easier it is to maintain constant pressure control (or pressure ratio control).

As inlet conditions change the operating point moves along the system resistance curve (assume that the system resistance curve is constant) as follows (see Figure 4-16):

- The operating point moves up along the system resistance curve to Point B as
  - Ambient pressure increases
  - Ambient temperature decreases, and the
  - Surge line moves to the right and closer to the operating point

- The operating point moves down along the system resistance curve to Point A as
  - Ambient pressure decreases
  - Ambient temperature increases, and the
  - Surge line moves to the left and further away from the operating point.

Whereas molecular weight and gas compressibility are not major

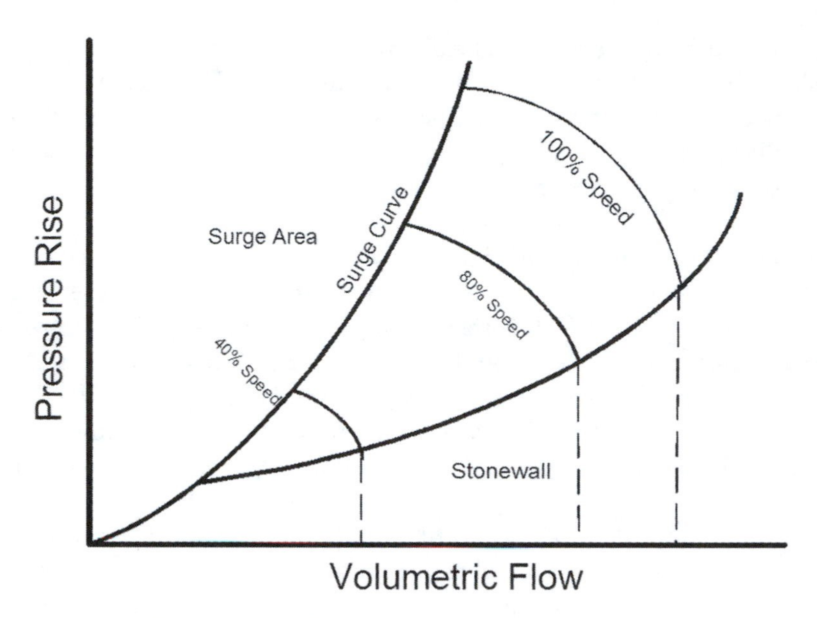

Figure 4-15a. Single-stage compressor map

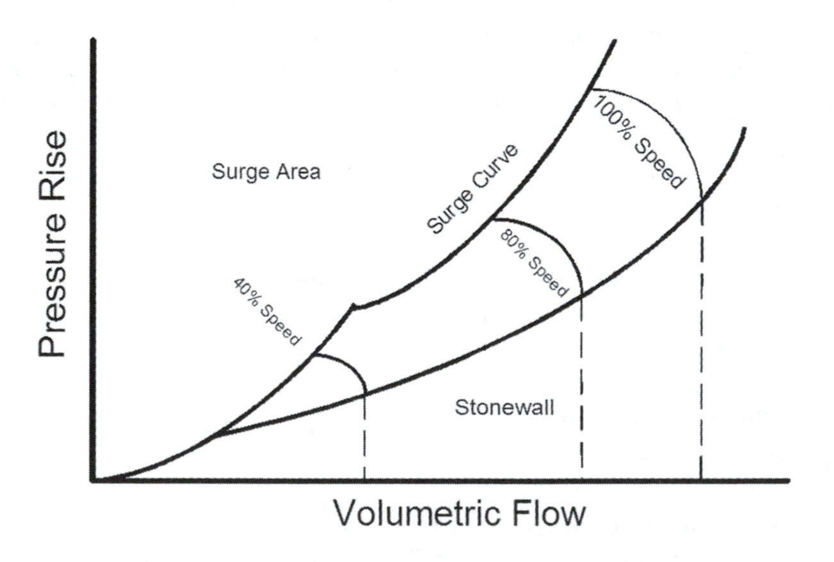

Figure 4-15b. Two-stage compressor map

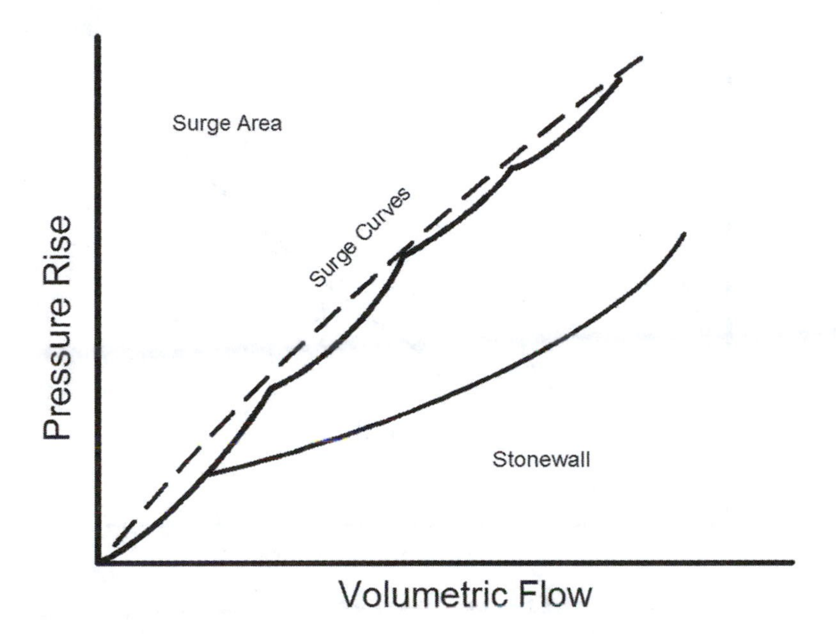

Figure 4-15c. Four-stage compressor map

considerations for air breathing gas turbines, inlet pressure and temperature do have a significant impact on controlling compressor operation. However, their impact can be minimized. This is accomplished by recognizing the similarities in the head and flow relationships.

The equations for head and flow[2] can be reduced as follows:

$$H_p = \left(\frac{Z_{ave} * T_s}{MW}\right) * \left(\frac{R_c^{\sigma} - 1}{\sigma}\right) \tag{4-26}$$

$$Q^2 = \left(\frac{Z_s * T_s}{MW}\right) * \left(\frac{\Delta P_{os}}{P_s}\right) \tag{4-27}$$

$$\text{since} \left(\frac{Z_s * T_s}{MW}\right) \approx \left(\frac{Z_{ave} * T_s}{MW}\right) \tag{4-28}$$

$$\text{then } H_{p'} \approx \left(\frac{R_c^{\sigma} - 1}{\sigma}\right) \tag{4-29}$$

$$Q^{2'}, \approx \left(\frac{\Delta P_{os}}{P_s}\right) \tag{4-30}$$

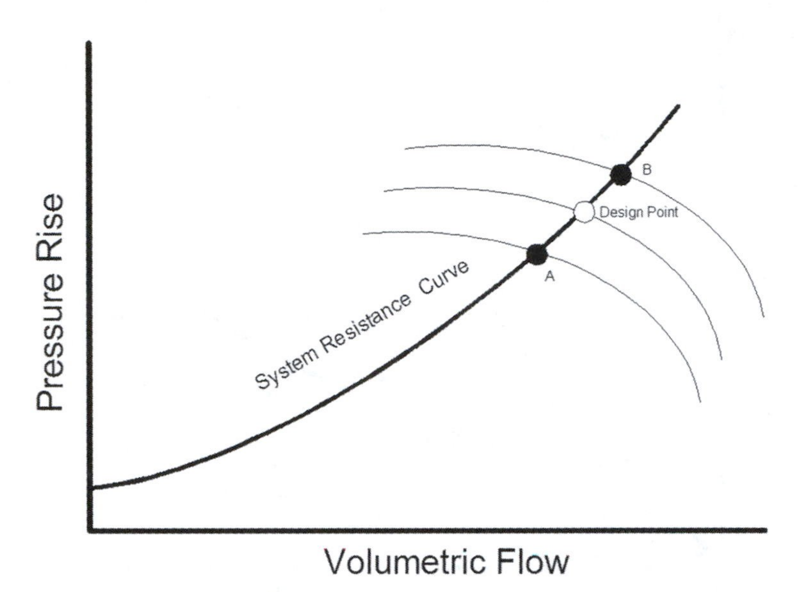

Figure 4-16. Effects of changing inlet conditions

Where

$H_p \equiv$ Polytropic Head, Ft

$Z_{ave} \equiv$ Compressibility Factor Average

$Z_s \equiv$ Compressibility Factor Suction Conditions

$T_s \equiv$ Suction Temperature, ^{0}R

$MW \equiv$ Molecular Weight

$Rc \equiv$ Compression Ratio

$\sigma, \equiv \left(\frac{k-1}{k\eta_p}\right)$; k ≡ Ratio of Specific Heats and η_p ≡ Polytropic Efficiency

$\Delta P_{os} \equiv$ Differential Pressure Across Flow Orifice in Suction Line

$P_s \equiv$Suction Pressure—psia

The reduced coordinates define a performance map (Figure 4-17) which does not vary with varying inlet conditions; Has one surge limit point for a given rotational speed and compressor geometry and permits or facilitates calculation of the operating point without obtaining molecular weight and compressibility measurements.

For a given rotational speed and compressor geometry, the operating point can be defined by a line from the origin to the operating

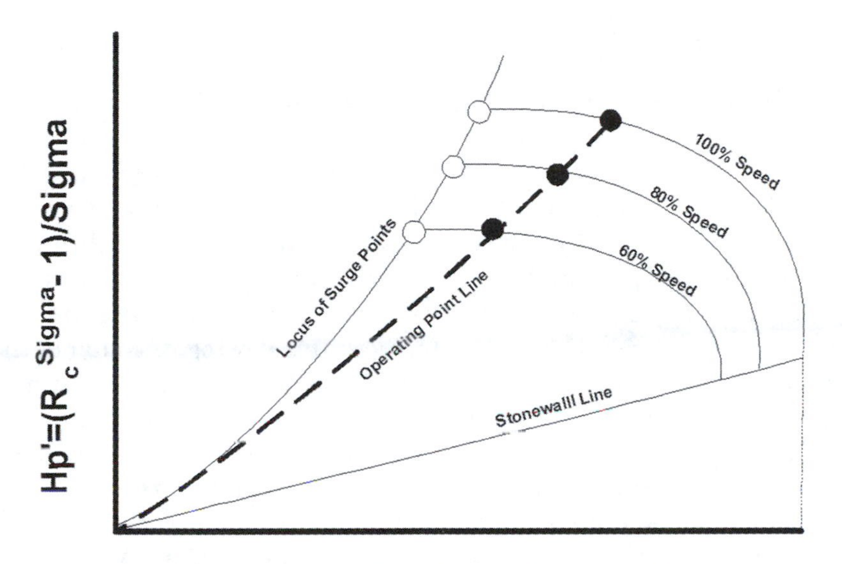

Figure 4-17. Simplified compressor map using pressures only

point. This line can be defined by its slope = Hp/Q_s^2. Each coordinate can be reduced by a common factor without changing the slope.

Thus the operating point can be precisely defined by: $(H_p/A)/(Q_s^2/A)$ or $Hp'/Q_s^{2'}$

Compressor Safety Limits

A. Performance limits (usually overspeed) are not uncommon.

B. Surge is a serious threat to all turbocompressors. Virtually every compressor is outfitted with some sort of antisurge protection (Bleed Valves, VGVs). The consequences are too severe to ignore.

1. If the surge limit is crossed, there will be severe high-speed oscillations accompanied by vibration and rising gas temperatures.
2. Surge is a high speed phenomenon—the drop from full flow to reverse flow can occur in 0.04 seconds.
3. In addition to the surge limit several other performance limits (pressure, speed, load, choke) have to be observed as well.

C. Usually choke will not be a problem.

BURNER SECTION

The burner section is made up of the diffuser duct, the combustor, fuel nozzle and the transition duct.

The velocity of the air leaving the compressor is decreased before it enters the combustor in order to reduce the burner pressure loss and the air velocity in the burner. According to Bernoulli's Law a decrease in velocity, and its resultant increase in static pressure, is achieved in the diffuser. [Bernoulli's Law states that at any point in a passageway through which a gas (or liquid) is flowing, the sum of the pressure energy, potential energy, and kinetic energy is a constant.] As the (subsonic) velocity of the air decreases with the expanding shape of the divergent duct, its static pressure increases, although the total pressure remains the same. (Note that total pressure is the sum of the static pressure and the velocity pressure; where velocity pressure is the pressure created by the movement of the air.)

The fraction of the velocity head that is converted to static pressure (diffuser efficiency) is a function of the area ratio and diffuser angle. Diffuser efficiency is defined as

$$\eta_D = \frac{1-\left(\dfrac{P_o}{P_i}\right)^{\sigma}}{1-\dfrac{T_o}{T_i}} \tag{4-31}$$

where subscripts i and o refer to the pressure and temperature entering and exiting the diffuser, respectively.[1] It must be noted that diffuser effectiveness is the designers problem and for most applications it can be considered part of the combustor efficiency.

The fuel nozzles function is to introduce the fuel into the burner in a form suitable for rapid and complete mixing with air. There are several types of fuel nozzles: gaseous, liquid, dual gas/liquid, and dual liquid/liquid. The liquid fuel nozzle must atomize the fuel into a fine, uniform spray. Furthermore, it must maintain this spray pattern throughout the range of operation from starting to maximum power output. To accomplish this, dual pressure features have been designed into the fuel nozzles. These are commonly referred to as duplex, dual orifice, multi-jet, and pintle fuel nozzles. The gaseous fuel nozzles do not share this problem, as the gas is already in a vapor form.

The combustor itself may be either a series of combustor cans in an annular configuration, an annual combustor, or a single "stand alone" design. Combustors typically have primary air holes at the dome for mixing with the fuel and providing the critical air ratio for ignition, and secondary air holes downstream for additional cooling (Figure 3-12). Combustor design continues to be an art and, although computer modeling of the air and fuel flow within the combustor is progressing, designs are developed and improved by trial and error testing.

Properties of a good burner are high combustion efficiency, stable combustion, low NO_x formation, freedom from blowout, uniform or controlled discharge temperature, low pressure loss, easy starting, long life, and (for liquid fuel operation) minimum carbon accumulation. Combustors must be able to withstand various conditions; namely, a wide range of airflow, fuel flow, and discharge temperature, rapid acceleration and deceleration, and variation in fuel properties.

Combustor efficiency[2] is defined as

$$\eta_B = \frac{W_a c_p (T_o - T_i)}{W_f Q_r} \qquad (4\text{-}32)$$

The pattern or profile of the gas temperature leaving the combustor has a direct impact on the operation and life of the turbine. Therefore, the design of the combustor and the transition duct are critical. As will be discussed later, the condition of the fuel nozzles, burner, and transition duct can have a deleterious effect on the temperature profile, and the turbine gas path components.

TURBINE

The turbine extracts kinetic energy from the expanding gases that flow from the combustion chamber, converting this energy into shaft horsepower to drive the compressor, the output turbine, and select accessories.

The axial-flow turbine is made up of stationary nozzles (vanes or diaphragms) and rotating blades (buckets) attached to a turbine wheel (disc). Turbines are divided into three types: "impulse," "reaction," and a combination of the two designs called "impulse-reaction." The energy drop to each stage is a function of the nozzle area and airfoil configuration. Turbine nozzle area is a critical part of the design: too small and the nozzles will have a tendency to "choke" under maximum flow conditions, too large and the turbine will not operate at its best efficiency. It is important to note that approximately 3/4 - 2/3 of the turbine work drives the compressor leaving approximately 1/4-1/3 for shaft horsepower (or thrust for the jet engine).

Impulse

In the impulse type turbine there is no net change in pressure between rotor inlet and rotor exit. Therefore, the blades Relative Discharge Velocity will be the same as its Relative Inlet Velocity. The nozzle guide vanes are shaped to form passages, which increase the velocity and reduce the pressure of the escaping gases.

Reaction

In the reaction turbine the nozzle guide vanes only alter the direction of flow. The decrease in pressure and increase in velocity of the gas is accomplished by the convergent shape of the passage between the rotor blades.

The differences between the impulse and reaction turbine may be depicted visually with the help of the velocity triangles (Figure 4-18). In the impulse turbine $W_1 = W_2$. In the reaction turbine $W_2 = C_1$, and $W_1 = C_2$.

There are two types of impulse turbines: a velocity compounded impulse turbine and a pressure compounded impulse turbine. The velocity compounded impulse turbine is often referred to as the Curtis Turbine and the pressure compounded impulse turbine is referred to as the Rateau Turbine.

Figure 4-19 summarizes, graphically, the pressure and velocity through the various types of turbines. Note that in the impulse turbine no pressure drop or expansion occurs across the moving rows. While in the reaction turbine the fixed nozzles perform the same function as in the impulse turbine.

Turbines may be either single- or multiple-stage. When the turbine has more than one stage, stationary vanes are located upstream of each rotor wheel. Therefore, each set of stationary vanes forms a

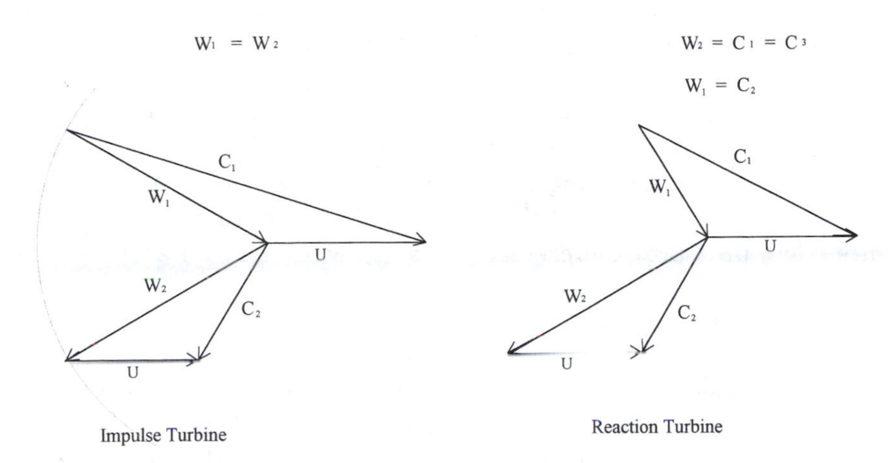

Figure 4-18. Comparison of impulse and reaction turbines.

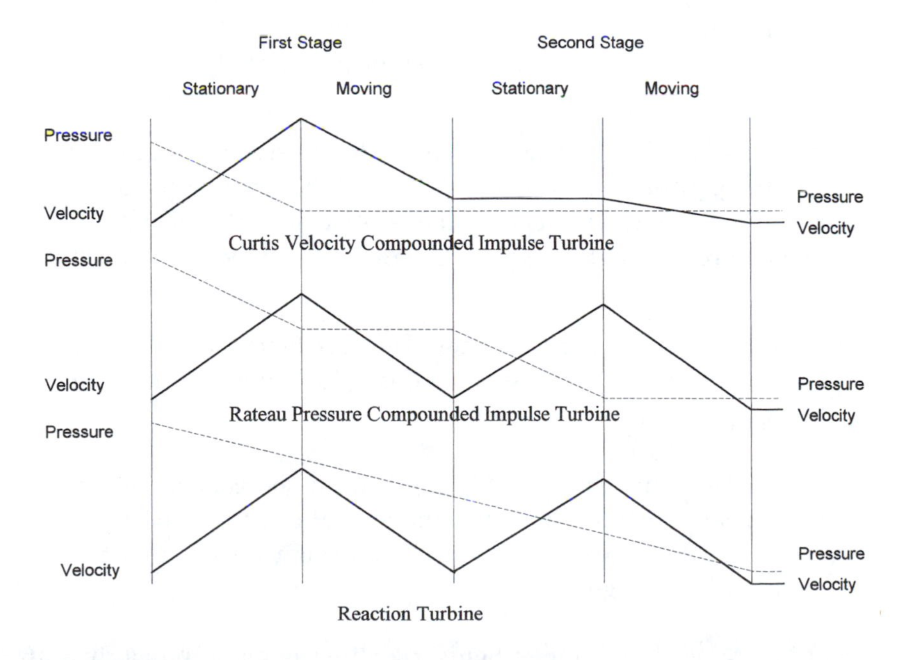

Figure 4-19. Inpulse and reaction turbine pressure and velocity profile.

nozzle vane assembly for the turbine wheel that follows. The rotor wheels may or may not operate independently of each other, depending upon the type of engine and the power requirements of the turbine.

Some gas turbines, based on their design objectives, incorporate two compressors and two turbines—with a shaft-in-a-shaft design. In this case the first (high pressure) turbine drives the last (high pressure) compressor, and the last (low pressure) turbine drives the first (low pressure) compressor. This is referred to as a split-shaft machine.

The turbine wheel, or turbine blade and disc assembly, consists of the turbine blades and the turbine disc. The blades are attached to the disc using a "fir-tree" design. This "fir-tree" configuration allows room for expansion (due to thermal growth) while still retaining the blades against centrifugal force. While almost all blades employ the fir-tree root design, the blades themselves may be solid or hollow (to provide for blade cooling), and with or without tip shrouds. Some blades, primarily in the lower pressure stages where the blade length can be very long, also include a damping wire through sets of blades. Each of these designs serves a specific purpose but they can be generalized as follows:

Hollow-air cooled—blades are used in the high pressure stages where gas temperatures exceed the limits of the specific blade material. Figure 4-20 shows the effects of cooling the airfoils. Solid blades are used where gas temperatures are below the critical limits of the blade material.

Tip Shrouds—form a band around the perimeter of the turbine blades, which serve to seal against tip leakage, and dampen blade vibration.

Squealer Cut—a change in blade length from the convex surface to the concave surface, thus allowing a smaller portion of the blade tip to rub into the tip shroud. This design also serves to seal against tip leakage.

Lacing Wire—used to dampen blade vibration is usually installed at or near the mid-point of the airfoil (Figure 3-7).

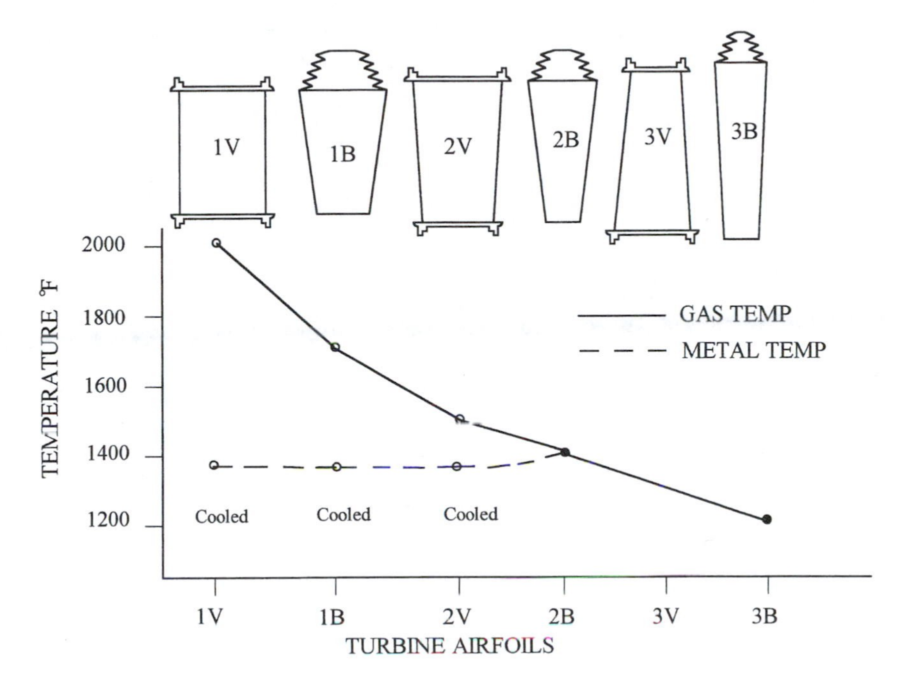

Figure 4-20. Turbine temperature profile.

Turbine blades and vanes tend to change pitch with continued use, straightening toward low pitch. Blades also undergo distortion and thinning and lengthening of the airfoil. This thinning and lengthening of the blade is known as "creep." This condition is cumulative. The rate of creep is determined by the load imposed on the turbine and the strength of the blade, which changes as a function of the temperature within the turbine. Creep also affects the stator vane. However, since the vane is held in place by the inner and outer platforms, the distortion takes the form of bowing. This is almost always accompanied by cracking of the airfoil.

References

1. *Gas Turbine Analysis and Practice*, Jennings and Rogers, McGraw-Hill, 1953.
2. *Gas Turbine Engine Parameter Interrelationships*, Louis Urban, Hamilton Standard, 1969.

Chapter 5

Gas Turbine Controls

The gas turbine is a highly responsive, high speed piece of machinery. In an aircraft application, the gas turbine can accelerate from idle to maximum take-off power in less than 60 seconds. In industrial gas turbines, the acceleration rate is limited by the mass moment of inertia of the driven equipment. This responsiveness does not come without a downside. Without a proper control system: the compressor can go into surge in less than 50 milliseconds; the turbine can exceed safe temperatures in less than a quarter of a second; and the power turbine can go into overspeed in less than two seconds. Furthermore, changes in ambient temperature and ambient pressure, deviations that may not even be noticed, can adversely affect the operation of the gas turbine.

As discussed in Chapters 2 and 3, the gas turbine can take many different forms (single shaft, dual shaft, hot end drive, cold end drive) depending on the application (turbojet, turboprop, generator drive, process compressor drive, process pump drive, etc.). Controlling the gas turbine in each of these different configurations and applications requires the interaction of several complex functions. Some of the complexity can be simplified by considering the gas turbine as a gas generator and a power-extraction-turbine, where the gas generator consists of the compressor, combustor, and compressor-turbine. The compressor-turbine is that part of the gas generator developing the shaft horsepower to drive the compressor; and the power-extraction-turbine is that part of the gas turbine developing the horsepower to drive the external load. The energy developed in the combustor, by burning fuel under pressure, is gas horsepower (GHP). On turbojets, the gas horsepower that is not used by the compressor-turbine to drive the compressor is converted to thrust. On turboprops, mechanical drive, and generator drive gas turbines this gas horsepower (designated GHP_{PT}) is used by the power-extraction-turbine to drive the external load. Note that this categorization applies whether the power extraction turbine is a free power turbine or integrally connected to the compressor-turbine shaft. Therefore, whether the gas horsepower is expanded through the remaining turbine stages (as on a single shaft machine), or through a free power turbine (as on a split shaft machine) this additional energy is converted into shaft horsepower (SHP). This can be represented as follows:

$$GHP_{PT} = \eta_{PT}\, SHP.$$

where η_{PT} = Efficiency of the power extraction turbine.

The control varies SHP of the gas generator turbine and power-extraction turbine by varying gas generator speed, which the control accomplishes by varying fuel flow (Figure 5-1). Control of the gas turbine in providing the shaft horsepower required by the operation or process is accomplished using parameters such as fuel flow, compressor inlet pressure, compressor discharge pressure, shaft speed, compressor inlet temperatures and turbine inlet or exhaust temperatures. At a constant gas generator speed, as ambient temperature decreases, turbine inlet temperature will decrease slightly and GHP will increase significantly (Figure 5-2). This increase in gas horsepower results from the increase in compressor pressure ratio and aerodynamic loading. Therefore, the control must protect the gas turbine on cold days from overloading the compressor airfoils and over-pressurizing the compressor cases. To get all the power possible on hot days it is necessary to control turbine inlet temperature to constant values, and allow gas generator speed to vary. The control senses ambient inlet temperature, compressor discharge pressure (also referred to as burner pressure-P_b), and gas generator speed. These three variables affect the amount of power that the engine will produce (Figure 5-3). Also, sensing ambient inlet temperature helps insure that engine internal pressures are not exceeded, and sensing turbine inlet temperature insures that the maximum allowable turbine temperatures are not exceeded. Sensing gas generator speed enables the control to accelerate through any critical speed points (gas turbines are typically flexible shaft machines and, therefore, have a low critical speed).

There are as many variations in controls as there are control manufacturers, gas turbines, and gas turbine applications. Controls can be divided into several groups: hydromechanical (pneumatic or hydraulic), electrical (hard wired relay logic), and computer (programmable logic controller or microprocessor)[1]. The hydromechanical type controls consist of cams, servos, speed (fly-ball) governors, sleeve and pilot valves, metering valves, temperature sensing bellows, etc. (Figure 5-4).

Electrical type controls consist of electrical amplifiers, relays, switches, solenoids, timers, tachometers, converters, thermocouples, etc. Computer controls incorporate many of the electrical functions such as amplifiers, relays, switches, and timers within the central processing unit (CPU). These functions are easily programmed in the CPU and they can be just as easily reprogrammed. This flexibility in modifying all or part of a program is especially useful to the user/operator in the field. Analog signals such as temperature, pressure, vibration, and speed are converted to digital signals before they are processed by the CPU. Also the output signals to the fuel valve, variable geometry actuator, bleed valve, anti-icing valve, etc. must be converted from digital to analog.

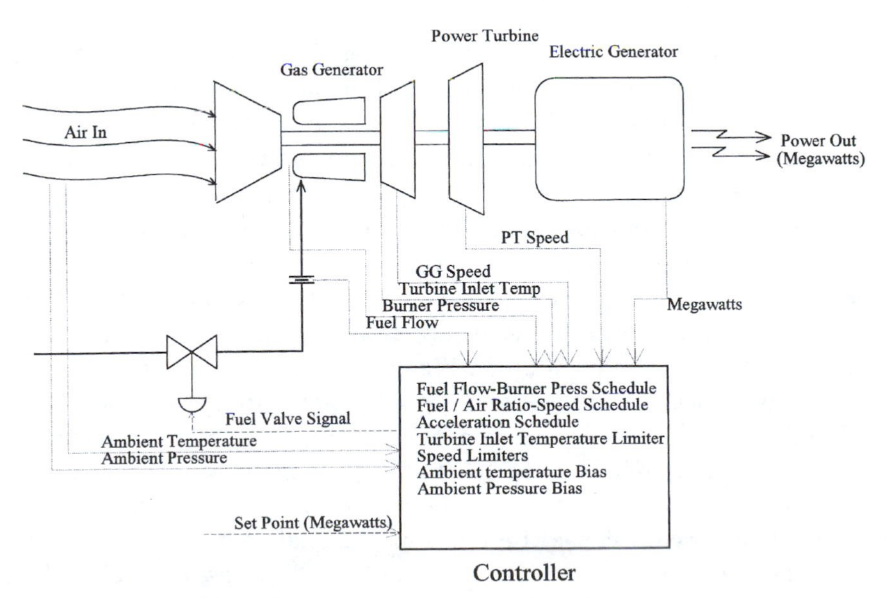

Figure 5-1. Simplified gas turbine-generator control system.

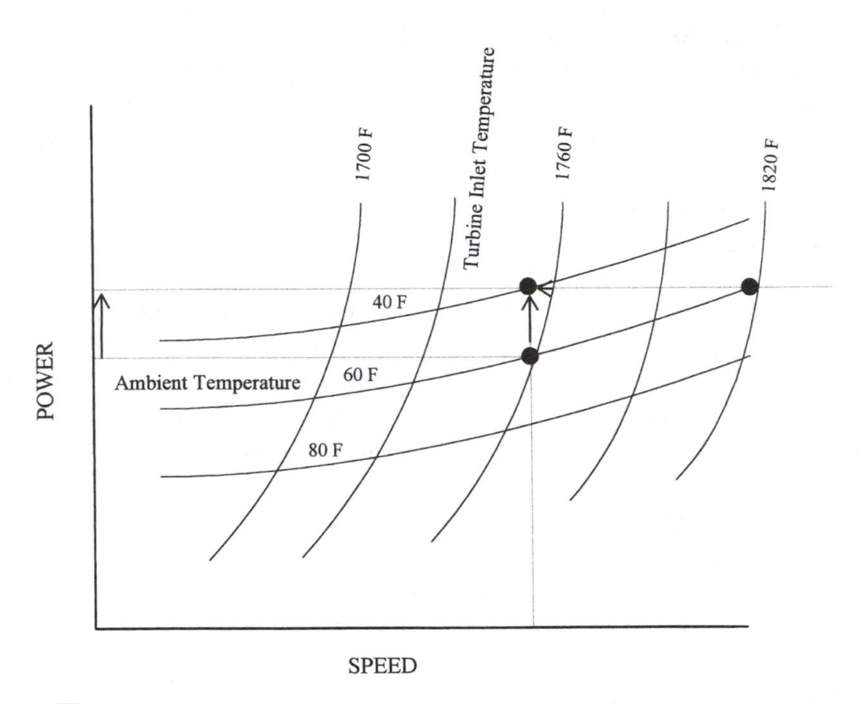

Figure 5-2. Temperature-power-speed interrelationships.

From the turn of the century through the late 1970s, control systems operated only in real time with no ability to store or retrieve data. Hydromechanical controls had to be calibrated frequently (weekly in some applications) and were subject to contamination and deterioration due to wear. A requirement for multiple outputs (i.e., fuel flow control and compressor bleed-air flow-control) required completely independent control loops. Coordinating the output of multiple loops, through cascade control, was a difficult task and often resulted in a compromise between accuracy and response time. In addition, many of the tasks had to be performed manually. For example, station valves, prelube pumps, and cooling water pumps were manually placed into the running position prior to starting the gas generator. Also protection devices were limited. The margin between temperature control set points and safe operating turbine temperatures was necessarily large because the hydromechanical controls could not react quick enough to limit high turbine temperatures, or to shutdown the gas generator, before damage would occur.

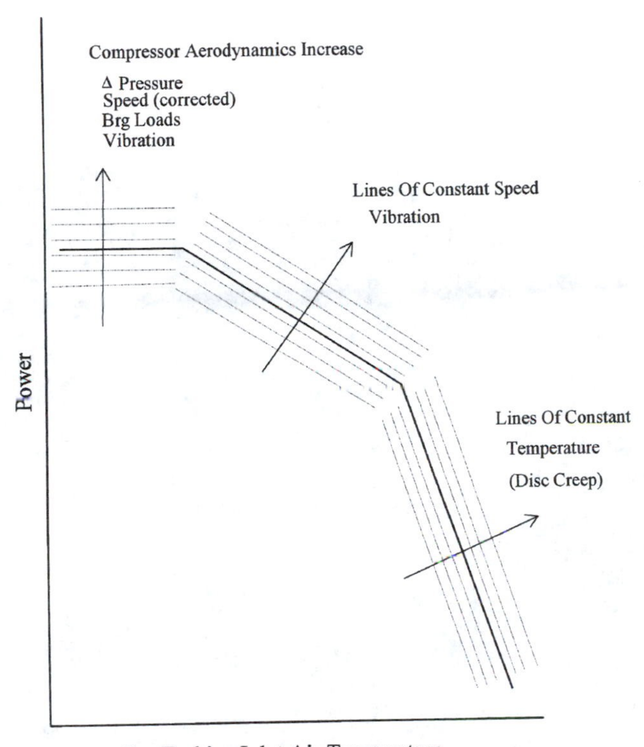

Figure 5-3. Shaft horsepower vs. inlet temperature.

Controls for a typical gas turbine generator or mechanical drive application in the early 1970s were electric controls and consisted of Station Control, a Process Control, and a Turbine Control[2]. In this type of control system all control functions (Start, Stop, Load, Unload, Speed, and Temperature) were generated, biased, and computed electrically. Output amplifiers were used to drive servo valves, employing high pressure hydraulics, to operate hydraulic actuators. These actuators were sometimes fitted with position sensors to provide electronic feedback.

The advent of programmable logic controllers and microprocessors (Figure 5-5) in the late 1970s eliminated these independent control loops, and facilitated multi-function control. Regardless of type, all controls must accomplish the same basic functions.

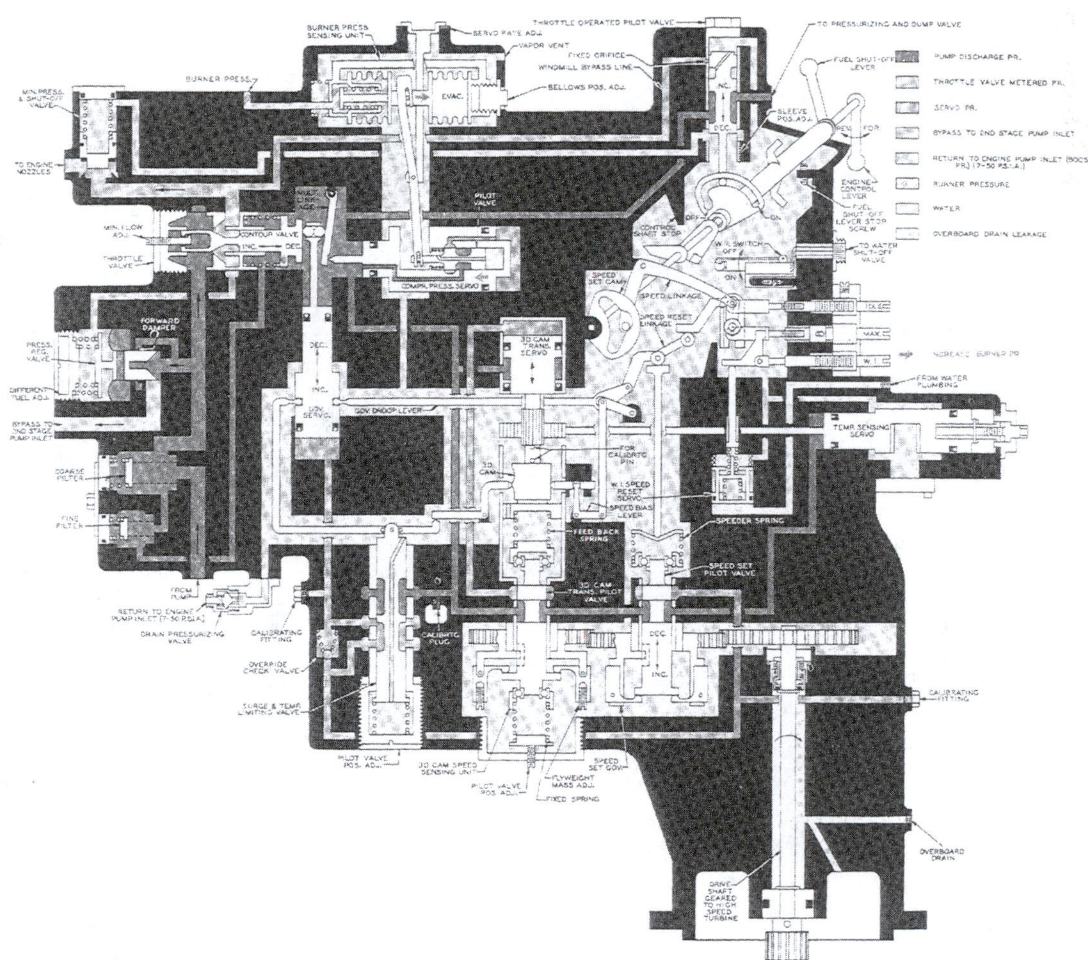

Figure 5-4. Courtesy of United Technologies Corporation, Pratt & Whitney Aircraft. Cutaway drawing of a typical hydromechanical control. This sketch shows the basic fuel-control functions. This type of control was state-of-the-art in the 1960s.

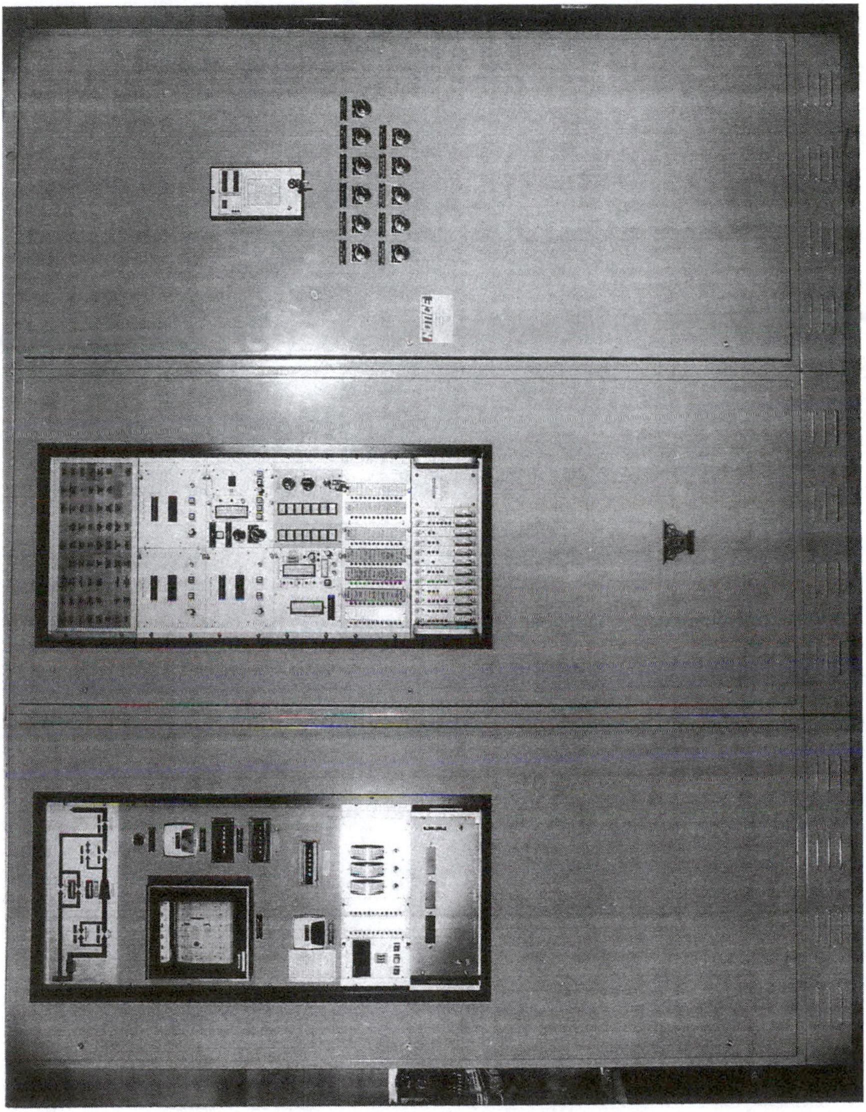

Figure 5-5. Courtesy of Cooper Cameron Corporation (formerly Cooper Bessemer Company). Solid state electrical control panels were a vast improvement over the mechanical, pneumatic, and hydraulic controls that they replaced. These controls came into wide use in the 1970s and 1980s. While still in use today, they are being replaced by Program Logic Controllers (PLCs) and microcomputers.

Table 5-A. Control Functions (Mechanical Type)

FUNCTION	DESCRIPTION
Sequencer	•Start/Stop Gas Turbine Accessories •Load/Unload Gas Turbine
Speed Control	•Constant Time-Based Schedule
Temperature Control	•Exhaust Control (Average of Thermocouples), •Override Control at Loading & Acceleration
Load Control	•Manual-Hand Set (Using Speed Control Function) •Driven Equipment Feedback (Discharge Pressure, Flow, Megawatts)
GT Compressor Surge Control	•Acceleration Schedule Temperature Biased
Driven Unit Output	•Process Flow •Process Pressure •Megawatts

While control system functions tend to overlap, they can be looked at as controlling three separate areas:

Sequencing Control
Routine Operation Control
Protection Control

Sequencing Control consists of the steps to Start, Load, Unload, and Stop the unit. Typical steps included in a normal start are:

- Check that all permissives are made
- Check that lube oil pumps are running

- Open the process compressor recycle valve (if applicable)
- Actuate the gas turbine starter
- Actuate the gas turbine starter time-out clock
- Energize the ignition system
- Actuate the ignition system time-out clock
- Open the fuel valve to preset point
- Stop the gas turbine starter when starter time-out clock expires
- De-energize the ignition system when the time-out clock expires.

When the start sequence is complete the gas turbine will have reached self-sustaining speed. At this time control can be, and usually is, handed over to the Routine Operation Controller. This controller will maintain stable operation until it receives an input (from the operator, or the process) to load the unit. Prior to initiating loading the unit, control is turned back over to the Sequence Controller to properly position inlet and discharge valves, electrical breakers, etc. On electric generator drives this is the point that the automatic synchronizer is activated to synchronize the unit to the electric grid. When these steps are completed, control is again turned back over to the Routine Operation Controller and the speed control governor, acceleration scheduler, temperature limit controller, and pressure limit controller all come into play. Controlling the gas turbine during steady state operation is a major function of the control system. However, while the control operates in this mode most of the time, it is the other modes of operation that are most critical. These are Starting, Stopping, Increasing Power and Decreasing Power. The direct result of varying fuel flow is higher or lower combustion temperatures. As fuel flow is increased, combustor heat and pressure increase and heat energy to the turbine is increased. Part of this increased energy is used by the compressor-turbine to increase speed which, in turn, causes the compressor to increase airflow and pressure. The remaining heat energy is used by the power extraction turbine to produce more shaft horsepower. This cycle continues until the desired shaft horsepower or preset parameter limit (temperature, speed, etc.) is reached.

Similarly, to reduce shaft horsepower, the control starts by reducing fuel flow. The lower fuel flow reduces combustion heat and pressure and reduces the heat energy available to the compressor-turbine. With less available energy the compressor-turbine slows down, thereby lowering the compressor speed as well as airflow and pressure. The downward spiral continues until the desired shaft horsepower is reached.

While it is possible to increase and decrease shaft horsepower by making minute changes in fuel flow, this method is time consuming. On the one hand, if the fuel flow is increased too rapidly than excessive combustor heat is generated and either the turbine inlet temperature will be exceeded or the increase in speed will drive the compressor into surge, or both. On the other hand if the fuel flow is decreased too rapidly then the reduction in fuel flow could be faster than the rate at which the compressor will reduce airflow and pressure and result in a flame-out or compressor surge (as speed decreases the compressor operating point moves closer to the surge line). High turbine inlet temperature will shorten the life of the turbine blades and nozzles, and compressor surge could severely damage the compressor blades and stators (and possibly the rest of the gas turbine). Flame-out as a result of power decease takes its toll in the long run because of the thermal stresses created with each shutdown and re-start.

The control must also guard against surge during rapid power changes, start-up, and periods of operation when compressor inlet temperature is low or drops rapidly. Note: the gas turbine is more susceptible to surge at low compressor inlet temperatures. Normally changes in ambient temperature are slow compared to the response time of the gas turbine control system. However, the temperature range from 28°F (–2.0°C) to 42°F (6.0°C) accompanied with high humidity is of major concern. Operation in this range has resulted in ice formation in the plenum upstream of the compressor. To address this problem, anti-icing schemes have been employed to increase the sensible heat by introducing hot air into the inlet. Anti-icing, therefore, becomes another control function that must address both the problem of ice formation, as well as the effect temperature changes have on the compressor relative to surge.

The acceleration schedule facilitates loading the unit as quickly as possible. Without the acceleration schedule it would be very time consuming to increase the load from the Idle-No-Load position to the Full-Load position. As prescribed by the acceleration schedule, the fuel valve is opened and loading is initiated. As load approaches its target set point the speed governor starts to override the acceleration schedule output and the fuel valve starts to approach its final running position. During this excursion the temperature limit controller and the pressure limit controller monitor temperatures and pressures to ensure that the preset levels are not exceeded. The temperature limit controller for turbine inlet temperature monitors the average of several thermocouples taking temperature measurements in the same plane. It is possible for the temperatures throughout this measurement plane to vary significantly while the average remains the same. This situation is dealt with in the Protection Controller, discussed later. Should the temperature (or pressure) reach its set point the limit controller will override the governor controller and maintain operation at a constant temperature (or pressure). These different controllers are always operating but the output signal to the fuel valve only reflects the controller signal requiring the minimum fuel flow.

Considering the control maps (Figures 5-6 and 5-7), the operating point will move along a locus of points that define the operating line for various load conditions. The number of operating lines and the number of governor lines are infinite. However, when the operating point intercepts a pressure, temperature, speed limit, or acceleration schedule limit condition, the gas turbine cannot operate beyond that point. Barring any other restrictions, the gas turbine would be able to operate at any load within the operating envelope.

The Protection Controller continuously checks speed, temperature, and vibration for levels that may be detrimental to the operation, the unit, and personnel. Usually two levels are set for each parameter, an alarm level and a shutdown level. When the alarm level is reached the system will provide an audible warning to the operator that there is a problem. If the transition from alarm to shutdown condition takes place so rapidly that operation response is not possible, the unit is automatically shutdown. Overspeed is one of the parameters monitored by the protection controller that does not include an alarm signal. Overspeed* of the compressor, compressor-turbine, or power extraction turbine could result from excessive fuel flow or loss of load (as a result of coupling failure).

---

*The turbine manufacturer has determined the yield and burst speed of each disc. Based on this information he has defined a speed above which the integrity of the disc is compromised. Whenever possible overspeed is set at a point below the disc yield point. When this is not possible, following an overspeed situation, the unit must be overhauled and the affected disc replaced.

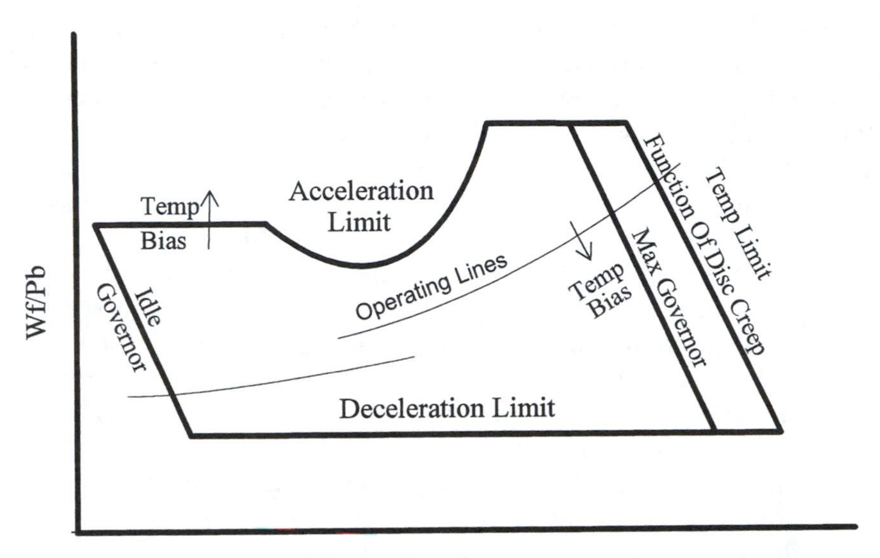

Figure 5-6. Control map.

Depicted is a typical control map for a dual-spool gas turbine. The parameters used for this map are fuel flow (Wf) divided by burner pressure (Pb), which represents the fuel-air ratio, and rotor speed (N). The upper portion of the map represents maximum allowable operating conditions to avoid compressor surge while attaining the maximum rate of acceleration. The extreme right line represents the exhaust gas temperature or turbine inlet temperature limit, which is a function of turbine disc creep temperature. On aircraft engines, this is trimmed in as a function of engine pressure ratio and exhaust gas temperature for standard ambient conditions. The "governor" lines indicate the infinite number of conditions between idle and maximum temperature limit at which the gas turbine can operate. On jet engines this is a function of power lever angle. On industrial engines it is a function of load. The lower line represents the gas turbine operating line for steady state conditions. The lowest line is the deceleration limit function. This function permits rapid deceleration to a lower power or idle without flame-out or loss of combustion. The type of governor depicted is a speed-droop governor and is in common use on aircraft engines and some split-shaft, mechanical drive gas turbines. An isochronous, or zero speed-droop, governor is used almost exclusively on power generating gas turbines.

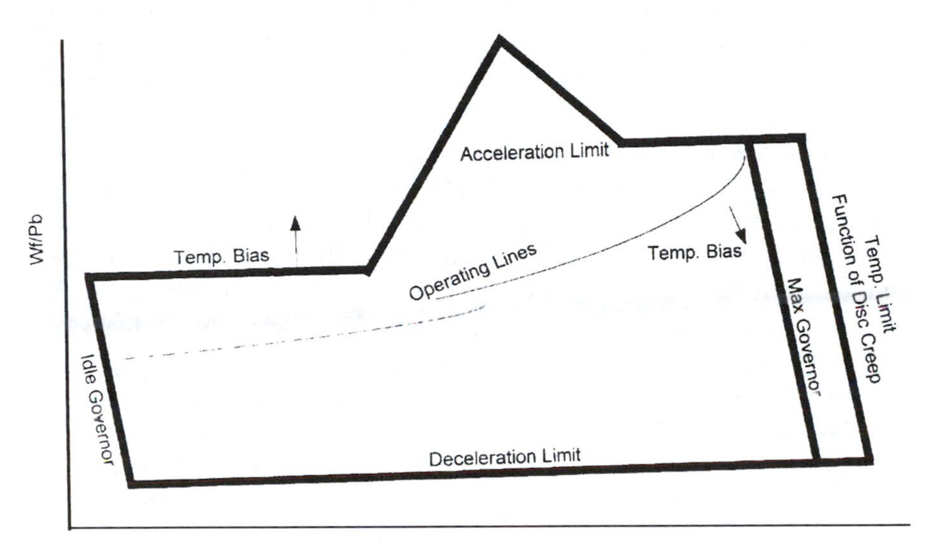

Figure 5-7 Control map.

Depicted is a typical control map for a single spool gas turbine. The parameters used for this map are fuel flow (Wf) divided by burner pressure (Pb), which represents the fuel-air ratio, and rotor speed (N). The upper portion of the map represents maximum allowable operating conditions to avoid compressor surge while attaining the maximum rate of acceleration. The extreme right line represents the exhaust gas temperature or turbine inlet temperature limit, which is a function of turbine disc creep temperature. On aircraft engines, this is trimmed in as a function of engine pressure ratio and exhaust gas temperature for standard ambient conditions. The "governor" lines indicate the infinite number of conditions between idle and maximum temperature limit at which the gas turbine can operate. On jet engines this is a function of thrust or power lever angle. On industrial engines it is a function of load. The lower split-lines represent the gas turbine operating line for steady state conditions. The lowest line is the deceleration limit function . This function permits rapid deceleration to a lower power or idle without flame-out or loss of combustion. The type of governor depicted is a speed-droop governor and is in common use on aircraft engines and some mechanical drive gas turbines. An isochronous, or zero speed-droop, governor is used almost exclusively on power generating gas turbines.

Temperature protection is provided to protect against excessively high temperatures, but it can also be used to protect against excessive temperature spread. The use of multiple temperature sensors, as in the case of turbine inlet temperature (intermediate turbine temperature or exhaust gas temperature), provides protection should the spread from minimum to maximum temperature (or minimum and maximum compared to the average) become excessive.

Because turbine inlet temperature is the most frequently activated limiting factor, one level is set for base load operation and a higher level is set for peaking operation. However, direct measurement of turbine inlet temperature is too risky, because a responsive temperature probe can not live in that environment and a robust probe would not be responsive. Therefore, temperatures are measured in a cooler part of the compressor-turbine, and sometimes between the compressor-turbine and the power extraction turbine. A predetermined temperature schedule is then prepared by the gas turbine manufacturer to ensure operation within safe temperature limits.

References

1. "Controls: The Old vs the New", Power Engineering, April 1996.
2. "Speedtronic* Mark II For Pipeline Applications," SOA-17-73, T.R. Chamberlin, General Electric.

Chapter 6

Accessories

Accessories include the starting system, ignition system, lubrication system, air inlet cooling system, water or steam injection system (for NO_x control or power augmentation), and the ammonia injection system (for NO_x control). The starting, ignition, and lubrication systems are covered in this chapter. The air inlet cooling system is discussed in Chapter 8, and the water or steam injection systems and the ammonia injection system are treated in Chapter 9.

Accessory systems are considered direct drive when they are connected directly to the shaft of the gas turbine (either the gas generator or power turbine). Usually only one or two accessories are direct connected. In most cases it is one of the lubrication pumps that is direct connected.

Indirect drives utilize electric, steam, or hydraulic motors for power. Using indirect drives also facilitates employing redundant systems, thus increasing the system and the gas turbine plant reliability.

Electric systems that are powered by a directly driven electric generator share the advantages of both the direct and indirect drive arrangements.

STARTING SYSTEM[1,2]

Starting systems fall into two categories: those that drive the gas generator directly and those that drive the gas generator through an intermediate gearbox. Starters may be diesel or gas engine, steam or gas turbine, electric, hydraulic, or pneumatic (air or gas). The starter satisfies two independent functions: the first is to rotate the gas generator until it reaches its self-sustaining speed, and the second is to drive the gas generator compressor to purge the gas generator and the exhaust duct of any volatile gases prior to initiating the ignition cycle. The starting sequence consists of the following:

- engage starter
- purge inlet and exhaust ducts
- energize ignitors
- switch fuel on.

The primary function of the starting system is to accelerate the gas generator from rest to a speed point just beyond the self-sustaining speed of the gas generator (Figure 6-1). To accomplish this the starter must develop enough torque to overcome the drag torque of the gas generator's compressor and turbine, any attached loads including accessories loads, and bearing resistance. The single shaft gas turbines with directly attached loads (such as electric generators) represent the highest starting torque as the driven load must also be accelerated from rest to a speed sufficiently above gas generator self-sustaining speed. Two shaft gas turbines (consisting of the gas generator and the driven load connected to the free power turbine) represent the lowest starting torque requirements. In this case only the gas generator is rotated.

Another function of the starting systems is to rotate the gas generator, after shutdown, to hasten cooling. The purge and cool-down functions have lead to utilization of two-speed starters. The low speed is used for purge and cooling and the high speed is used to start the unit. When sizing the starter, the designer should keep in mind that the gas generator must move 3 to 5 times the volume of the exhaust stack to insure purging any residual gas from that area. Also control system programmers and operators should be aware that the purge time, within the start cycle, is necessary for safe operation.

Gas generators are started by rotating the compressor. This is accomplished in a number of ways:

- starter directly connected to the compressor shaft
- starter indirectly connected to the compressor shaft via the accessory gearbox
- impingement air directed into the compressor or compressor-turbine.

Devices used to start gas generators include electric (alternating current and direct current) motors, pneumatic motors, hydraulic motors, diesel motors, and small gas turbines.

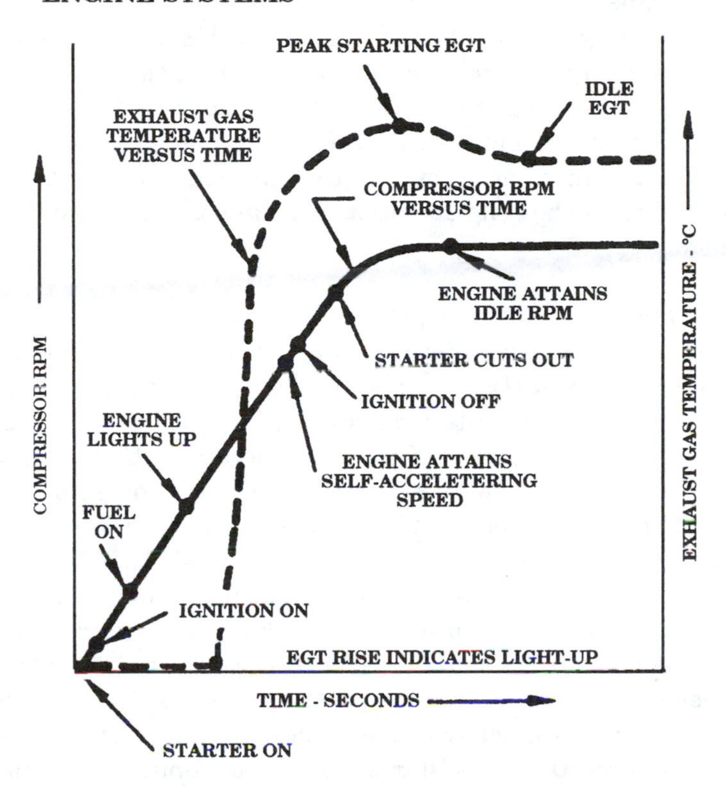

Figure 6-1. Courtesy of United Technologies Corporation, Pratt & Whitney Aircraft. A typical starting cycle for a gas turbine. Note that the time from starter 'on' to engine 'idle' is measured in seconds. This could be as little as 10 seconds for an aero-derivative gas turbine to 30 seconds for a heavy frame industrial gas turbine.

Electric Motors

Alternating Current

Where alternating current (AC) power is available, three-phase induction type motors are the preferred choice for starter drivers. In general, the induction motor is directly connected to the compressor shaft or the starter pad of the accessory gearbox. On some engine models the starter pad mount is a tight fit due to the size of an electric motor and the configuration of the accessory gearbox. Also, as these accessory gearboxes are located under the gas generator, it is a hostile (temperature) environment for an electric motor.

Once the gas generator has reached self-sustaining speed, the motor is de-energized and mechanically disengaged through a clutch mechanism. In some applications a clutch mechanism is not included and the motor is simply de-energized. In applications where the clutch mechanism is not provided, the gas generator must carry the motor load throughout its operation. This imposes additional wear on the motor.

Direct Current

Where AC power is not available, such as black start applications, direct current (DC) motors may be used. The source of power for the DC motor is a battery bank of sufficient capacity to carry the cranking and starting loads of the gas generator. DC starting motors are more commonly used with small gas turbines and aeroderivative-type gas turbines (where stating torque is relatively low). As in the AC starter motor application, the DC starter motor may be configured with a clutch mechanism to dis-engage the motor from the gas generator. Another approach is to convert the DC motor (electrically) into a electric generator to charge the battery system. This is a convenient arrangement where large battery packs are also used to provide direct power for other systems (controls, motorized valve operators, etc.). Battery powered DC motor starters are predominately used in small, self-contained, gas turbines under 500 brake horsepower (BHP).

Electric motors require explosion proof housings and connectors and must be rated for the area classification in which they are installed. Typically this is Class I, Division 2, Group D. See Appendix C-4 for more details on area classifications.

Pneumatics Motors

Pneumatic starter motors may be either impulse-turbine or vane pump type. These motors utilize air or gas as the driving (motive) force, and are coupled to the turbine accessory drive gear with an overriding clutch. The overriding clutch mechanism disengages when the drive torque reverses (that is when the gas turbine self-accelerates faster than the starter) and the air supply is shutoff. The housing of this type starter must be sufficiently robust to sustain the high gas generator speeds in the event the mechanism fails to disengage.

Air or gas must be available at approximately 100 psig and in sufficient quantity to sustain starter operation until the gas generator exceeds self-sustaining speeds. Where a continuous source of air or gas is not available, banks of high and low pressure receivers and a small positive displacement compressor can provide sufficient air for a limited number of start attempts. As a rule of thumb, the starting system should be capable of three successive start attempts before the air supply system must be recharged.

In gas pipeline applications, the pneumatic starter can use pipeline gas as the source of power. In these applications it is critical that the starter seals are leakproof and the area is well ventilated.

Hydraulic Motors

Hydraulic pumps often provide the power (motive force) to drive hydraulic motors or hydraulic impulse turbine (Pelton Wheel) starters. Hydraulic systems are often used with aeroderivative gas turbines as they are easily adaptable to the existing hydraulic systems. Hydraulic systems offer many advantages such as small size, light weight, and high time between overhaul.

Diesel Motors

Due to their large mass moment of inertia, heavy frame (25,000 SHP and above) gas turbines require high torque, high time starting systems. Since many of these units are single shaft machines, the starting torque must be sufficient to overcome the mass of the gas turbine and the driven load. Diesel motors are the starters of choice for these large gas turbines. Since diesel motors cannot operate at gas turbine speeds, a speed increaser gearbox is necessary to boost diesel motor starter speed to gas turbine speed. Diesel starters are almost always connected to the compressor shaft. Besides the speed increaser gearbox, a clutch mechanism must be installed to insure that the diesel motor starter can be disengaged from the gas turbine. Advantages of the diesel motors are that they are highly reliable and they can run on the same fuel as the gas turbine, eliminating the need for separate fuel supplies.

Small Gas Turbines

Small gas turbines are used to provide the power to drive either pneumatic or hydraulic starters. In the aircraft industry a combustion starter, essentially a small gas turbine, is used to start the gas turbine in remote locations. They are not used in industrial applications.

Impingement Starting

Impingement starting utilizes jets of compressed air piped to the inside of the compressor or turbine to rotate the gas generator. The pneumatic power source required for impingement starting is similar to air starters.

Regardless of the starter type it must be properly sized to provide sufficient torque for the purge time and the acceleration time from zero speed to gas generator self-sustaining speed.

IGNITION SYSTEM

Ignition is one part of the gas turbine system that is often taken for granted. That is until a problem develops. Even when a problem does develop the ignition system is not the first item checked. One reason is that this system has developed into one of the most reliable systems in a gas turbine package. The other reason is that ignition is only required during start-up. Once the unit has accelerated to self-sustaining speed the ignition system can, and usually is, de-energized. Experience has shown that the ignition system should not be energized until the gas generator has reached cranking speed, and remained at that speed long enough to purge any volatile gases from the engine and exhaust duct. As soon as the igniters are energized, fuel can be admitted into the combustor. These two distinct functions are often implemented simultaneously, and are commonly referred to as "pressurization."

To add to their dependability, it is standard practice to install two igniters in each engine. An igniter is installed on each side of the engine, although not normally 180 degrees apart. This design approach holds for all combustor designs (i.e. the annular design, can-annular design, or single "stand-alone" design combustor). In the can-annular combustor design the flame propagates from "combustor can to combustor can" via interconnecting flame tubes (also referred to as cross-over tubes). Also, for redundancy, the ignition system consists of two identical, independent ignition units (exciters) with a common electrical power source. During the start cycle each igniter discharges about 1.5 - 2 times per second with an electrical energy pulse of 4 to 30 joules (depending on combustor size, fuel, etc.).[1]

"One joule is the unit of work or energy expanded in one second by an electric current of one ampere in a resistance of one ohm. One joule/second equals one watt."

Once the gas generator has been started the function of the igniter is fulfilled and any further exposure to the hot gases of combustion shortens its life. To eliminate this unnecessary exposure, some igniter designs include a spring loaded retracting mechanism that permits the igniter to move out of the gas path as combustion pressure increases.

To achieve ignition consistently a high voltage and a high heat intensity spark are required. Ignition systems include both high and low energy systems, and energy sources comprise both AC and DC sources. Typical electrical potential at the spark plug is approximately 25,000 volts. Due to this high voltage the electrical wiring (also referred to as the ignition harness) to each igniter plug is shielded and the ignition exciter is hermetically sealed.

Several types of ignition systems have been developed: they are the capacitive AC and DC, high and low tension systems and inductive AC and DC systems. Of these the capacitive systems generate the hottest spark.

The capacitive AC high tension ignition system (Figure 6-2) consists of the power source, on/off switch, radio interference filter, low tension AC transformer, rectifier, storage capacitor, high frequency/high tension transformer, high frequency capacitor, and spark plug. The power source may be either AC or DC. The function of the ignition system components is as follows:

- The radio interference filter eliminates ignition energy from affecting local radio wave signals
- The AC transformer (or transistorized chopper circuit transformer) boosts the voltage to approximately 2,000 volts
- The rectifier is an electrical one-way valve allowing the flow of current into the storage capacitor but preventing return flow
- The high tension transformer charges the high tension capacitor.

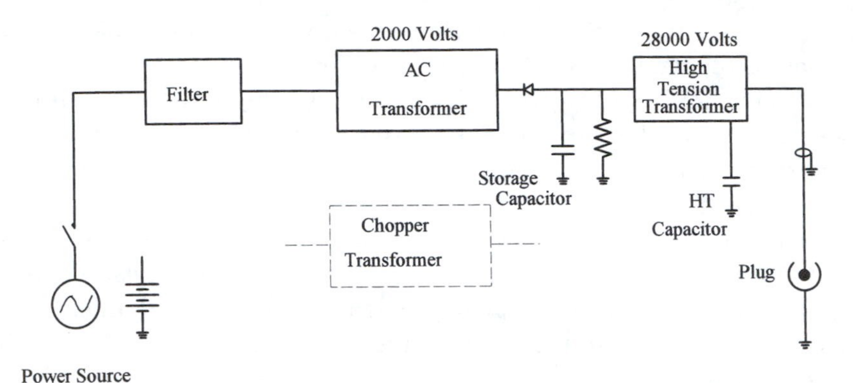

Figure 6-2. Typical high energy capacitor-type ignition system.

The low voltage charge in the storage capacitor is not sufficient to bridge the gap across the spark plug electrodes until the path is provided by the very high voltage, high tension discharge. Therefore the high tension capacitor discharges first to bridge the gap across the electrodes of the spark plug and reduce the resistance to the low tension discharge. The low tension capacitor then discharges providing a long, hot spark. One of the drawbacks to this system is that the leads from the ignition exciter to the spark plugs must be shielded, heavy duty, leads.

By contrast the low tension ignition system does not have a high tension transformer, high tension capacitor, or high voltage leads. However, it does have a semiconductive surface in direct contact with the center electrode and ground shell of the igniter plug. This semiconductor is shown schematically as the resistor in Figure 6-3. The initial flow of current bridges the gap between the center electrode and the ground shell to reduce resistance to the low tension discharge. Due to its lower voltages, the low tension system can use smaller, lighter, ignition cables.

The inductive ignition system (Figure 6-4) utilizes a rapid variation in magnetic flux in an inductive coil to generate the high energy required for the spark. While this system produces a high frequency, high voltage spark, the spark energy is relatively low. This system is small, light, and less expensive, but it is only suitable for use with easily ignitable fuels. And, the spark igniter is more prone to fouling from moisture and carbon build-up.

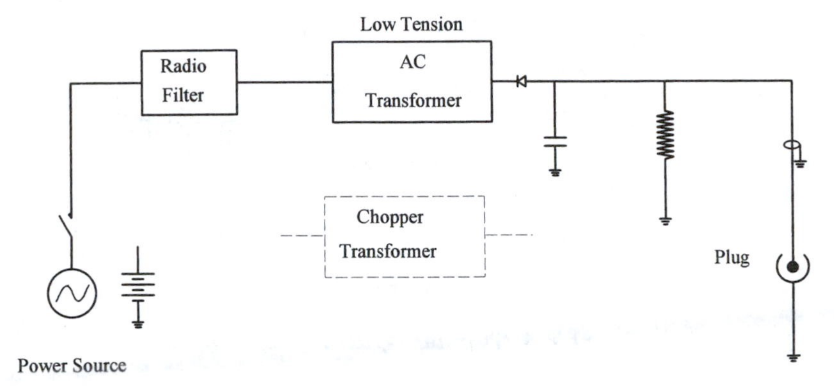

Figure 6-3. Typical low energy capacitor-type ignition system.

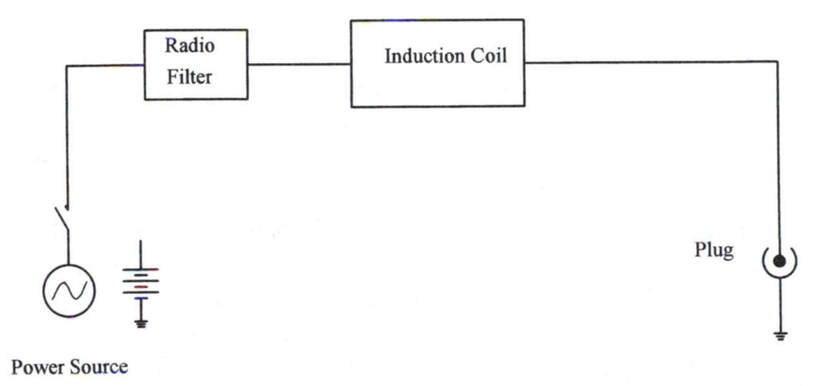

Figure 6-4. DC inductive-type ignition system.

Spark Plug

Because the high energy current causes rapid erosion of the igniter plug, it is advisable not to operate the ignition system any longer than necessary to insure ignition. However, this same high energy helps eliminate fouling. Ignitor plugs are of the annular-gap type and the constrained-gap type (Figure 6-5). The annular-gap plug projects slightly into the combustor, while the constrained-gap plug can be positioned in the plane of the combustor liner, and therefore operate in a cooler environment. There is more flexibility in positioning the constrained-gap plug because its spark jumps in an arc from the electrode. This carries the spark into the combustion chamber.

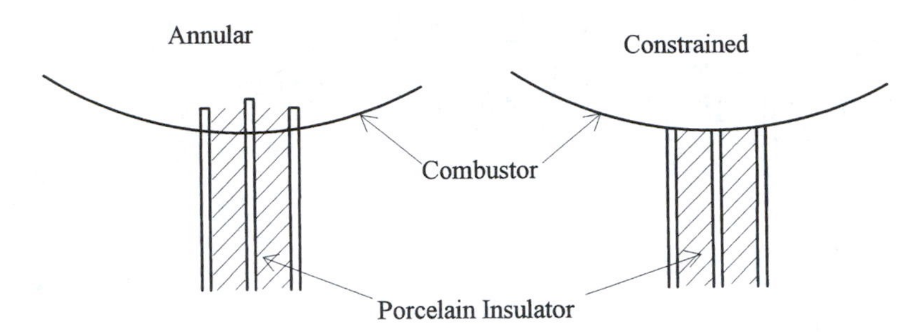

Figure 6-5. Typical ignitor plug types.

LUBRICATION SYSTEMS

General

Lubrication systems in gas turbines must address two requirements: one is to provide lubrication between the rotating and stationary bearing surfaces, and the other is to remove heat (cooling effect) from those surfaces. Lube oil system designs must consider bearing types, rpm, load, temperature, and oil viscosity. Bearing types fall into two main categories: hydrodynamic bearings and anti-friction bearings. Lubrication in a hydrodynamic bearing converts sliding friction into fluid friction. While fluid friction is considerably lower, it still consumes power and produces heat. Therefore, besides providing the fluid wedge that separates the surfaces, lubrication oil also removes the heat from that area.

Anti-friction bearings work on the principal of rolling friction. The shaft load is directly supported by the rolling elements and races in metal-to-metal contact. Furthermore, the high unit pressures between the rolling element and the inner and outer races prohibit the formation of an unbroken oil film. Therefore, the role of the lubrication oil is slightly different. Lubrication oil is still required to remove heat from the bearing area, and it reduces the sliding friction between the rolling elements and the cage that maintains the position of the rolling elements. The engine service and the type of bearing selected by the engine designer, dictates the type of lubrication oil used in the system.

Most heavy frame gas turbines, which incorporate hydrodynamic bearings, use mineral oil while the aeroderivative gas turbines, which incorporate anti-friction bearings, use a synthetic oil. Mineral oils are manufactured (distilled) from petroleum crude oils and are generally less expensive than synthetic lubricants. Synthetic lubricants do not occur naturally but are "built up" by reacting various organic chemicals, e.g., alcohol's, ethylene, etc., with other elements. Synthetic lubricants are used primarily in very high temperature applications (less than 350°F and 175°C) or where its fire-resistant qualities are required.

Lubrication Oil Application

Bearings in gas turbines are lubricated, almost without exception, by a pressure circulating system. Pressure circulating lubrication consist of a reservoir, pump, regulator, filter, and cooler. Oil in the reservoir is pumped under pressure through a filter and oil cooler to the bearings and then returned to the reservoir for re-use (Figure 6-6).

Reservoir

The reservoir is used to store the oil during periods of unit shutdown, and to deaerate the oil during operation. The reservoir must be large enough to contain all the oil used in the gas turbine, piping, and

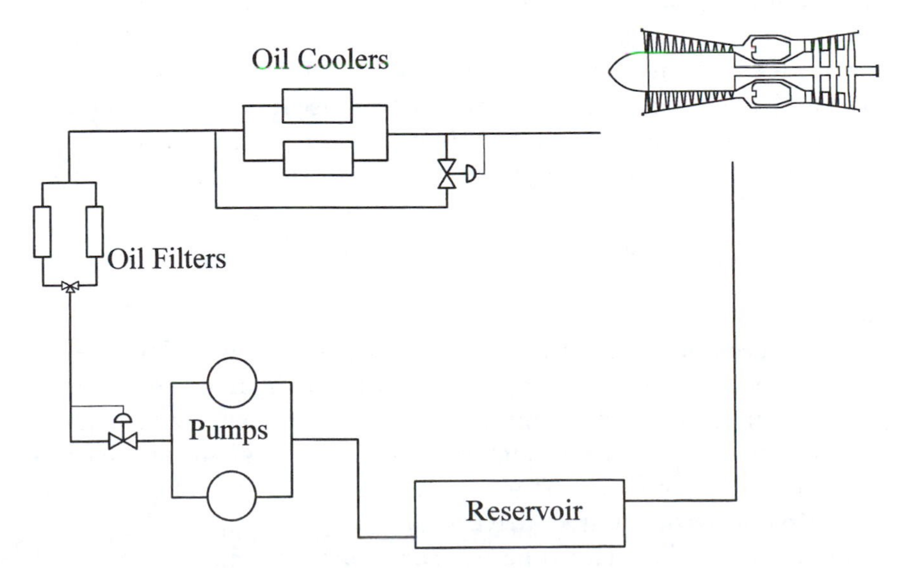

Figure 6-6. Typical lubrication oil schematic.

any accessories. To protect the gas turbine, in the event of a lubrication oil leak, the reservoir is sized such that the holding capacity, or volume, between the "low alarm level" and "pump suction" provides enough time for the unit to be shutdown before the oil supply is depleted. In cold climates a heater, built into the reservoir, warms the oil prior to start-up. Care must be taken in sizing the heater. An oversized heater, while it will heat the oil faster, may result in carborizing the oil around the heater element. This will lead to deterioration of the oil and failure of the heater. A heater that is undersized will be ineffective at low ambient temperatures. Normally natural convection is sufficient to circulate the oil as it is heated. When it is necessary to heat the oil quickly, a circulating pump should be installed within the reservoir to keep the oil in motion.

During operation the reservoir also serves as a deaerator. As the lubrication oil circulates through the bearings it may entrap air in the oil. This is commonly referred to as oil "foaming." The foam must be removed before the oil is returned to the pump or the air bubbles will result in pump cavitation. To deaerate the oil the reservoir surface area is made as large as possible and screens, baffles, or both are built into the reservoir. Anti-foaming agents can also be added to the oil. Oil reservoirs usually operate at a positive pressure of 1 to 2 inches water gauge above atmospheric pressure. Depending on the location and the application, the oil reservoir is either vented locally or is vented to a safe area. In either case as the vented oil mist cools it condenses and leaves an oil film. If left to accumulate, this oil film can become a hazard. Care should be taken to collect this oil and route it to a safe waste dump. The inside of reservoirs should never be painted. Where corrosion is a concern the reservoir should be of stainless steel construction.

Pump

In a pressure circulating system, oil pressure and flow is provided by the oil pump. A pump failure will result in severe damage to the bearings in the gas turbine, and possibly secondary damage to the compressor and turbine blades. Over the years two approaches have evolved to minimize the effects of a pump failure: one is to utilize a direct-drive or indirect-drive shaft driven pump, the other is to install a backup (redundant) pump. Direct-drive pumps are mounted on the accessory gearbox. Depending on the gas turbine type the accessory gearbox may be located at the front of the gas generator (some heavy and lightweight industrial machines), under the gas generator (some aeroderivative units), or on the power turbine accessory gearbox (some aeroderivative units). In any of the above locations, these pumps cannot be serviced, removed, or replaced while the gas turbine is in operation. However, a second (redundant) pump may be installed, usually electric motor driven (hydraulic and steam driven systems have also been used). This second pump also serves as the pre- and post-lube pump. In cold climates the second pump may be used to circulate oil in order to maintain heat in the oil and the gas turbine.

Indirect-drive pump systems rely on a hydraulic pump or an electrical generator mounted on the accessory gearbox. The hydraulic pump (or electrical generator) provides the motive force to drive the lube oil pump through a separately mounted hydraulic motor (or electric motor). This configuration allows either pump to be the primary unit. Through judicious control techniques, the secondary pump can replace the primary pump, in the event of its failure, without disrupting the operation of the gas turbine. The redundant lube oil pump can then be serviced with the gas turbine in operation.

Lube oil pumps may be either centrifugal or positive displacement type pumps. The positive replacement pumps are usually rotary (gear, lobe, screw or vane) pumps[3], which are self-priming and can be used in various suction lift applications. These pumps lend themselves to being directly mounted on the accessory gearbox, which invariably will be some distance from the reservoir.

The redundant (alternate, standby) pump, either centrifugal or positive displacement type may be mounted within the reservoir. This type of installation eliminates the concern about the pump's net positive suction head (NPSH). However, servicing a pump in this type of installation presents another set of problems. Monitoring pump operation, temperature, and vibration is further complicated by its installation in the reservoir, and the area above the reservoir must be clear to enable removal of the pump.

The arrangement most widely used is to install both pumps at the same level as the reservoir and as close as possible to the reservoir. In this arrangement the pumps are identical and both pumps are electric motor driven. The only difference is that the primary pump motor is powered by the shaft driven electrical generator.

Filters

Because the lube oil system is basically a closed system, the function of the filters is to remove pump and gas turbine bearing wear particles from the oil. However, it must be realized that this system was not always closed, and will not always remain closed. Therefore, during initial start-up and after major overhauls the "starting" filter elements should be installed in place of the "running" filter elements. Where 5 or 10 micron filters are satisfactory for "running" conditions, 1 to 3 micron filters should be installed during oil flushing prior to start-up (initial or after overhaul).

Redundant oil filters should be installed along with a three-way-transfer valve. In the event the primary filter clogs, the transfer valve can easily switch over to the clean filter. During initial installation and replacement of filter elements care should be taken to remove all air from the filter housing. Failure to do so can result in temporary loss of lubrication to one or more bearings.

Wear particles from the pump and gas turbine bearings will accumulate in the filter element. Also temperature related oil degradation and oil additives can create a sludge that will accumulate in the filter. As the filter clogs, the differential pressure across the filter will increase. Instrumentation normally included is a pressure differential gauge for local readout, and a differential pressure transducer for remote readout and alarm. A typical differential pressure alarm setting is 5 psig. Integrity of the filter element is usually greater than 2 times the operating pressure of the lube oil system. Therefore, filter failure due to differential pressure should never be a concern. However, pressure pulses in the oil, caused by a positive displacement pump, may weaken the filter element over a period of time. Therefore, filter elements should be replaced when the gas turbine is overhauled or at least once a year.

Regulators

Regulators are used to maintain a constant pressure level in the lube system regardless of which pump is running or even if both pumps are running. Regulators make it possible for maintenance personnel to operate the secondary pump on a preventive maintenance schedule. While regulators are an absolute necessity for positive displacement pumps, they are recommended and should also be used with centrifugal pumps.

Coolers

Lube oil coolers are required to remove heat from the oil before it is re-introduced into the gas turbine. The degree of cooling required is a function of the friction heat generated in each bearing, the heat transferred from the gas turbine to the oil by convection and conduction, and heat transferred from the hot gas path through seal leakage. The oil must be cooled to within acceptable limits, normally 120°F-140°F (50°C-60°C). Cooler design and size is based on the viscosity of the oil at the bearing operating temperature, the maximum allowable bearing metal temperature, and the temperature and available flow rate of the cooling media. To maximize heat transfer, fins are installed on the outside of each tube and turbulators are placed inside each cooling tube. The turbulators help transfer heat from the hot oil to the inner wall of the cooling tubes and the fins help dissipate this heat. When turbulators are used it is necessary to provide sufficient maintenance pull space to facilitate removal and replacement of each turbulator. The cross section design of the turbulators is a function of oil flow rate and viscosity. Tubes with turbulators tend to foul faster than tubes without turbulators. Tube fin construction may be either rolled in, tension wound, or welded "L." To be effective the fin must be in continuous contact with the tube. U-bend tubes should not be used as there is no way to effectively clean them.

The cooling media may be either air or a water/glycol mix. The cooling media selection is primarily a function of location and the availability of adequate utilities. For example air/oil coolers are widely used in desert regions, while tube and shell coolers can be found in the arctic regions and most coastal regions.

Air/Oil Coolers

Air/oil coolers utilize ambient air as the cooling media. Cooling coils are arranged horizontally in the length and number of passes necessary to satisfy the footprint limits and the cooling load (plus a margin of safety of approximately 15% to 20%). Cooling fans are usually electric motor driven, often with two speed motors. This arrangement allows for high and low cooling flows. To closely match the cooling flow to the required heat load, changeable pitch fan blades can be provided. If the heat load changes over a period of time the blades can be adjusted in the field to meet the new heat flow requirements. When variable cooling control is necessary, an oil bypass control loop and a temperature control valve can be installed. Air/oil coolers may also include top louvers to protect the cooling coils from hail. These louvers are not effective for temperature control. Appendix C-5 includes a checklist that summarizes the requirements of an air/oil cooler.

Tube And Shell Coolers

Tube and shell coolers usually use a water glycol mix as the cooling media when it is available in the area. It would not be economical to use water/glycol to cool the lube oil and then cool the water/glycol with an air cooler. To prevent contamination of the oil, in case of cooler failure, the oil side operating pressure should be higher than the water side operating pressure and water/glycol should be on the tube side of the cooler. Also, it is important that vent and drain connections are provided on both the water and oil sides of the cooler. Oil temperature control is best accomplished with a cooler bypass control loop and a temperature control valve. Schematically this is similar to the control loop around the air/oil cooler in Figure 6-6.

CHARACTERISTICS OF LUBE OILS

Physical and chemical properties of lube oils are determined by a number of standard laboratory tests. The more common physical and chemical tests are discussed in this section.[4]

Specific Gravity

Specific gravity, ρ, is defined as follows:

$$\rho = \frac{Mo}{Mw}$$

where,

Mo is the mass of a given volume of oil
Mw is the mass of the same volume of water
both measured at the same temperature, usually 60°F.

Specific gravity is not a significant indicator of oil quality and, therefore, is not used in the selection of lube oils. Specific gravity is an important "first indicator" of oil contamination.

Fire Point

The lowest temperature at which the vapor will continue to burn when a flame is applied.

Flash Point

The temperature at which the vapor from an oil will flash when a flame is applied under standard conditions. The flash point minimums are set to meet safety limits for products lighter than lube oil. Flash point is useful in detecting contamination of the lube oil by fuel oils. Typical values for flash point are listed in Table 6-1:

Table 6-1.

Product	Flash Point
Gasoline	>32°F (0°C)
Diesel Fuel	150°F (65°C) to 200°F (93°C)
Lube Oil	300°F (149°C) to 650°F (343°C)

Pour Point

The lowest temperature at which an oil will flow when cooled under standard conditions.

Cloud Point

The temperature at which a cloud of wax crystals appear when oil is cooled under certain standard conditions.

Flock Point

The temperature at which wax separates as a flock when a mixture of 10% oil and 90% refrigerant is cooled under certain conditions.

These temperatures define the flow properties of oil under low temperature conditions (pour and cloud points are for non-miscible refrigerants and flock point is for miscible refrigerants). This property is significant to refrigeration and air conditioning plant compressor operators. Pour points of –30°F (–34°C) to –40°F (–40°C) or flock points of –60°F (51°C) to –70°F (–57°C) are necessary.

Viscosity

This is a measure of the resistance of a liquid to flow. Viscosity is determined by measuring the time for a quantity of oil to flow through a specified orifice at a certain temperature. Two basic measurements are:

- Absolute or Dynamic Viscosity - measured in Poises (P) or Centipoises (cP)

- Kinematic Viscosity = Absolute Viscosity/Density. Measured in Stokes (St.) or Centistokes (cST.)

Table 6-2 lists several instruments which measure the time for a quantity of oil to flow through an orifice:

Table 6-2.

INSTRUMENT	EXPRESSED IN	SYMBOL	WHERE USED	MEAS. TEMP
Saybolt Universal	seconds	SUS	USA	100°F, 130°F, 210°F 37.8°C, 54.4°C, 98.9°C
Redwood No 1	seconds	secs R.I.	UK	70°F, 140°F, 212°F 21.1°C, 60°C, 100°C
Engler	degrees	°E		20°C, 50°C, 100°C 68°F, 122°F, 212°F

Note: a viscosity figure must always state the units and the temperature of measurement, otherwise it has no meaning.

Viscosity is one of the most important properties of a lube oil. Oil viscosity must be high enough to maintain an unbroken fluid film at the operating temperature of the oil in the bearing without causing excessive fluid friction in the oil itself. An increase in viscosity indicates deterioration of the oil or contamination with a heavier grade. A decrease in viscosity indicates fuel dilution or contamination with a lighter grade.

Carbon Residue

Measures the amount of residue left after evaporation and burning of a sample of oil under standard conditions. Distillate oils leave less carbon residue than residual oils and most naphthenic oils leave less carbon residue than paraffin oils.

Ash Content

Measures the ash that remains after burning a sample of oil. Clean, straight mineral oils have zero ash content. This test is useful for non-additive oils to detect contamination.

Emulsion Characteristics

Measures the ability of oil to separate from water. This is important in lube oil circulating systems where contamination by water may occur.

Oxidation Characteristics

Measures the ability of the oil to resist oxidation. All oils oxidize and deteriorate in service. Therefore, good resistance to oxidation should be built into the oil selected for the lube oil service.

Neutralization Number

An increase in the Neutralization Number (NV) indicates an increase in acidity due to oxidation of the oil. To determine the NV a used oil value must be compared with the value for the new oil.

Effective Additive

This test compares the total base number (TBN) of a used oil sample with the TBN of the new oil. This test is used to determine the effectiveness of the remaining additives in the oil.

Water Content

This test measures the actual water content of the oil. However, a visual inspection for cloudiness or an opaque yellow appearance will usually suffice to indicate the presence of water.

References

1. "The Aircraft Gas Turbine Engine And Its Operation," Pratt & Whitney Aircraft, Division of United Technology, PWA Oper Instr 200.
2. "Sawyers Gas Turbine Engineering Handbook," Vol III, Third Edition, 1985.
3. "API 614 Second Edition," January 1984, Lubrication, Shaft Sealing, and Control-Oil Systems for Special-Purpose Applications.
4. "ARAMCO Lubrication Manual," July 1972.

Chapter 7

Parameter Characteristics

In controlling, monitoring, or analyzing gas turbines the basic objective is to maximize performance, reduce maintenance, and reduce unit downtime. To accomplish this, the indicators of gas turbine health (or condition) must be identified and understood. In general, a gas turbine may be viewed as consisting of accessory equipment (controls, lube oil pumps, ignition system, etc.), rotational mechanical equipment (rotor bearings), and thermodynamic gas path elements (gas containment path, compressor, combustor, and turbines).

A controlling or monitoring system must pay attention to the three areas of the gas turbine system:

- The thermodynamic gas path
- Vibrations of bearings, rotors, and gearboxes
- Lubrication, control, and other accessory subsystems

Furthermore, the information gathered from these three areas may often be used in a cross complementary fashion to verify diagnoses and more precisely isolate faults.

Table 7-1 shows the key engine parameters and their symbols used in controlling, monitoring and analyzing gas turbine operation.

VIBRATION

Every piece of equipment, regardless of size or configuration, has a natural or resonant frequency. Gas turbines are no exception. If the resonant frequency is below the operating frequency or speed, the unit is considered to have a flexible shaft. If the resonant frequency is above the operating speed, the unit is said to have a stiff shaft. Almost all gas turbines are considered to have a flexible shaft; that is, the normal op-

Table 7-1.

PARAMETER	SYMBOL
Ambient Air Temperature	T_{am}
Ambient Air Pressure (Barometer)	P_{am}
Inlet Air Pressure	P_1
Exhaust Total Pressure	P_{t5} OR P_{t7}
N_1 (RPM)	N_1
N_2 (RPM) a dual spool	N_2
N_3 (RPM) power turbine	N_3
Exhaust Gas Temperature	EGT
Fuel Flow	W_f
Power Output (Electric Generator)	MW
Low Pressure Compressor Out Pressure (static)	P_{S3}
High Pressure Compressor Out Pressure (static)	P_{S4}
Vibration	V

erating speed is above the resonant frequency. The gas turbine as a dynamic, fuel consuming, power producing entity has a specific natural frequency, which is a combination of the frequencies of its components. Where the natural frequency of the shaft connecting the compressor and turbine may be at 250 Hz, and the natural frequency of the assembled compressor and turbine rotor may be at 25 Hz, the gas turbine, as an entity, may be at 1.5 Hz.

Vibration measurements usually consist of amplitude (inches), velocity (inches per second) and acceleration (inches per second per second or "**g**"). These measurements are taken with probes, such as seismic probes that measure amplitude, proximity probes that measure primarily velocity, and acceleration probes or accelerometers that measure acceleration. By the use of integrating or differentiating circuits, amplitude, velocity or acceleration readouts are possible from any of the three probe types. However, more accurate information can be obtained from direct measurements using the proximity probe and accelerometer.

Typical sources of vibration that are usually found in a gas turbine package are depicted on a real time spectrographic plot (Figure 7-1). Note that rotor imbalance is seen predominantly at the running speed, whereas misalignment can be seen at two times the running speed and is seen on readouts from both the horizontal or vertical probe and the axial probe. Looseness due to excessive clearances in the bearing or bearing support is also seen at two times speed, except the axial probe does not show any excursions. Hydrodynamic journal bearing oil whirl is seen at 0.42 running speed and is both violent and sporadic. Very close to that, at 0.5 running speed, is baseplate resonance. Note that baseplate resonance is picked up predominantly on the vertical probe. Gear noise is seen at running speed times the number of gear teeth. Amplitude is not the best criteria to judge excessive or acceptable vibration since a high amplitude for a low rpm machine might be acceptable whereas that same amplitude for a high rpm machine would not be acceptable.

These same guidelines apply to the lube oil and fuel supply systems. Pumps and motors exhibit the same relative indicators of unbalance. Even the piping of the lube oil and fuel forwarding systems have their own unique natural frequencies.

VIBRATION MEASUREMENT

A seismic probe, which consists of a spring mass system, is primarily used where the frequency measured is well below the machine's natural frequency and that of the spring-mass system.

A proximity probe, which is mounted on the equipment case, measures the relative movement of the shaft to the case. Proximity probes are used where the frequency measured is well below the machine's natural frequency.

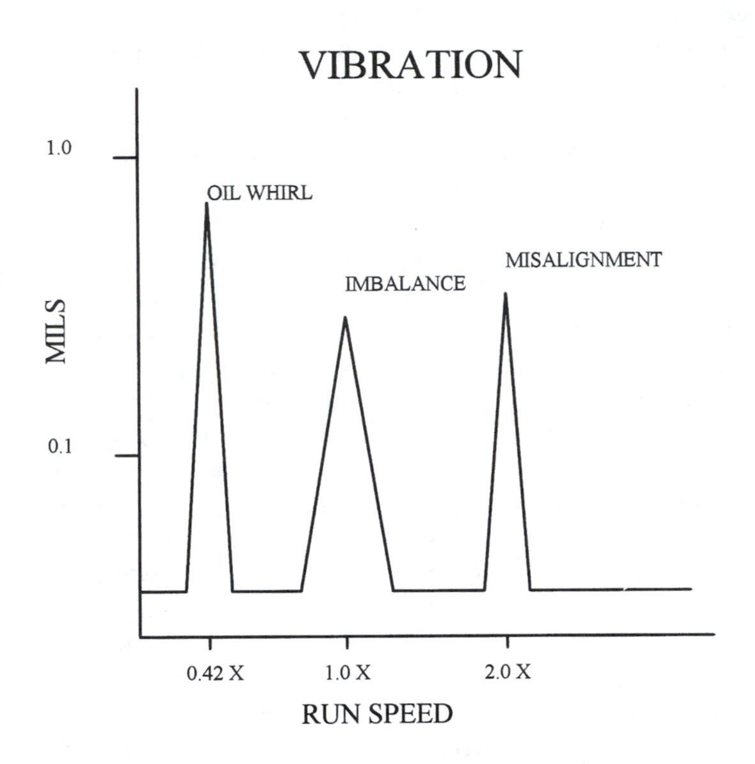

1 X SPEED	ROTOR IMBALANCE	
2 X SPEED	MISALIGNMENT OR LOOSENESS	AXIAL HIGH AXIAL LOW
0.42 X SPEED	OIL WHIRL - VIOLENT BUT SPORADIC	
0.5 X SPEED	BASEPLATE RESONANCE - VERTICAL MORE SENSITIVE	
RPM X GEAR TEETH	GEAR NOISE	

Figure 7-1. Real time spectrographic plot

An accelerometer, which consists of a piezoelectric cell or transducer, measures frequencies that are higher than the natural frequency of the machine. Accelerometers are usually mounted internally on the bearing support, but they may be mounted on the outer case at the bearing support.

A velocity, acceleration and amplitude map is shown in Figure 7-2.

Note that one mil amplitude is considered smooth at 10 cycles per second and rough at 100 cycles per second. The most uniform measurement is velocity, which is always smooth below .02 inches per second and is always rough above 0.11 inches per second. Velocity can be obtained from amplitude and running speed by multiplying amplitude in mils, times π, times rpm, divided by 60,000.

$$V = MILS \bullet \pi \bullet RPM/60{,}000$$

EXHAUST GAS TEMPERATURE (EGT)

Exhaust gas temperature is one of the most critical parameters in a gas turbine, in view of the fact that excessive turbine temperatures result in decreased life or catastrophic failure. In older machines, when temperatures were not as high, turbine inlet temperature was measured directly. In the current generation of machines, temperatures at the combustor discharge are too high for the type of instrumentation available and, so, intermediate stage or exhaust gas temperature is used as an indication of turbine inlet temperature.

The quality of the temperature measured is a function of the number of probes, their positions, and their sensitivity. Since the temperature profile out of any gas turbine is not uniform, the greater the number of probes, the better the representation of the exhaust gas average temperature and profile. A greater number of probes will help pinpoint disturbances or malfunctions in the gas turbine by highlighting shifts in the temperature profile.

As metal temperature increases, creep also increases. At any given power, exhaust gas temperature increases with the increase in ambient temperature. Therefore, to remain within the safe temperature envelope as outside temperature increases, output power must decrease (Figure 7-3).

ROTOR SPEED

Rotor speeds are commonly used as control functions on almost all gas turbines. At any constant power, rotor speeds will increase with an increase in outside air temperature. Therefore, most control functions

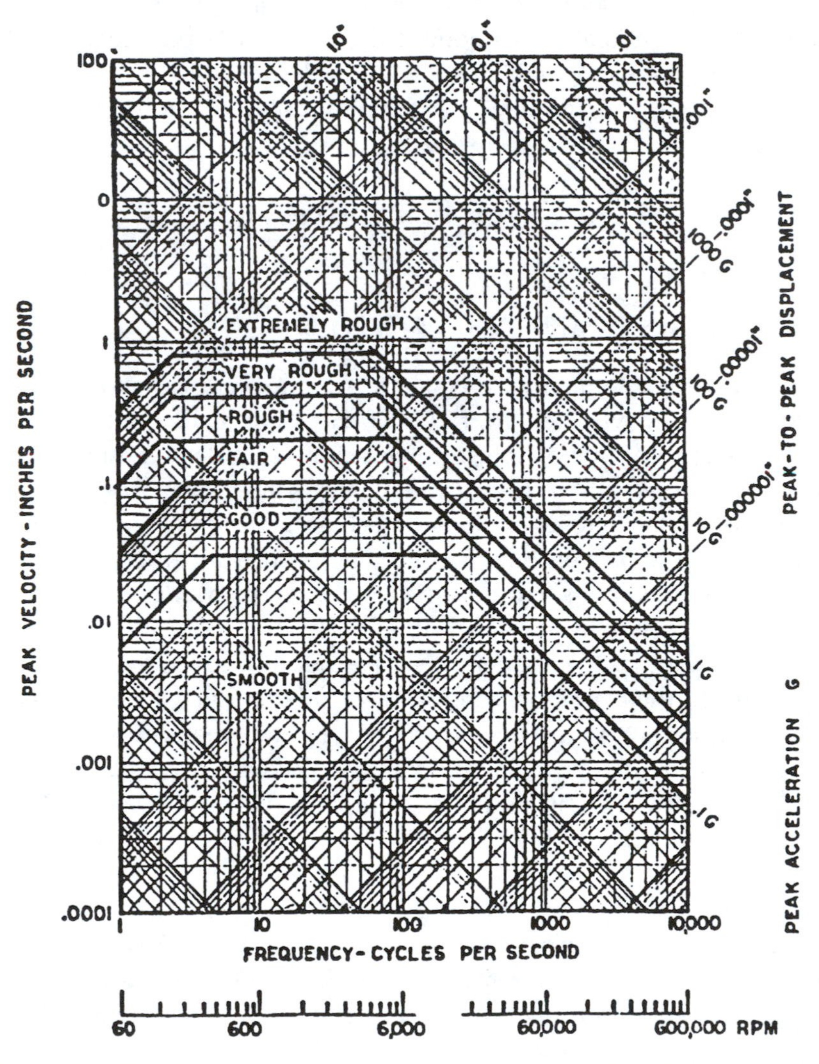

Figure 7-2. Machine vibration as shown on a combined acceleration, amplitude, and velocity plot. This approach to viewing vibration shows the relative importance of machine speed, either rpm or Hertz (Hz).

utilizing rotor speeds are biased by temperature (Figure 7-4). On split-shaft or dual-spool engines, rotor speeds vary as a function of engine match changes between the compressor and the turbine. The complex interplay of these components, as seen by the interaction between N_1 (low speed) and N_2 (high speed), is very subtle and difficult to define without a detailed gas path analysis. On single-spool engines driving generators, rotor speed is one of the controlled functions and in itself will not change.

OIL PRESSURE AND TEMPERATURE

Measurement Methods

Oil pressure is indicative of the pressure drop across the filters and external and internal leaks. External leaks will eventually become obvious, but internal leaks can be difficult to detect. Internal leaks can result in oil leaking into the hot gas path and may or may not be evidenced by exhaust smoke. Care should be taken when the unit is down to determine if oil is leaking within the engine by inspecting the engine inlet, the exhaust duct and where possible a spot check (boroscope inspection) of the compressor discharge or bleed air discharge.

Exhaust Gas Temperature (EGT)

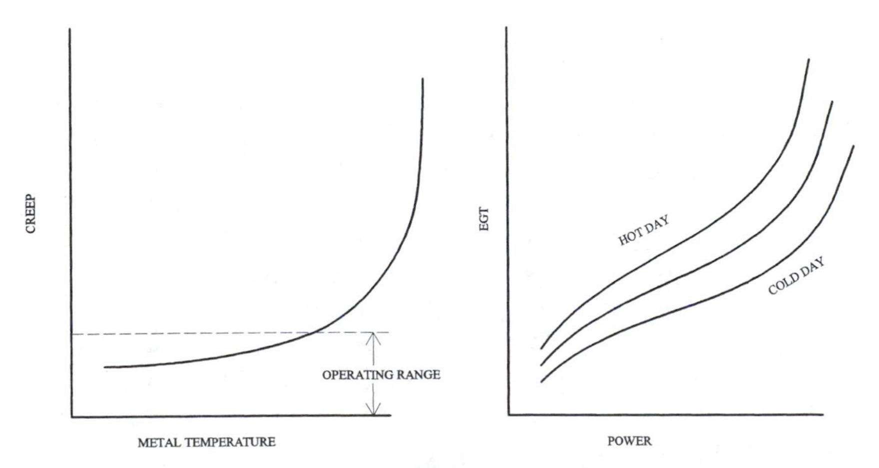

Figure 7-3. Temperature measurement is a function of the number of probes, their position and their sensitivity.

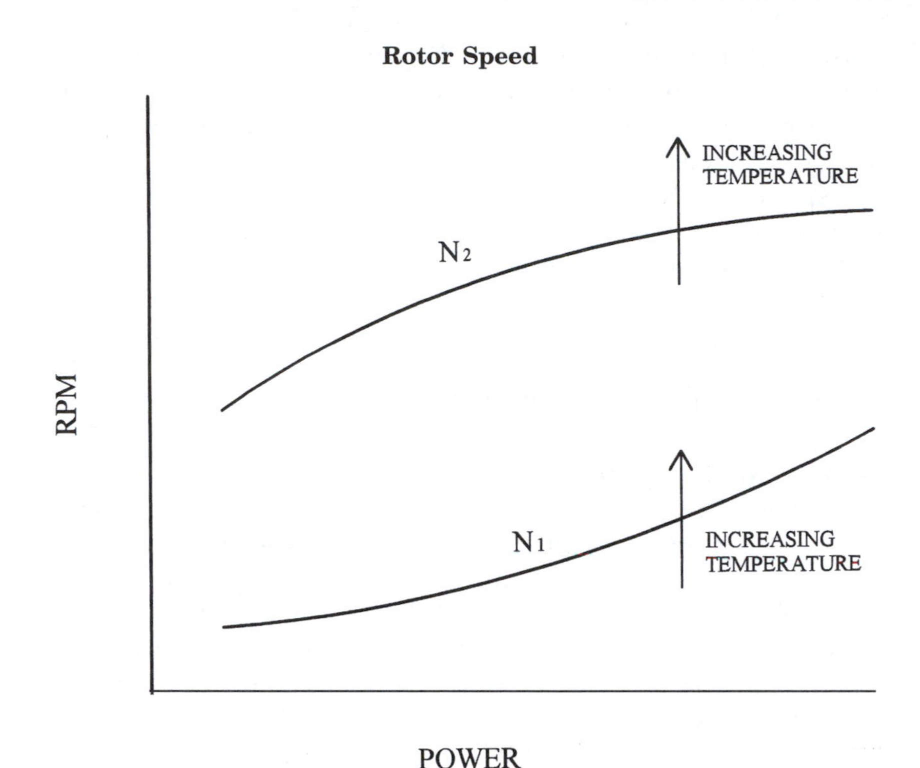

Figure 7-4. Rotor speed is a control function biased by temperature.

Three measurement methods are commonly used to monitor oil temperature: oil throw-off temperature from the bearing, immersed oil temperature (actual oil temperature), and bearing metal temperature. Since, the object of measuring oil temperatures is to determine the bearing metal temperature, that is the most direct method to use. This usually implies contact thermocouples or resistance temperature detectors, commonly referred to as RTD's. RTD's are accurate in low temperature applications up to 500°F. RTD's are used in measuring throw-off or immersion oil temperature. The other significance of oil temperature is determining cooler capacity, cooler efficiency and seal leaks, i.e., hot gases leaking into the oil.

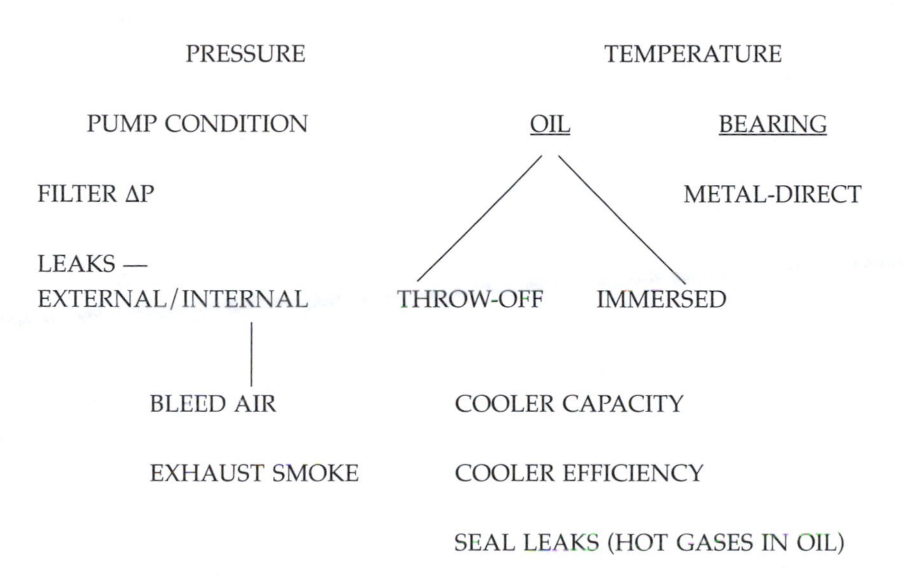
OIL PRESSURE AND TEMPERATURE
PRESSURE
TEMPERATURE
PUMP CONDITION
OIL
BEARING
FILTER ΔP
METAL-DIRECT
LEAKS —
EXTERNAL/INTERNAL
THROW-OFF
IMMERSED
BLEED AIR
COOLER CAPACITY
EXHAUST SMOKE
COOLER EFFICIENCY
SEAL LEAKS (HOT GASES IN OIL)

Chapter 8

Gas Turbine Inlet Treatment

THE ENVIRONMENT

Gas turbines are installed in many environments from desert to coastal, tropical, arctic, agricultural, oil fields, etc. The weather conditions, temperature range, and type, concentration, and particle size of airborne contaminates can be defined well in advance of equipment purchase and installation. Chart 8-1 summarizes the dust concentration and particle size in various environments in which gas turbines operate. Each environment brings with it specific atmospheric contaminants. For example, sand, salt, and dust are the problem contaminants in the desert; salt is the major concern in coastal areas; dust, pollen, and chemicals (fertilizers) are the impediments in agricultural locations; and snow and ice must be contended with in the arctic regions. These contaminants are the cause of erosion, corrosion and fouling. Some contaminants like salt, when combined with high temperatures, attack turbine blade and nozzle material. This is commonly referred to as sulfidation attack and can seriously and rapidly reduce performance and shorten turbine life. A general overview of the characteristics of particles and particle size analysis methods can be found in Appendix C-1.

In the effort to improve performance and lower fuel consumption and emissions, secondary problems are often created. Two examples are: the injection of steam or water into the combustor to increase power output and decrease fuel consumption, and the installation of evaporative coolers or fogger systems in the gas turbine inlet to reduce temperature and increase power output. For example, if potable (drinking) water were injected into the combustor the mineral deposits that would form on the hot turbine airfoils would render them completely useless in less than a year. Another example, one of overtreatment, would be the use of demineralized/deionized (DI)

Chart 8-1 Gas Turbine Environments

ENVIRONMENT	RURAL COUNTRY SIDE	COASTAL AND PLATFORM	LARGE CITES (POWER STATIONS CHEMICAL PLANTS)	INDUSTRIAL AREAS (STEEL WORKS, PETRO-CHEM. MINING)	DESERTS (SAND STORMS, DUSTY GROUND)	TROPICAL	ARCTIC	MOBILE INSTALLATIONS
WEATHER CONDITIONS	DRY AND SUNNY, RAIN, SNOW, FOG	DRY AND SUNNY, RAIN, SNOW, SEA MIST, FREEZING FOG IN WINTER	DRY AND SUNNY, RAIN SHOW, HAIL STONES, SMOG	DRY AND SUNNY, RAIN SNOW, HAIL STONES, SMOG	LONG, DRY, SUNNY, SPELLS; HIGH WINDS; SAND AND DUST STORMS; SOME RAIN	HIGH HUMIDITY TROPICAL RAIN INSECT AND MOSQUITO SWARMS	HEAVY SNOW HIGH WINDS, ICING CONDITIONS INSECT SWARMS IN SUMMERTIME IN SOME AREAS	All POSSIBLE WEATHER CONDITIONS
TEMPERATURE RANGE °F	-4 to +86	–4 to +77	–4 to +95	–4 to +95	+23 to +113	+41 to +113	–40 to +41	–22 to +113
TYPES OF DUST	DRY, NON-EROSIVE	DRY, NON-EROSIVE BUT SALT PARTI-CLES EXIST; CORROSIVE MIST	SOOTY-OILY; MAY BE EROSIVE, ALSO CORROSIVE	SOOTY-OILY; EROSIVE; MAY BE CORROSIVE	DRY, EROSIVE IN SAND STORM AREAS; FINE TALC-LIKE IN AREAS OF NON-SAND STORMS, BUT DUSTY GROUND	NON-EROSIVE MAY CAUSE FOULING	NON-EROSIVE	DRY, EROSIVE, SOOTY-OILY, CORROSIVE
DUST CONCENTRATION gr/1000 ft.^{3}	0.004-0.0436	0.004-0.0436	0.01-0.13	0.043-4.36	0.04-306	0.004-0.10	0.004-0.10	0.04-306
PARTICLE SIZE	0.01 - 3.0	.01 - 3.0 Salt 5.0	.01 - 10.0	.01 - (50)*	1 - (500)**	.01 - 10.0	.01 - 10.0	.01 - (500)***
EFFECT ON G.T.	MINIMAL	CORROSION	FOULING SOMETIMES CORROSION AND FOULING	EROSION SOMETIMES CORROSION AND FOULING	EROSION CORROSION	FOULING	PLUGGING OF AIR INTAKE SYSTEM WITH SNOW AND ICE	FOULING EROSION CORROSION

* IN EMISSION AREAS OF CHIMNEYS
** DURING SEVERE SAND STORMS
*** AT TRACK LEVEL AND/OR DURING DUST STORMS

water to wet the evaporative cooler pads. In this case, the DI water would break down the bond in the evaporative cooler pads and the filter (assuming that the filters are downstream of the cooling flow) and reduce them to a mush-like substance. On the other hand, using water with a high calcium (C_a) or calcium carbonate (C_aCO_3) content would result in calcium deposits throughout the compressor and the cooling passages of the turbine.

Also, contaminant particle size varies within each area. For example, in a desert environment 50% of the contaminant consists of particles smaller than 1.5 microns (μm). The following details the size of contaminant and the major components:

>5.0 μm dust, rain, fog, chemicals, minerals and metals

1-2 μm dust, salts, fog, soot, chemicals, minerals and metals

0.3-0.5 μm hydrocarbon emissions, smog.

INLET AIR FILTERS

Filter types and configurations should be selected on the basis of the environment in which they are expected to operate, the amount of maintenance available, and the degree of protection expected. **Inertial Separators** are available with and without dust removal capability (Figure 8-1). They are useful in areas where there is a significant level of contaminants above 25 microns. **Prefilters** provide a similar function as the inertial separator and their initial cost is considerably lower. **Intermediate Filters**, with and without viscous coating, help to extend the life of the high efficiency filter. Intermediate filters must be capable of capturing 30% or more of the target micron size of the problematic contamination in the environment. **High Efficiency Filters** must be capable of capturing in excess of 95% of the problematic contamination down to 1.0 micron. Prefilters, intermediate filters, and high efficiency filters are commonly available in the barrier or pad type configuration. These are usually 24 inch × 24 inch square with depths ranging from 6 inches to 24 inches (Figure 8-2). This configuration lends itself to stacked filters. That is the intermediate filter and the high efficiency filter elements may be mounted back to back. This is convenient and economical.

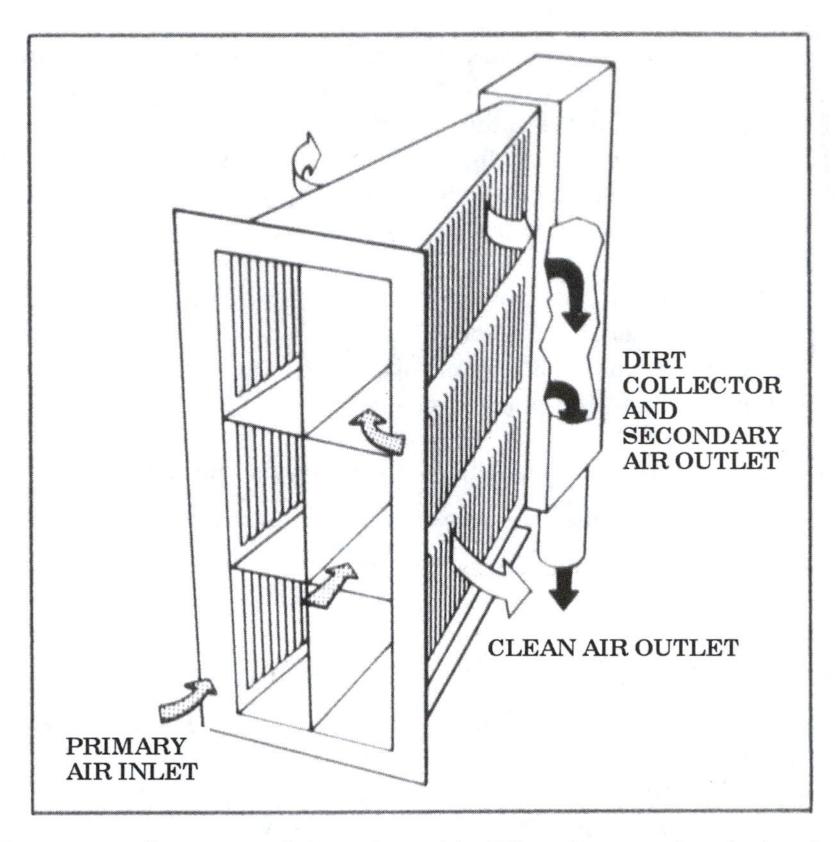

Figure 8-1. Courtesy of American Air Filter International. A schematic view of an inertial separator showing the paths taken by the clean air flow through the separator and the dirty air out of the separator.

Another type of filter is the cannister filter (Figure 8-3). This filter is provided with and without a self-clean feature, and with and without prefilters. Unlike barrier type filters, the cannister filters do not include an intermediate filter stage (although it has been proposed to install the cannister filter as the intermediate filter stage immediately upstream of a high efficiency barrier filter). The "self-clean" cannister filter design uses high pressure compressor air to backflow or reverse flow each filter cartridge. The reverse flow is designed to blow excess dust from the element. In principle these filters would accumulate a dust load to increase efficiency up to a target level. At the target differential pressure the reverse flow, self-clean, feature would

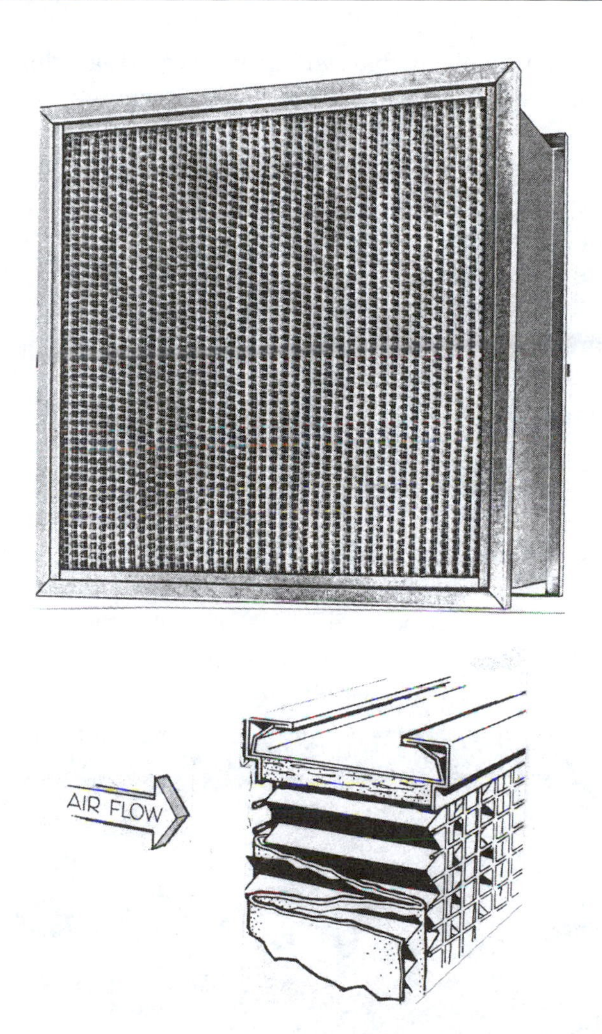

Figure 8-2. Courtesy of American Air Filter International. A typical 24" x 24" barrier filter. Note the insert sketch that details the internal labyrinth, which provides a large media surface area while maintaining a constant filter frontal area.

maintain the filter efficiency level. These filters evolved for desert and arctic regions where blowing sand or snow would quickly overload and clog the barrier type filter. Cannister filters are very effective at and above 5.0 microns. However, in areas where the major contamination is from particulate at and below 2.0 microns, cannister filters alone are inadequate. Attempts to improve the overall efficiency by installing a high efficiency barrier filter downstream have not been widely accepted due to the initial cost of that configuration.

Since site conditions may change during the life of the gas turbine plant, it is advisable to periodically test the environment and the effectiveness of the inlet filter system. This is particularly appropriate if major changes in surroundings are evident. Contaminants such as salts can be present in sufficient quantities and in the range where filter effectiveness is inadequate. This situation will, in time, result in significant sulfidation corrosion of turbine hot gas path parts (primarily, the turbine blades and turbine nozzles). A properly functioning filter system will protect the gas turbine from erosion, corrosion, and fouling; and help achieve performance, efficiency, and life expectations. However, a filter system that is not functioning properly exposes the gas turbine to these undesirable elements resulting in shorter time-between-overhauls (TBO), reduced power output, and increased fuel consumption.

Operators may first become aware of problems when their units require more frequent compressor cleaning to maintain power output. Or they may find evidence of erosion, corrosion, or contaminant

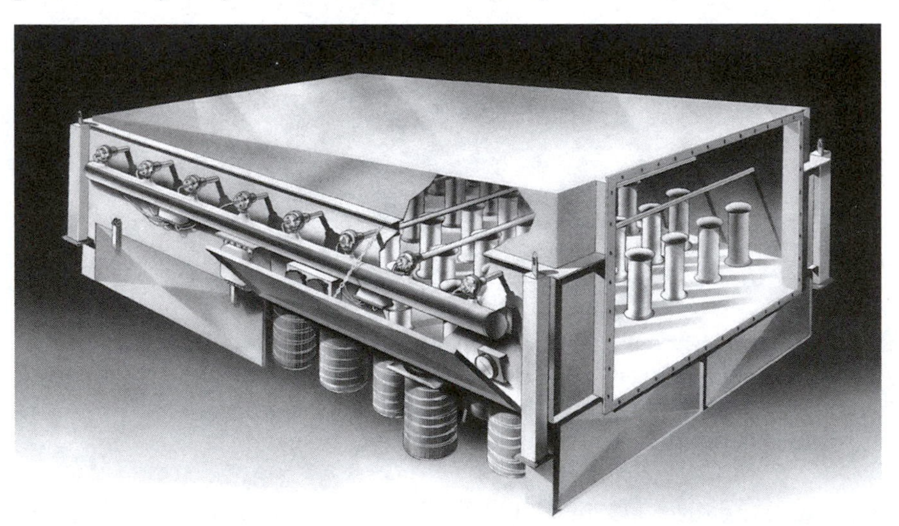

Figure 8-3. Courtesy of Farr Filter Company. The cartridge or canister filter proved to be an extremely effective filter for desert and arctic environments. Part of its success is attributed to the self-clean feature designed into the module.

buildup on the compressor blades and stators, or the turbine blades and nozzles during an inspection of the gas turbine (boroscope or physical teardown).

The mere presence of a filter is no guarantee that the required filtration is being achieved. To ensure full filter efficiency, the following items must be checked:

- The filter elements have been properly installed.
- The correct type of filter media has been installed.
- The filter element is being properly maintained.
- The inlet plenum is being properly maintained.
- The operating environment has not changed since the unit was installed.

A decrease in gas turbine power output, a decrease in compressor efficiency, or an increase in the frequency of cleaning the gas turbine compressor are all signs that the filter system is not functioning properly. Filters become more efficient as they load up with contaminants. The dust laden filter is significantly more effective than a new (unloaded) filter. However, as the dust load increases so does the pressure drop across the filter and this adversely affects gas turbine performance and power output. Therefore, the operator must determine the amount of power degradation his operation can "tolerate," and allow the filters to load to that level. Too frequent cleaning of the self-clean filter will reduce its efficiency and could possibly damage the filter element.

Once a filter system has been installed the operator or maintenance personnel have the task of maintaining or even improving the filter system. For the barrier type filter configuration there are new, more efficient, filter elements in the same 24-inch × 24-inch size that will fit into the existing rack. If prefilters were not part of the initial design, they can be added with a simple, inexpensive, modification to the main filter rack. Considering the cannister type filter, a prefilter wrap can be installed after the target differential pressure is reached. This will extend the time at this level of efficiency. Operators should establish a target filter differential pressure based on their operating requirements, and they should allow the filter to load to that differential pressure target. For most installations it is advisable to rely solely on the automated mechanism to clean the filters when the target differential pressure is reached. Manual activation of the cleaning mechanism should be considered only after the unit has reached its target differential pressure and the interval between cleaning cycles becomes noticeably short.

Other factors such as velocity of the air through the filter, the filter media's ability to tolerate moisture, and air leaks around the doors and windows also influence filter performance. Deteriorated seals around the filter elements and air leaks around the doors are especially detrimental to the effectiveness of inertial separator type filters (as they reduce the velocity through the separator).

FILTER & GAS TURBINE MATCH

The filter selection should be matched to the site environment, the gas turbine type, and economics. This selection should be based on the contaminants at the site location, the air flow into and through the gas turbine, and the level of support located on site. For example, a self-clean cannister type filter system may be well suited for the desert environment where sand storms will load it up quickly, and the self-clean feature will maintain its efficiency and pressure drop at an acceptable level. However, it is ill suited for installations where the element takes too long to load up, and where it does not filter a majority of the contaminants that are detrimental to the gas turbine. Figure 8-4 provides a comparison of the effectiveness of various filter types.

INLET AIR COOLING

Rating Curve

Gas turbines are rated for a power output level (horsepower or megawatts) at a temperature and altitude condition. Two standards used are the International Standards Organization (ISO) and the National Electrical Manufacturers Association (NEMA). ISO conditions are defined as 59°F inlet temperature, 14.7 psia inlet pressure, and 60% relative humidity, and NEMA conditions are defined as 80°F inlet temperature and 1,000-foot elevation. For surface applications, a rating curve showing power vs. inlet temperature is widely used (Fig-

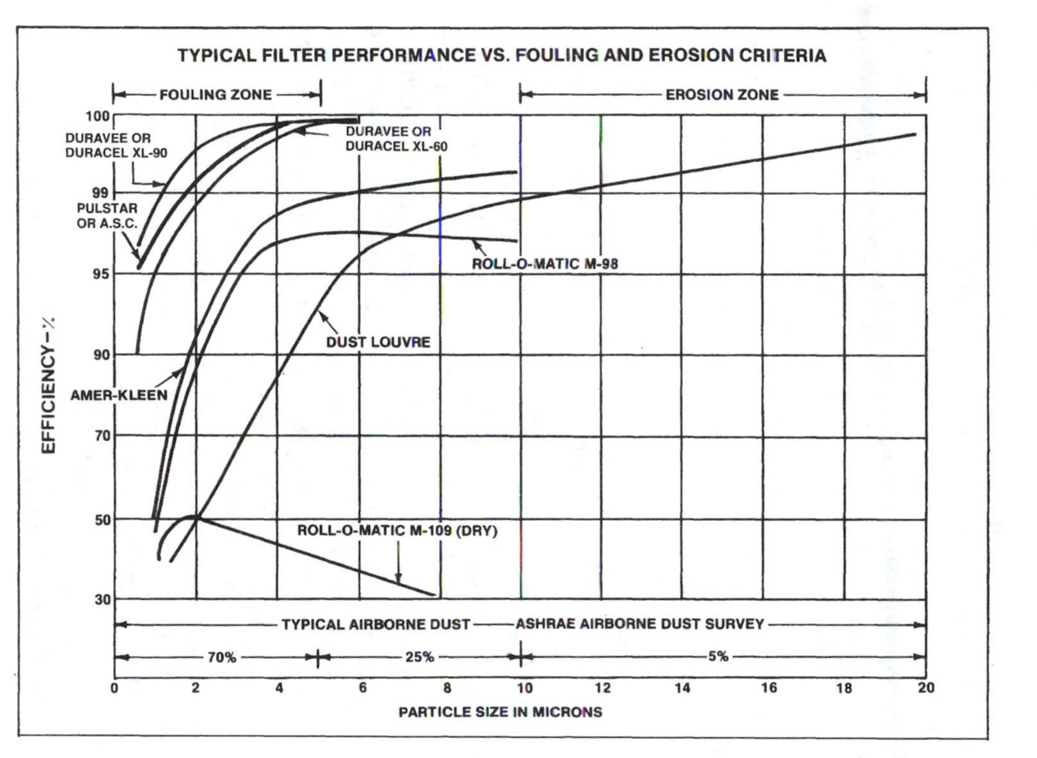

Figure 8-4. Courtesy of American Air Filter International. This filter performance curve shows the particle size distribution attributable to erosion and fouling. Manufacturers had found it relatively easy to filter out the large 'erosion causing' particles.

ure 5-3). This power curve is influenced by four distinct factors: inlet air temperature, turbine inlet temperature (i.e., combustor discharge temperature), rotor speed(s), and compressor aerodynamics.

Turbine inlet temperature is a limiting factor in all gas turbines, regardless of design configuration. This limit has been extended through utilization of internal turbine airfoil cooling. Limitations due to speed and compressor aerodynamics are distinguished as break points or slope changes in the power curve. Often, these limitations are of little consequence since they are small relative to the turbine inlet temperature limitations. The effects of increasing ambient temperature on gas turbine output clearly indicate the advantages of cooling the compressor inlet, especially in hot climates. Lowering the compressor inlet temperature can be accomplished by the installation of an evaporative cooler, a fogger system, or a chiller in the inlet ducting.

Evaporative Cooling

The evaporative cooler is a cost-effective way to recover capacity during periods of high temperature and low or moderate relative humidity. The biggest gains are realized in hot, low humidity climates. However, evaporative cooler effectiveness is limited to ambient temperatures of 50°F (10°C) to 60°F (16°C) and above. Below these temperatures, parameters other than turbine temperature will limit gas turbine operation. Also, as the inlet air temperature drops the potential for ice formation in the wet inlet increases.

Evaporative cooler effectiveness is a measure of how close the cooler exit temperature approaches the ambient wet bulb temperature. For most applications coolers having an effectiveness of 80%-90% provide the most economic benefit. The actual temperature drop realized is a function of both the equipment design and atmospheric conditions. The design controls the effectiveness of the cooler, defined as follows:

$$\textbf{cooler effectiveness} = (T_{1DB} - T_{2DB})/(T_{1DB} - T_{2WB}) \qquad (8\text{-}1)$$

where: T_{1DB} is the dry bulb temperature upstream of the cooler
T_{2DB} is the dry bulb temperature downstream of the cooler
T_{2WB} is the wet bulb temperature downstream of the cooler.

As an example, assume that the ambient temperature is 100°F and the relative humidity is 20%. Referring to the Psychometric Chart[1] (Figure 8-5), the corresponding wet bulb temperature is 70°F. For an 80% effective design cooler, the temperature drop through the cooler should be:

$$\Delta T_{DB} = 0.8(T_{1DB} - T_{2WB}), \text{ or } 24F. \tag{8-2}$$

The actual cooler effectiveness can be determined by measuring dry-bulb temperatures and either relative humidity or wet-bulb temperatures before and after the evaporative cooler. This testing determines if the design target is being met and if there is margin for improving the evaporative cooler effectiveness. The exact increase in available power attributable to inlet air cooling depends upon the machine model, ambient pressure and temperature, and relative humidity. Likewise, the decrease in heat rate attributable to inlet air cooling depends upon the machine model, ambient pressure and temperature, and relative humidity. Figure 8-6[1] can be used as a rough

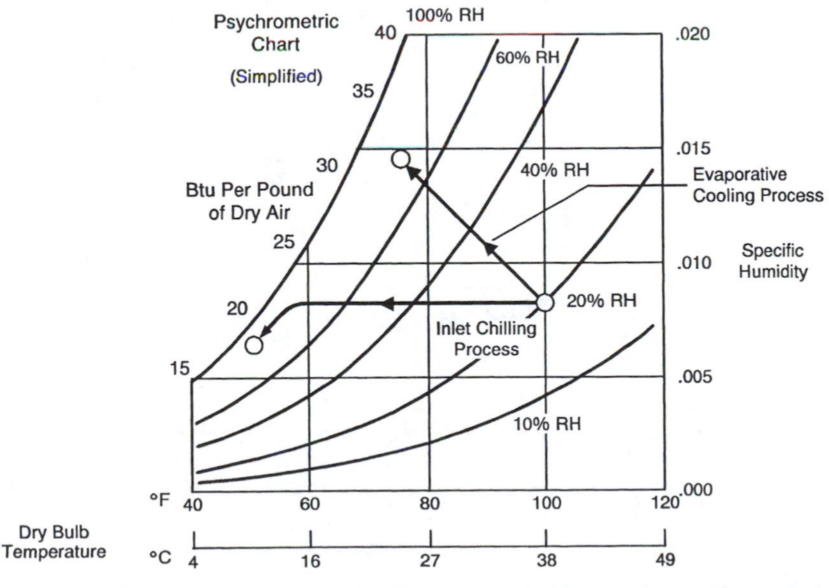

Figure 8-5. Courtesy of General Electric Company. This psychrometric chart shows the path taken by an evaporative cooling process and an inlet air chilling process.

estimate of the effect of inlet cooling. As expected the improvement is greatest in hot dry weather. The addition of an evaporative cooler is economically justified when the value of the increased output exceeds the initial and operating costs, and appropriate climatic conditions permit effective utilization of the equipment.

Water quality is a concern. The water must be treated to remove contaminants such as salt, calcium, magnesium, aluminum, etc. When wetted (100% humidity) these contaminants function as electrolytes and can result in severe corrosion of materials. Careful application of this system is necessary, as condensation or water carryover can intensify compressor fouling and degrade performance. The higher the levels of dissolved solids and salts in the water the greater the effort must be to avoid water carryover. Besides control-

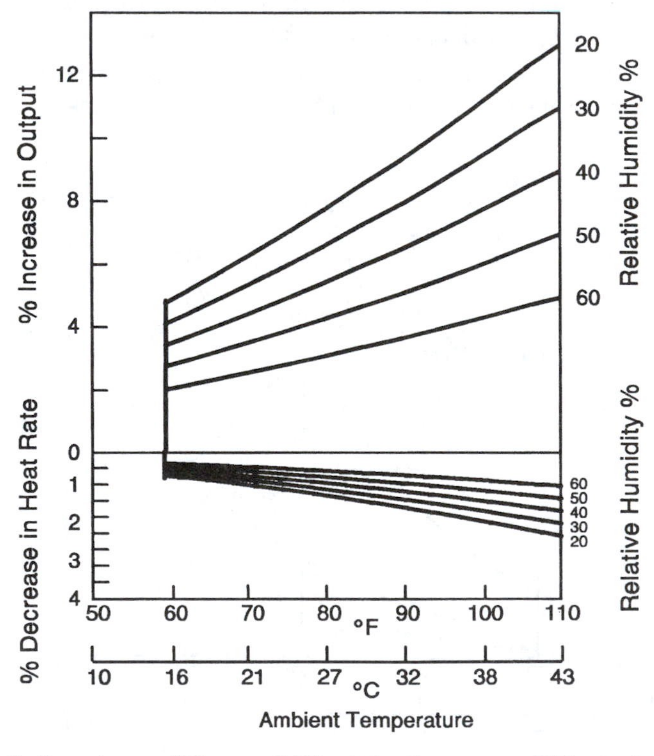

Figure 8-6. Courtesy of General Electric Company. Effect of evaporative cooling on power output and heat rate as a function of ambient air temperature and relative humidity.

ling water quality, controlling water flow rate is also important (too little is ineffective and too much may be detrimental to machine life). There are two types of evaporative coolers: pad and fogger.

The pad-type cooler consists of wettable porous media pads that are kept moist by water sprayed or dripped continuously on their upper edges. Air flows horizontally through the wetted media where the water wicks through the media wall and into the airstream. Pad-type coolers require low (less than 100 psig) pressure water flow and can tolerate moderate levels of calcium carbonate (less than 150 ppm). Calcium carbonate levels between 50 ppm and 150 ppm require additional water flow to wash contaminants from the media. The media pads have approximately the same capture efficiency as a pre-filter. Media pad life is a function of clogging, sagging, decay, and bacterial action. To maximize the life and effectiveness of the media pads, these coolers should be installed downstream of the inlet air filter system. The media pads should be sufficiently dried and removed when not in use for prolonged periods of time.

A fogger-type cooler consists of water flow nozzles placed across the face of the gas turbine inlet and a coalescer stage. These nozzles distribute a fine mist of water into the air stream and the coalescer stage eliminates non-evaporated water carry over. The quantity of fogger nozzles is a function of nozzle orifice size, spray angle, cross sectional area of the gas turbine inlet, and air flow velocity. Nozzle orifice size is a function of the residence time the droplet has to evaporate and the amount of contaminate in the water. Residence time is defined as the water droplet size (measured in microns) and the duct length from the plane of the fogger nozzles to the coalescer vanes. Typical nozzle sizes range from 0.006 inches to 0.010 inches (producing droplets in the 25- to 75-micron size range) and water pressure ranges from 300 psig to 1,000 psig. Due to the small nozzle sizes even moderate levels of calcium carbonate in the water present a problem. Therefore, it is recommended that water hardness (C_aCO_3) be reduced to less than 10 ppm. Operation between 10 ppm and 50 ppm is not impossible but will incur increased maintenance with smaller nozzle sizes. Fogger-type evaporative cooler nozzles can be placed either before or after the air filters. (However, due to the "foreign object" hazard created by the number and size of the nozzles immediately upstream of the gas turbine compressor inlet, it is recommended that this system only be installed upstream of the air filters.)

WET COMPRESSION

As an extension of the fogger-type cooling approach, water is allowed to enter the compressor and evaporation takes place within the compressor. When water droplets enter the compressor the process is referred to as "overspray" or "wet compression."

Evaporation of the water droplets inside the compressor provides continuous cooling of the air thus leading to a reduction in the compressor work and compressor discharge temperature for a given pressure ratio, and a change in the stage work distribution. Since 65% to 75% of the turbine work is used to drive the compressor, a reduction in compressor work results in an increase in shaft output power. While more fuel is required to bring the vaporized water/air mixture up to a given turbine inlet temperature, the percentage increase in power output is greater than the percentage increase in fuel consumed resulting in a net decrease in overall heat rate.

Figure 8-7. Fogger spray upstream of the compressor inlet.

Wet compression results in higher compressor airflow at a given speed and pressure ratio, which tends to unload the first few compressor stages and to increase loading on the last few stages. The maximum desirable ratio of water-to-air flow is limited by compressor surge or stall and combustor efficiency. A coating of liquid on the airfoil surfaces will change the blade path geometry and the related position of the surge line. The injection of untreated water will result in contaminants being deposited on the airfoil surfaces. This, in turn, leads to a change in airfoil geometry and the position of the surge line. A more general impact is a rapid decrease in compressor performance. Therefore, it is essential that the water be treated as discussed earlier in this chapter. The combustor and turbine performance is also affected by wet compression but the affect is small for the range of water-to-air flow ratios encountered.

The mass flow capability of some compressors is limited as evidenced by the decrease in available power on cold days *(Figure 5-3)*. Therefore, when increasing the mass flow care must be

Figure 8-8. Reverse osmosis water treatment prior to installation.

taken not to exceed the structural capability of the compressor.

A second concern is the formation of ice in the gas turbine inlet if the wet compression system is operated at or below 50 degrees F. Without the protection of the inlet air filters ice could be ingested directly into the gas turbine compressor. This could result in severe damage to the compressor and possibly the entire gas turbine.

Another concern is foreign object damage (FOD) caused by a loose or broken nozzle head. Some fog system manufacturers recognize this potential and have provided a "tie" (made of plastic or stainless steel wire) to retain the nozzle if it should separate from the supply line. Each time the nozzle is removed and replaced this "tie" must be cut off and a new one installed. Considering that the inlet of the large gas turbines could have several hundred of these nozzles and "ties," the possibility of just one piece being left loose in the inlet increases exponentially. A loose nozzle head would do an enormous amount of damage to the gas turbine compressor components and possibly also the turbine components. Other system manufacturers have placed the wet compression nozzles far enough up stream of the compressor inlet such that if a nozzle should break loose it would fall to the bottom of the duct and come to rest before getting close to the compressor inlet.

An ideal application would install an inlet fogger evaporation system upstream of the inlet air filters and a wet compression system upstream of the compressor inlet (either upstream or down stream of the acoustic silencers, but downstream of the filters). The fogger evaporation system would reduce the inlet temperature and the wet compression would provide compressor intercooling. An example is a study sponsored by the Electric Power Research Institute (EPRI) and reported in 1995.[2] In this study a General Electric MS7001 was used and a fogger evaporative cooling system injected 23 gpm upstream of the inlet air filters. This alone boosted power 10% (from 61 MW to 67.1 MW).

A wet compression system installed downstream of the acoustic silencers injected 23 gpm of water directly into the compressor inlet. The wet compression system increased power an additional 3.8% to 69.4 MW. Placing the fogger evaporative cooling nozzles upstream of the inlet air filters eliminated the need for highly treated water as the filters would "catch" any contaminants in the water-air stream. This left only the wet compression water to be highly treated.

A guide for estimating the affects of water fogging is provided in detail in Appendix C-11.

CHILLERS

Chillers, unlike evaporative coolers, are not limited by the ambient wet bulb temperature. The power increase achievable is limited by the machine, the capacity of the chilling device to produce coolant, and the ability of the coils to transfer heat.

Cooling initially follows a line of constant humidity ratio (Specific Humidity). As saturation is approached, water begins to condense from the air. Further heat transfer cools the condensate and air, and causes more condensation (Figure 8-5 and 8-9). Because of the relative high heat of vaporization of water, most of the cooling energy in this regime goes to condensation and little to further temperature reduction. Therefore, since little benefit is achieved as the cooled temperature approaches the dew line, chillers should be designed to avoid forming excessive condensate.

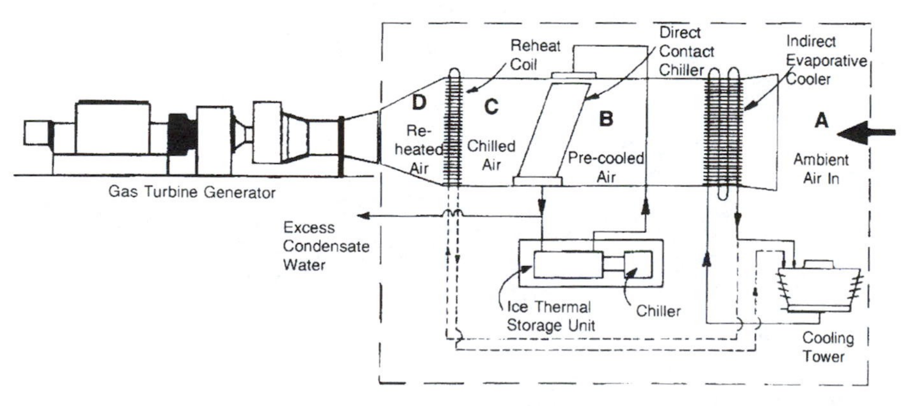

Figure 8-9. Courtesy of Baltimore Aircoil Company. This gas turbine inlet air conditioner combines three steps of air conditioning: Air Pre-Cooling, Air Chilling, and Air Reheat.

References

1. "GE Gas Turbine Performance Characteristics," Reference No. GER-3567C, by F.J. Brooks, GE Industrial & Power Systems, Schenectady, NY, August 1992.
2. Gas Turbine and Combined Cycle Capacity Enhancement, Second Interim Report, EPRI TR-104612, January 1995.

Chapter 9

Gas Turbine Exhaust Treatment

To one degree or another we all live with air pollution. The wind and rain continuously dilute and remove these polluting contaminants (although to some extent this merely transforms them from air pollution to water pollution). Awareness of air pollution is greater around commercial and industrial fossil fuel burning plants and in our large cities worldwide. Air pollution is a problem because, in sufficient quantities, it is detrimental to our health and that of our children. We have identified the components of air pollution as hydrocarbons, carbon monoxide, sulfur oxides, and nitrogen oxides. We have also determined that, on a worldwide basis, nature produces more pollution than man. As shown in Table 9-1 nature produces more than three times the amount of sulfur oxides and ten times the amount of nitrogen oxides than does man. This table is a compilation of the annual emission rates of hydrocarbons, carbon monoxide, sulfur compounds, and nitrogen oxides for which the world-wide studies of sources, concentrations, and sinks have been made.

While this table shows that natural sources exceed man-made sources, locally—where most man-made pollutants are produced—the picture may be entirely different. An example of this is in our large cities where the automobile makes a large contribution to the amount of pollution present. When we look at the amount of contaminant generated by each fossil fuel burning piece of equipment, and the number and location of such equipment, we begin to understand that we can and must improve our own environment.

The products of combustion from gas turbines burning hydrocarbon fuels are detailed in Table 9-2.

Table 9-1.

NATURAL & MAN-MADE SOURCES OF ATMOSPHERIC CONTAMINANTS

Emission	Natural-kg/yr	Man-Made - kg/yr	Ratio Natural/ Man Made
Hydrocarbons[1]	159×10^9	24×10^9	6
Carbon Monoxide[2]	3.2×10^{12}	0.24×10^{12}	13
Sulfur Oxides[1,3,4]	415×10^9	139×10^9	3
Nitrogen Oxides[5,6]	45×10^{10}	4.5×10^{10}	10

Table 9-2.
GAS TURBINE EXHAUST PRODUCTS FROM HYDROCARBON FUEL COMBUSTION IN DRY AIR

CONSTITUENT	% BY WEIGHT	REMARKS
N_2 - Nitrogen	74.16	Mostly inert, from atmosphere
O_2 - Oxygen	16.47	From excess air
CO_2 - Carbon Dioxide	5.47	Product of complete combustion
H_2O - Water	2.34	Product of complete combustion
A - Argon	1.26	Inert, from atmosphere
UHC - Unburned Hydrocarbons	trace	Product of incomplete combustion
CO - Carbon Monoxide	trace	Product of incomplete combustion
NO_x - Oxides of Nitrogen	trace	
• Thermal		From fixation of atmospheric N_2
• Organic		From fuel bound nitrogen
SO_x - Oxides of Sulfur		From Sulfur in fuel

A look at how pollution is created in the combustion process leads us to an understanding of how to control or eliminate it at the source.

Carbon Monoxide (CO)

Carbon monoxide emissions are a function of the combustor design, specifically the combustor's primary reaction zone. New combustor designs are being evaluated to reduce these emissions. In the interim, CO can be effectively treated with a catalytic converter.

Oxides of Nitrogen (NO_x)

Oxides of nitrogen are produced primarily as nitric oxide (NO) in the hotter regions of the combustion reaction zone of the combustor. Nitrogen oxides (NO_x) found in the exhaust are the products of the combustion of hydrocarbon fuels in air. In this process nitrogen oxides are formed by two mechanisms: Thermal NO and Organic NO. The predominant mechanism in the formation of NO_x in gas turbine combustors depends on such conditions as the reaction temperature, the resident time at high temperature, the fuel/air ratio in and after the combustion reaction zone, the fuel composition [the fuel bound nitrogen (FBN) content], the combustor geometry, and the mixing pattern inside the combustor.

- Thermal NO is extremely temperature dependent, and therefore, is produced in the hottest regions of the combustor. Water and steam injection contribute to reducing the combustion temperature at a given load.

- Organic NO is formed during combustion by chemical combination of the nitrogen atoms, which are part of the fuel molecule, and the oxygen in the air. The amount of Organic NO produced is affected by the nitrogen content of the fuel, a yield factor (a measure of FBN), the fuel/air ratio, and abatement techniques such as water and steam injection.

NO_x emissions from burning various fuels, compared to burning methane (CH_4) gas, can be estimated as shown in Figure 9-1 for gaseous fuels and Figure 9-2 for liquid fuels.[7] As shown in these figures, NO_x emissions for most liquid fuels (except liquid methanol) are higher than methane gas. While the NO_x emissions from most gaseous fuel corresponds to that of methane. This is directly related to the amount of fuel bound nitrogen in the fuel. Fuel gas treatment, such as water scrubbing, can remove some of the nitrogen-bearing compounds (such as hydrogen cyanide). The amount of fuel bound nitrogen in petroleum liquid fuels depends on the source crude oil and the refining process. Properties of liquid and gaseous fuels are detailed in Appendix C7 and C8.

In general, nitrogen oxide emissions increase in direct proportion to the increase in combustion temperature as shown in Figure 9-3. Modifications to the combustor design of a number of gas turbine models have already demonstrated that significant reductions in NO_x are achievable.

Another factor influencing the degree of pollutant level in the atmosphere is the time necessary for each constituent to be consumed by the various mechanisms available—called sinks. As Table 9-3 shows, nitrogen oxides last up to 5 days while carbon monoxide can last up to 3 years.

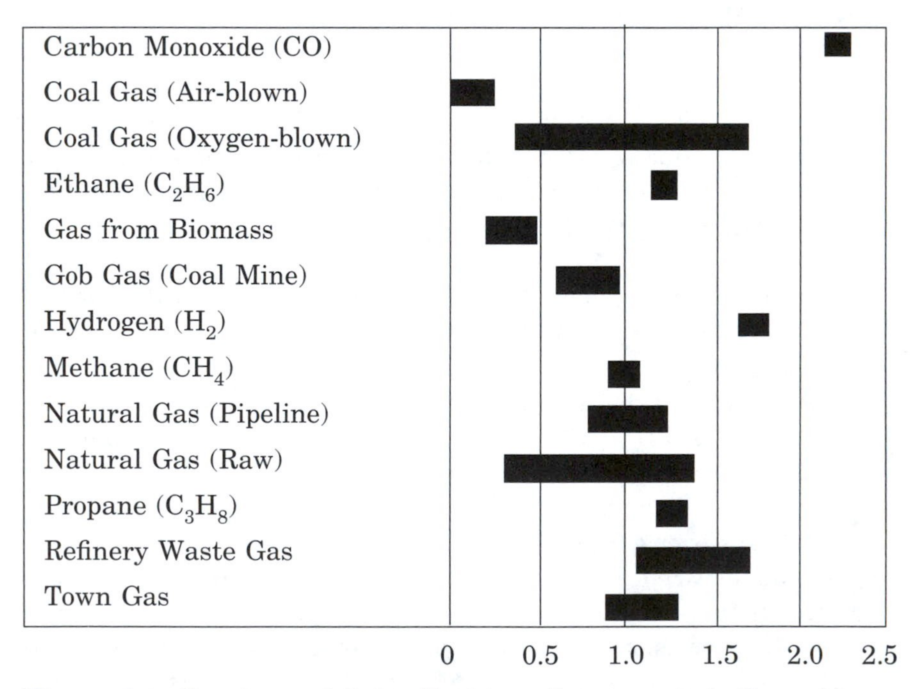

Figure 9-1. Courtesy of Solar Turbines Incorporated. Comparison to methane of the relative Thermal NO_x produced by various fuel gases.

In addition to low NO_x combustor designs, the methods currently in use to reduce emissions are water injection and steam injection into the combustor, and a selective catalytic reactor (SCR) in the turbine exhaust.

Table. 9-3. Residence Time of Atmospheric Trace Elements.

EMISSIONS	LIFE
CO[8]	1-3 years
NO[5]	1/2 - 1 day
NO_2[5]	3-5 days
HC (as CH_4)[8]	1.5 years
SO_x[5]	3 - 4 days
P-M[9]	hours to days (troposphere) days to years (stratosphere)

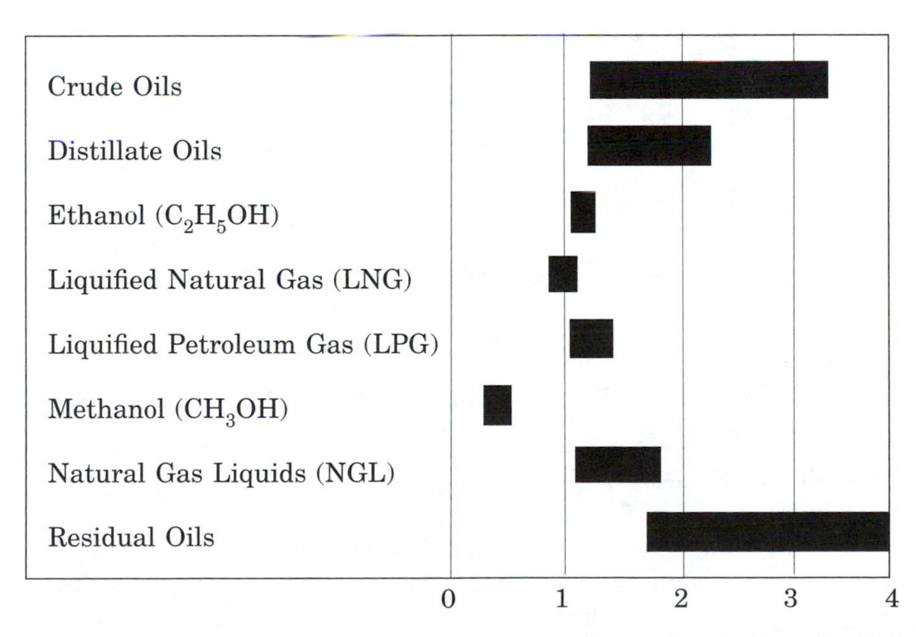

Figure 9-2. Courtesy of Solar Turbines Incorporated. Comparison to methane of the relative Thermal NO_x produced by various liquid fuels.

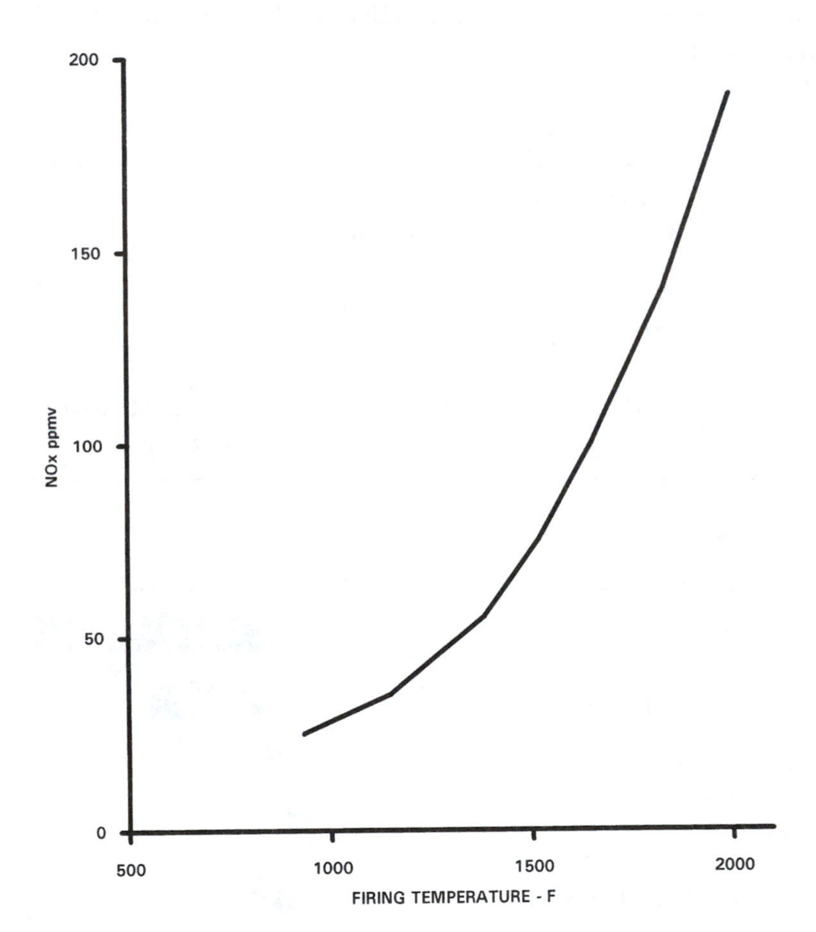

Figure 9-3. Typical NO_x emissions as a function of temperature.

WATER OR STEAM INJECTION

Power augmentation is obtained by cooling the air entering the gas turbine combustor by means of vaporization of water introduced into the airstream. Cooling the air increases the air-density and the mass flow. More air and cooler air to the combustor permits more fuel to be burned before reaching the turbine inlet temperature limit. To obtain effective cooling a liquid with a high heat of vaporization is preferred and water has a high heat of vaporization. Both water and steam injection increase the power output of the gas turbine by adding to the mass flow through the turbine.

WATER INJECTION

During the early development of gas turbines, water was injected into the compressor, diffuser, or combustor to increase power output. This technique was applied to both the aircraft "jet" gas turbine engine to increase takeoff thrust, and the stationary heavy industrial gas turbines for peaking power. Today, in order to control the formation of organic NO, demineralized/deionized water is injected directly into the combustion zones of the gas turbine thereby influencing the chemical reaction of the combustion process. In addition to lowering the flame and gas temperatures, vaporized water also increases the mass flow through the engine. As a result, at a constant power output, combustion and turbine temperatures are reduced. The combination of the reduced combustion temperatures and changes in the chemical reaction can reduce NO_x formation up to 80%. The amount of water necessary to accomplish this reduction in NO_x is a function of the diffuser, combustor, and fuel nozzle design. Water injection rates are generally quoted as a water-to-fuel ratio or as a percentage of compressor inlet air flow. The water injection rate for a typical 80 MW heavy frame gas turbine would be 0.6 water-to-fuel ratio or 1.15 % of total air flow. The amount of water injected into the diffuser or the combustor is limited by several factors. Water injection moves the operating line closer to the compressor surge line (Figure 9-4). Also, too much water will quench the combustion flame, resulting in flame-out.

The water used for NO_x control is demineralized and deionized to eliminate deposits from forming on the hot metal surfaces of the combustor, turbine nozzles, and turbine blades. When handling demineralized/deionized water, care must be taken to select materials that are resistant to its highly reactive attack. Therefore, piping should be AISI 304L and valves and pumps should be 316L stainless steel.

OPERATING LINE SHIFT
WITH WATER INJECTION

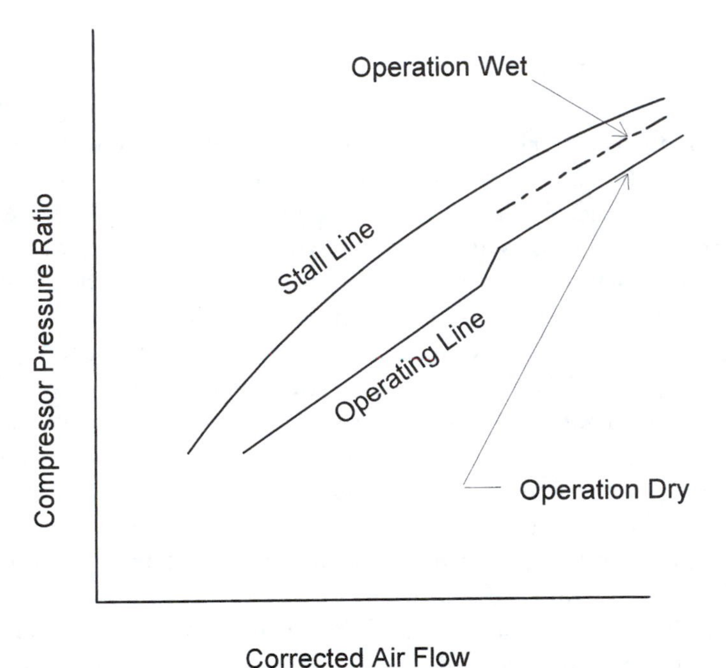

Figure 9-4. Operating line shift with water injection.

STEAM INJECTION

For cogeneration and combined cycle applications, steam is an excellent choice for controlling emissions from gas turbines. This is due not only to the availability of steam at appropriate conditions, but also to the substantial improvement that can be achieved in gas turbine heat rate. Unlike water, the steam entering the combustor is already vaporized. Therefore, the energy to vaporize the steam is conserved. However, because the steam is 300°F (150°C) to 500°F (260°C) degrees hotter than water, its flame quenching capabilities are reduced, and hence more steam is required to accomplish the same amount of NO_x reduction. Steam flow requirements are nominally 1.5-2.0 times those required for water injection. The steam injection rate for a typical aeroderivative gas turbine operating at 25 megawatts is 3.3% of the compressor inlet air flow. As a function of mass flow (steam to fuel), ratios of up to 2.4 by weight have been used[10].

However, there is also a limit to the amount of steam that may be injected into the gas turbine in general, and into the primary combustor zone in particular. The maximum total steam injection into the gas turbine is between 5% and 20% of existing air flow. The allowable amount of steam to be injected into the combustor primary zone is limited by the flameout characteristics of the combustor. Also the allowable amount of steam injected for power augmentation is set by mechanical considerations and compressor pressure ratio limitations.

On a per-pound basis, steam contains more expansion energy than air; the Cp of steam is approximately twice the Cp of air. The power output gain from steam injection is about 4% for each 1% of steam injected (where % of steam injected refers to main turbine flow). The following graph (Figure 9-5) shows the effect of a constant 5% steam injection on the load carrying ability of a typical gas turbine. For comparative purposes, the output variation with ambient temperature is shown for the same turbine without steam injection.

SELECTIVE CATALYTIC REDUCTION

Selective catalytic reduction (SCR) is a process in which NO_x is removed from the exhaust gas stream by the injection of ammonia (NH_3) into the stream and the subsequent chemical reaction in the presence of a catalyst. For a given gas condition (temperature, gas composition, etc.) the performance of the SCR is a function of the catalyst type and geometry, the residence time of the gas in the reactor, and the amount of ammonia injected upstream of the reactor. Selection of the catalyst is specific to the temperature in which it is expected to operate. The ammonia utilized in the process may be either anhydrous or aqueous. The injection systems differ slightly depending on the type of ammonia injected. The basic chemical reactions[11] are:

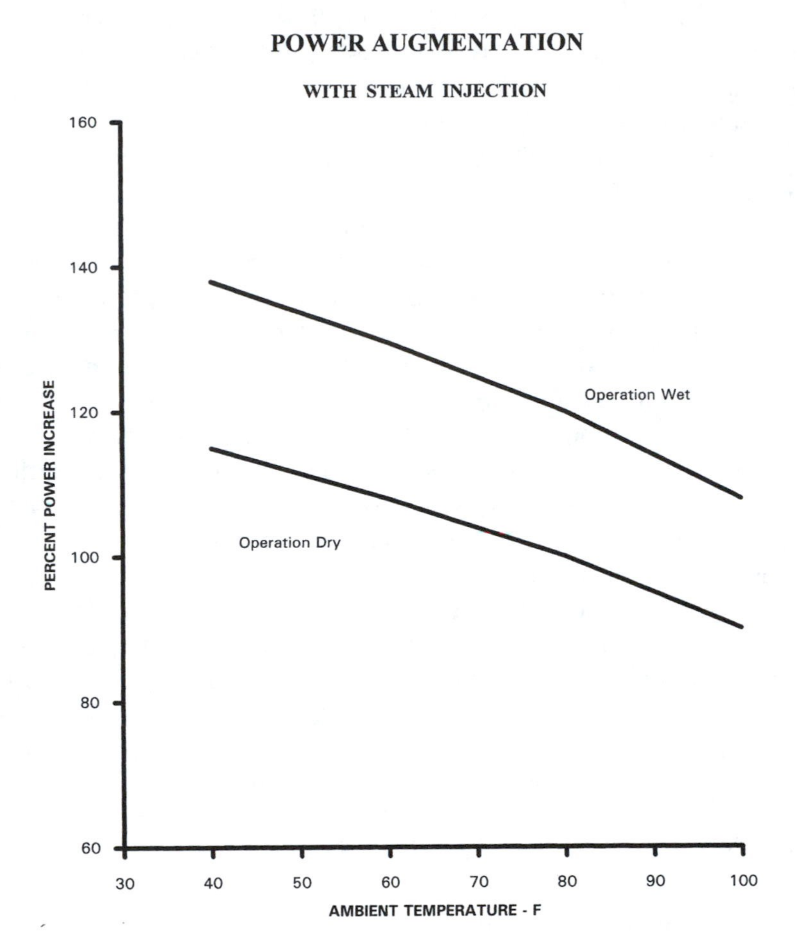

Figure 9-5. Power augmentation with steam injection.

$$4NO + 4NH_3 + O_2 \Rightarrow 4N_2 + 6H_2O \quad (9\text{-}1)$$
$$6NO + 4NH_3 \Rightarrow 5N_2 + 6H_2O \quad (9\text{-}2)$$
$$2NO_2 + 4NH_3 + O_2 \Rightarrow 3N2 + 6H_2O \quad (9\text{-}3)$$
$$6NO_2 + 8NH_3 \Rightarrow 7N_2 + 12H_2O \quad (9\text{-}4)$$
$$NO + NO_2 + 2NH_3 \Rightarrow 2N_2 + 3H_2O \quad (9\text{-}5)$$

SCR catalysts are made of titanium, tungsten, and vanadium elements. The factors that affect the SCR design are type of fuel, dust loading, exhaust gas conditions, and the required NO_x removal efficiency. The efficiency of the SCR can be defined as the quantity of NO_x removed divided by the quantity of NO_x present in the inlet stream.

The SCR reactor consists of the catalysts, housing, and ammonia injection system (Figure 9-6). In most applications the reactor size is larger than the duct size and therefore, diverging/converging transition ducts are used before and after the reactor, respectively. Care should be taken in preparing the physical design of the SCR reactor, particularly with regard to the gas-stream pressure drop. Gas-stream pressure drop in the order of 2 inches (50mm) to 4 inches (100mm) water column are normally acceptable.

Anhydrous ammonia is undiluted, pure, liquid ammonia. In this form it is toxic and hazardous. The liquid anhydrous ammonia is expanded through a heater, mixed with air, and injected into the gas turbine exhaust (Figure 9-7).

Aqueous ammonia (NH_4OH) is a mixture of ammonia (30%) and water (70%). Since it is diluted it is less hazardous than anhydrous ammonia. Injecting aqueous ammonia is only slightly more complicated than injecting anhydrous ammonia. A pump is required to move the aqueous ammonia to a vaporizer tank where it is mixed with hot [575°F (300°C) to 850°F (455°C)] air. There the ammonia/water

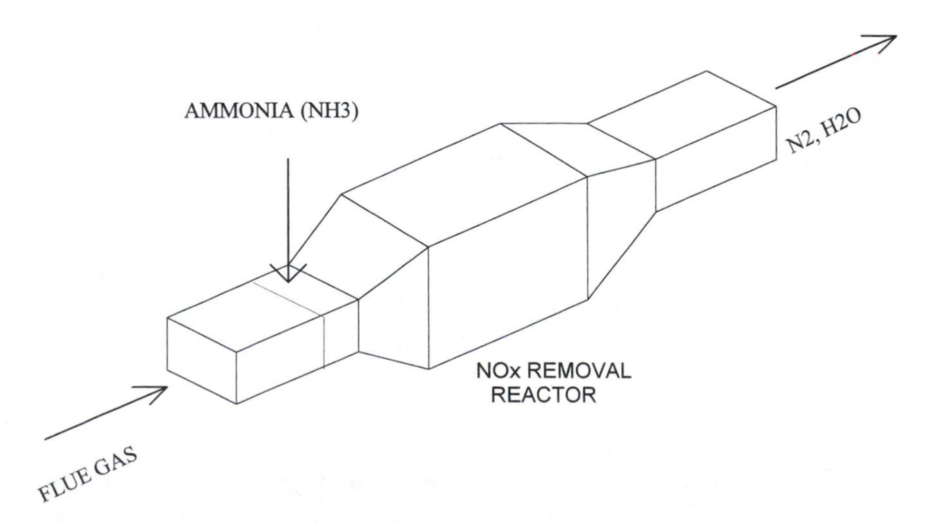

Figure 9-6. Selective catalytic reduction.

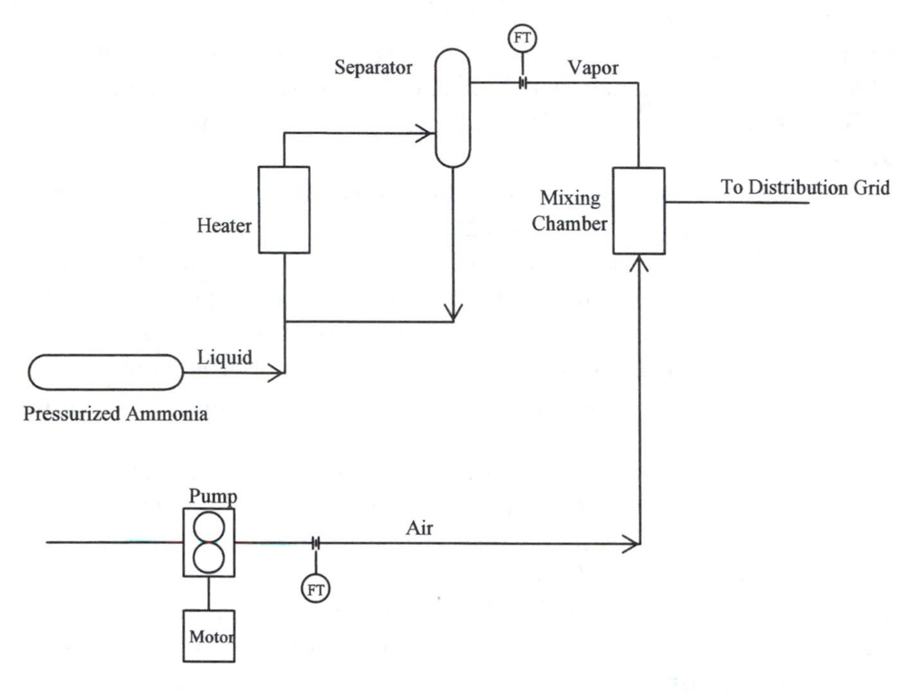

Figure 9-7. Anhydrous ammonia injection system.

mixture is vaporized, and the ammonia vapors and water vapors are forwarded to the injection manifold for injection into the gas turbine exhaust.

EFFECTS OF OPERATING PARAMETERS

Gas Temperature

NO_x removal efficiency peaks at a different temperature for different catalysts. Therefore, the choice of catalyst and location in the exhaust stream depends on the gas turbine's operating profile and temperature.

Water Vapor Concentration

The water vapor content in the exhaust gas has an adverse effect on the NH_3 removal efficiency. Therefore, increases in water vapor content result in lower SCR efficiency.

Aging Effect

The catalyst's performance tends to deteriorate with time. The rate of deterioration is high at the beginning of operation and becomes moderate after initial settlement.

Ammonia Slip

In theory the amount of ammonia injected into the gas stream should equal the amount of NO_x to be removed from the gas stream. However, because the ammonia does not completely mix with the NO_x, more ammonia must be injected. Ammonia slip is the excess residual ammonia in the downstream gas. Also, based on the catalyst selected, SO_2 can convert to SO_3. When the ammonia, water vapor, and SO_3 combine ammonia sulfates are formed as follows[12]:

$$SO_3 + 2NH_3 + H_2O \Rightarrow (NH_4)_2HSO_4 \tag{9-6}$$

$$SO_3 + NH_3 + H_2O \Rightarrow (NH_4)HSO_4 \tag{9-7}$$

Ammonia sulfates are sticky substances that will deposit on the equipment downstream of the SCR. These deposits will result in reducing the cross-sectional area and increasing backpressure. For each inch water column increase in backpressure power decreases by 0.25% and heat rate increases 0.25%. Ammonia sulfate starts forming when ammonia slip is greater than 10 ppm and the SO_3 concentration is greater than 5 ppm. This problem is less likely to occur when exhaust gas temperatures are maintained above 400°F (205°C) and natural gas or low sulfur fuel is used.

References

1. "Atmospheric Chemistry: Trace Gases and Particulates," W.H. Fisher (1972).
2. "The Sulfur Cycle," W.W. Kellogg (1972).
3. "Ground Level Concentrations of Gas Turbine Emissions," H.L. Hamilton, E.W. Zeltmen (1974).
4. "Air Quality Standards National and International," S. Yanagisawa, (1973).
5. "Emissions, Concentrations, And Fate Of Gaseous Pollutants," R. Robinson, R.C. Robbins (1971).
6. "Abatement of Nitrogen Oxides Emission for Stationary Sources," National Academy of Engineers (1972).
7. "Gas Turbine Fuels," W.S.Y. Hung, Ph.D., Solar Turbines, Inc, 1989.

8. "Carbon Monoxide: Natural Sources Dwarf Man's Output," T.N. Maugh II (1972).
9. "Physical Climatology," W.D. Sellers (1965).
10. "Examine Full Impact Of Injecting Steam Into Gas-Turbine Systems" Elizabeth A. Bretz, *Power Magazine*, June 1989.
11. "Design And Operating Experience of Selective Catalytic Reduction Systems for NO_x Control In Gas Turbine Systems," S.M. Cho and A.H. Seltzer, Foster Wheeler Energy Corp., and Z. Tsutsui, Ishikawajima-Harima Heavy Ind.
12. "Design Experience of Selective Catalytic Reduction Systems For Denitrification of Flue Gas," S.M. Cho, Foster Wheeler Energy Corp.

Chapter 10

Combustion Turbine Acoustics and Noise Control

COMBUSTION TURBINES—LOTS OF POWER IN A LITTLE SPACE

A very attractive attribute about combustion turbines is the modular packaging of these units for fairly rapid delivery and installation. Some OEMs or other packagers package the units to produce no more than a certain sound level at 400 feet (or at some *standard* distance) from the site, but multiple units co-located and combined with other sources of noise can cause some difficulty in determining the total sound level at a location. The complexity of combustion turbine systems and their ancillary equipment generally requires an acoustical engineer well experienced with such systems to fully analyze the acoustical emissions and make a determination whether the expected sound levels meet regulatory requirements, are in balance with the environment or community, or need some level of noise mitigation. Industrial combustion turbines are built as compact high power delivery systems and the only place available to reduce the noise is by adding on duct-silencing equipment to the front and rear of the turbine system, which requires real estate and hardware that increase cost. Technological advances over the last two decades have given designers more tools and materials to develop some significant noise control breakthroughs but these engines still produce significant noise levels that must be mitigated or reduced to make these engines compatible with either environmental or occupational regulatory noise requirements. Combustion turbines typically have inlet (compressor end) and exhaust (turbine end) total sound power levels ranging from 120 decibels (dB) to over 155 dB. A reduction of 20-30 dB is fairly easy, 30-50 is typical of most silencer installations, and reducing levels over 60 dB require careful analysis to include duct breakout noise, flanking noise, and silencer self-flow noise to be considered.

Noise from combustion turbines can cause a dramatic change to the environment, particularly in rural settings and communities if adequate noise control or mitigation is not incorporated. This can cause negative community responses when future expansions are needed or new industrial plants are planned. Potential environmental noise impacts[1] must be considered at the earliest planning stages of a project. After a project is completed, post installation of noise control equipment frequently proves to be very expensive and may also negatively affect turbine performance. Another area that is sometimes overlooked is low frequency noise or infra-sound[2-4] that can cause lightweight structures to vibrate from the low frequency airborne noise coupling with the building or structure. This can have a dramatic impact upon residential communities, hospitals, offices, schools, and nearby industrial/commercial enterprises that operate high precision equipment. The low frequency noises we are addressing are usually associated with simple cycle exhaust systems (turbine exhausts directly to the environment) that produce low frequency tones (less than 40 Hz) that can propagate over large distances; low frequency noise problems have occurred over one-mile (1.6 km) from sites.

Recent development of microturbines for local power production and other applications has gained interests over the last several years. These units have the same physical properties as the larger industrial turbines except these units typically operate at much higher speeds. This has the advantage of producing high frequency sound which is easier to attenuate; however, care is necessary if the turbine blade tip speed is supersonic because it creates a sonic shock wave causing excessively high sound levels and what is frequently termed as "buzz-saw noise" (and can occur on all turbines). Good packaging and silencing treatment can have these units in a backyard with little concern over noise.

When it comes to noise control there are no simple closed form equations that exist for "simply finding the answer." In noise control the process is complex and the variables quite numerous; for instance, just in designing silencer panels there are upwards of 19 variables to be considered. Then the placement of the silencers in the duct system must be considered as that affects performance too. Typically, a series of iterations are performed before arriving at "one" potential solution. Then the entire system design must be reevaluated to ensure meeting any other performance requirements such as allowable system pressure loss or drop. Thus, this chapter is limited to the principal elements of noise control that need to be considered in the design process and is not intended as a solution manual.

NOISE SOURCES

There are three major sources of noise that contribute to the combustion turbine's noise emissions; they are casing, inlet and exhaust emissions. The casing emission is the noise radiation from the body of the turbine itself and through its ancillary equipment, piping, and foundation. In most instances the turbine is in an enclosure or inside a building that greatly reduces this source of noise. The inlet and exhaust noises are principally aerodynamic sources. The inlet, or compressor section, generally produces high frequency tones generated by the blade passing frequency (bpf) and typically has several harmonics as illustrated in Figure 10-1.

The fundamental bpf is calculated by counting the number of inlet compressor blades and multiply by the turbine's rotating speed.

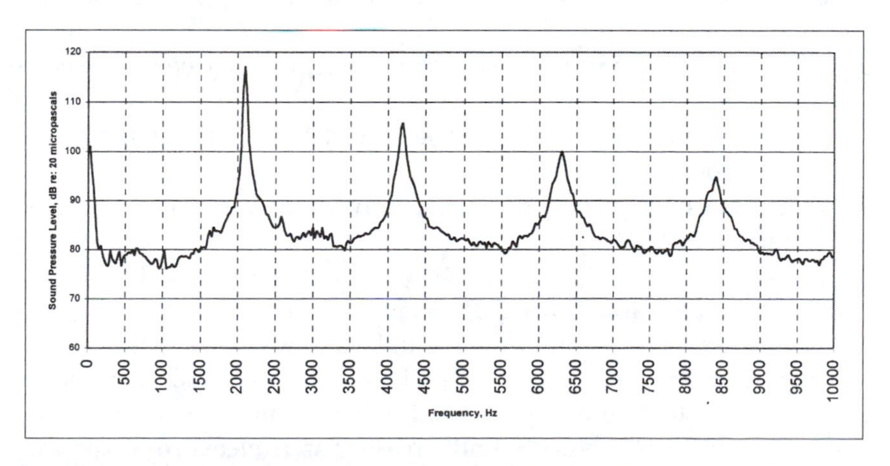

Figure 10-1. Spectrum of a Turbine Inlet Tone and Harmonics

For example, an inlet having 35 blades spinning at 60 Hz will have a fundamental bpf of 2,100 Hz with harmonics as illustrated.

Inlets have the advantage that usually there is some type of air conditioning or filtration system that is necessary and can provide a great reduction to the inlet noise, frequently up to a 20 decibel reduction. These include, evaporative coolers, moisture separators, chiller coils, and an inlet filter system, a barrier or self-cleaning (pulse type) filter.

The exhaust also produces tones but because of the amount of silencing needed to reduce the high aerodynamic turbulence and combustion noises the tones are seldom an issue. This principally broadband noise is the main concern and passive silencers generally work well.

SOME FUNDAMENTALS OF ACOUSTICS

Acoustical Notations

The following are some common sound descriptors used in acoustics.

L_P or L_p is the symbol used to describe a spectral sound pressure level in decibels (dB).

L_A is the symbol used to describe the overall A-weighted sound level in decibels.

L_W is the symbol used to describe the spectral sound power level in decibels.

L_{WA} is the symbol used to describe the overall A-weighted sound power level in decibels.

dB(A) or dBA is not promoted in usage anymore (or any other suffixes attached to "dB").

Sound Pressure and Sound Power

There are two elements of sound that we work with: sound pressure level and sound power level as measured and reported in decibel levels. Sound pressure is what is heard and what is measured using a sound level meter. Sound pressure is a measure of the distribution of sound power in direction and distance from a source of noise. Sound pressure level (SPL) is a decibel scale where:

$$SPL = Lp = 10Log\left(\frac{P}{Pr}\right)^2 = 20Log\left(\frac{P}{Pr}\right)dB \qquad (10\text{-}1)$$

where

P = sound pressure measured in Pascals (N/m^2), and

Pr = the reference pressure of 20 x 10^{-6} Pa or 20μPa.

In these equations the logarithm is to the base 10.

Sound power is the rate at which acoustic energy is emitted. Sound power level (PWL) is described by:

$$PWL = Lw = 10Log\left(\frac{W}{Wr}\right)dB \qquad (10\text{-}2)$$

where

W = calculated acoustic power in watts, and

Wr = the reference power of 1 x 10^{-12} watts.

In general noise control applications we work strictly with the decibel levels but one must be aware that on occasion, listed sound power levels may use an old reference value of 1 x 10^{-13} watts, and when converting to current standard, results in a ten-decibel reduction. Therefore, when specifying equipment sound power levels be sure to use the correct reference level, otherwise you can have a ten decibel deficiency in the design.

Sound power can only be calculated from a set of sound pressure level measurements where a microphone measures the sound pressure over a given area that completely envelops the equipment being measured. For instance, a piece of equipment setting on the ground, measures 1m x 1m x 1m (5 square meters exposed surface area) but measuring the sound pressure level on a rectangular grid one meter away results in an area of 33 square meters. If the area averaged adjusted sound pressure level is 85 dB, then the sound power is:

$$Lw = Lp + 10\ LOG\ (Area) = 85 + 10\ LOG(33) = 100\ dB \qquad (10\text{-}3)$$

This is a simplified demonstration but shows that when a supplier states his equipment does 85 dB at one meter that there is more to it to determine the total sound power a device emits.

So why is sound power so important? It is primarily used for the calculation of sound level and is used in modeling turbine noise emissions as well as all other types of equipment. The sound level at a particular distance is calculated by[5]:

$$Lp = Lw + D_C - A\ dB \quad (10\text{-}4)$$

where:

D_C = the directivity correction based on distance and angle between the source of noise (equipment) and the receiver (where the sound level is desired to be known).

A = the attenuation of the sound between the source and the receiver.

As an analogy, think of a light bulb, which has a power level of watts (L_W) but the brightness of the light (Lp) is determined by how far away you are (D_C) and whether an object blocks it or if the light is absorbed (A) by a dark surface. Sound behaves in a similar manner.

Decibels

Decibels (dB) are a convenient way of handling very large or very small scalar values and as described above, are a logarithmic function that requires a special method for combining decibels. In certain applications the sound levels, in decibels, L1, L2, etc., of two or more sources are summed. However, pressure and energy can be summed, not decibels, so the decibel levels are converted to energy levels, summed and then put back into decibel form:

$$Summation = 10\ Log\ [10^{(L1/10)} + 10^{(L2/10)} + 10^{(L3/10)} + \ldots 10^{(Ln/10)}]\ dB \quad (10\text{-}5)$$

Similarly, an average of several measurements is calculated by:

$$Average = 10\ Log\ \{1/n[10^{(L1/10)} + 10^{(L2/10)} + 10^{(L3/10)} + \ldots 10^{(Ln/10)}]\}\ dB \quad (10\text{-}6)$$

To extract the ambient or a contaminating noise level (L_1) from a total operating noise level (L_2) to determine what the source is (and L2 should be at least 3 dB higher than L1):

$$Source = 10\ Log\ [10^{(L2/10)} - 10^{(L1/10)}]\ dB \qquad (10\text{-}7)$$

Reporting or calculating sound levels in excess of one decimal point is unwarranted.

Frequency and Bandwidth

When we speak of acoustics most think of music, stereo systems and other elements of the science of producing sound in a pleasurable manner. Unwanted sound is called noise. It is this unwanted sound with which we are concerned. Sound or noise emanates at particular frequencies (discrete) or groups of frequencies (broad-band) that we can hear or sometimes feel. The unit of frequency is the hertz (Hz) having units of 1/sec (cycles per second is no longer used). The noise signature or emissions of equipment can span thousands of discrete frequencies, and thus, can be very challenging to analyze (in fact, the cited equations used in this chapter are all a function of frequency but for simplicity are not annotated as such). Industry Standards grouped these discrete frequencies into bands of frequencies known as octaves, which means a doubling and is a geometric-logarithmic function where 1,000 Hz has been defined as the datum point; that is, an octave above 1k Hz is 2k Hz, an octave down is 500 Hz, etc. By grouping these frequencies into just nine octave bands (31.5 – 8k Hz) makes noise control work much easier and somewhat more manageable. Other frequency bandwidths are used too, such as one-third octave band, and what is commonly known as narrowband analysis where the frequency spectrum can be measured in fractions of a hertz. This is typically the approach when identifying the source of a particular sound. Table 10-1 lists one-third and octave bands that are commonly used for sound measurement and analyses.

The Weighted Sound Level

The human ear receives sound at many frequencies, at many different amplitudes and all at the same time. It acts as a filter and it shifts sensitivity based on the amplitude and frequency of the sound. It performs poorly at low frequencies, very well at middle

Table 10-1. Acoustical Bandwidths and Weighting Values

Frequency Bandwidths, Hz	1/3 Octave Band Center Freq. Hz	Octave Band Center Freq. Hz	A Weighting Adjustment, dB	C Weighting Adjustment, dB
11.2 - 14.1	12.5		-63.4	-11.2
14.1 - 17.8	16	16	-56.7	-8.5
17.8 - 22.4	20		-50.4	-6.2
22.4 - 28.2	25		-44.7	-4.4
28.2 - 35.5	31.5	31.5	-39.4	-3.0
35.5 - 44.7	40		-34.6	-2.0
44.7 - 56.2	50		-30.2	-1.3
56.2 - 70.8	63	63	-26.2	-0.8
70.8 - 89.1	80		-22.5	-0.5
89.1 - 112	100		-19.1	-0.3
112 - 141	125	125	-16.1	-0.2
141 - 178	160		-13.4	-0.1
178 - 224	200		-10.9	0
224 - 282	250	250	-8.6	0
282 - 355	315		-6.6	0
355 - 447	400		-4.8	0
447 - 562	500	500	-3.2	0
562 - 708	630		-1.9	0
708 - 891	800		-0.8	0
891 - 1,122	1,000	1,000	0	0
1,122 - 1,413	1,250		+0.6	0
1,413 - 1,778	1,600		+1.0	-0.1
1,778 - 2,239	2,000	2,000	+1.2	-0.2
2,239 - 2,818	2,500		+1.3	-0.3
2,818 - 3,548	3,150		+1.2	-0.5
3,548 - 4,467	4,000	4,000	+1.0	-0.8
4,467 - 5,623	5,000		+0.5	-1.3
5,623 - 7,079	6,300		-0.1	-2.0
7,079 - 8,913	8,000	8,000	-1.1	-3.0
8,913 - 11,220	10,000		-2.5	-4.4

frequencies and begins to fade at high frequencies. Researchers developed electronic weighting curves or filters for sound level meters to process the sound as the ear would hear it. These weighting curves are commonly known as A, B, C, D, and E and are applied depending upon what is being measured. A and C weightings (or filtering) are the most widely used in environmental acoustics. A sound level meter set to measure A-weighted sound levels captures all the sound energy from 10 Hz to about 12,000 Hz or more and processes the measurement to arrive at a single sound level value. The following figure illustrates the weighting curves as applied to the sound spectra from 10 Hz through 10,000 Hz and the actual weighting values are listed in Table 10-1. An example of performing an overall A-weighted calculation is given in Appendix C-12.

Applying the A-weighting curve shown in Figure 10-2 to the noise signature shown in Figure 10-1 results in an overall A-weighted sound level of 120.5 dB. This is a much easier number to deal with than having to always produce a spectrum and show all the discrete sound pressure levels. To continue this simplification approach, the

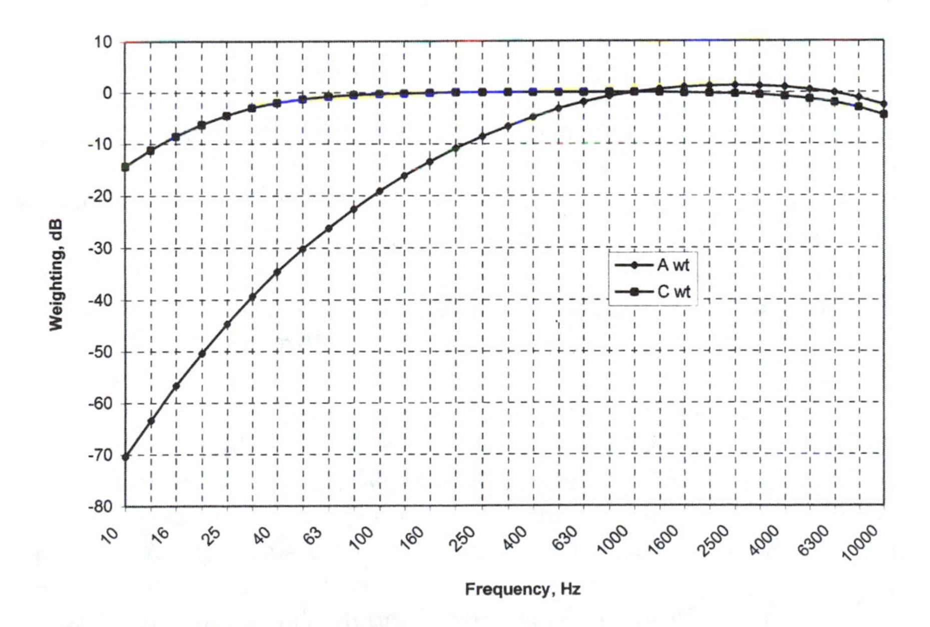

Figure 10-2. Frequency response curves for A and C weighting measurements

overall A-weighted sound level or sound power level is typically used for reporting sound levels. The A-weighted level is commonly denoted simply as the sound level (or power level) given as a decibel level without the suffix (A) added to dB. All sound levels are now implicitly understood to be A-weighted unless it is defined as something else. This simplification is acceptable for generally making a determination as to the amount of acoustical energy a device has near the equipment and out to a moderate distance but is limiting in determining equipment sound levels at large distances. This is because the atmosphere greatly attenuates the middle and high frequencies leaving the low frequencies as the main source of noise. So, at a few hundred meters the overall A-weighted sound level is essentially controlled by the low frequency noise from the equipment. This is most apparent when having to meet very low sound level requirements, say less than 45 dB in a remote area. For this reason, octave band sound power (preferred) or sound pressure levels are required for any detailed noise control modeling and analysis.

NOISE CRITERION

Before applying noise control or mitigation attributes to the turbine system we first need to know how quiet does it have to be? Environmental sound or noise considerations in the United States is a collection of state, county, and local laws, ordinances, and regulations that range from well written, to well meaning, to woefully inadequate. In many states and communities there are no environmental noise limits and most only have a general nuisance ordinance. The federal government (mainly through the EPA) has provided guidelines and recommendations but there are no federal laws regarding community noise levels as caused by land based industrial sites except for the Federal Energy Regulatory Commission (FERC) which regulates interstate energy transport (mainly oil and gas pumping stations). Frequently, it is left for the site developer to negotiate with the local permitting board or a state agency what is acceptable for noise emissions and, in a cost driven environment, the minimal is usually done. There are a number of organizations that promulgate environmental noise limits[6-11] and there are two national standards that are readily available; they are the American National

Standards Institute, ANSI B133.8—1977 (R2000) *Gas Turbine Installation Sound Emissions,*[6] and ANSI S12.9-1998/Part 5, *Quantities and Procedures for Description and Measurement of Environmental Sound—Part 5: Sound Level Descriptors for Determination of Compatible Land Use.*[7] ANSI B133.8 offers guidelines for developing a criterion and ANSI S12.9 provides guidelines based on land use.

One criterion that is most familiar is of course the Occupational Safety and Health Act of 1970 (OSHA) [12] that limits employee noise exposure. Frequently, many assume that the sound level limit is 85 dB, which can cause unneeded expenses. OSHA applies a time weighted average such that the total exposure an employee has does not exceed a certain limit; thus, near field sound levels can easily be in excess of 85 or 90 dB without penalty so long as the exposure time is controlled. The following table presents the maximum limits of selected allowable exposure times, but the total employee time weighted average (TWA) level must also be considered.

Table 10-2. Maximum Noise Exposure Limits (without hearing protection)

Sound Level, dB	80	85	90	92	95	97	100	102	105	110	115
Exposure Time, hours	32	16	8	6	4	3	2	1.5	1	0.5	0.25

The OSHA *action level* is one-half the allowable time exposure so when an employee is exposed to 85 dB for 8 hours or more (TWA $\geq$ 85 dB) then the employer shall administer a continuing, effective hearing conservation program including providing hearing protection at no cost to employees. For convenience, the threshold *action levels* are listed in the following table.

Table 10-3. OSHA TWA "*Action Level*" Thresholds

Sound Level, L_A (dB)	80	85	90	92	95	97	100	102	105	110	115
Exposure Time, T (hrs.)	16	8	4	3	2	1.5	1	.75	0.5	0.25	0.125

It is always recommended to wear hearing protection around any moderately loud equipment, even at home. The interested reader is directed to Appendix C-13 for more information.

NOISE CONTROL

Noise control or mitigation involves several steps. First of course is, "does it need to be quiet and if so, how much?" The amount of noise reduction is driven by having to meet an environmental noise limit or some regulatory limit as discussed. Frequently, more silencing is needed than what can be described as the bare minimum in order to account for noise from other equipment or sources that all combine to create a total sound level (see equation, 10-5); thus, a *balance of plant* noise analysis must be performed. There is also a certain amount of uncertainty, particularly with turbine sound power levels, and there may be some design margin incorporated, perhaps 3 decibels or more depending on the acoustical performance risks and guarantees that must be met. The following figure illustrates the classical approach to noise control.

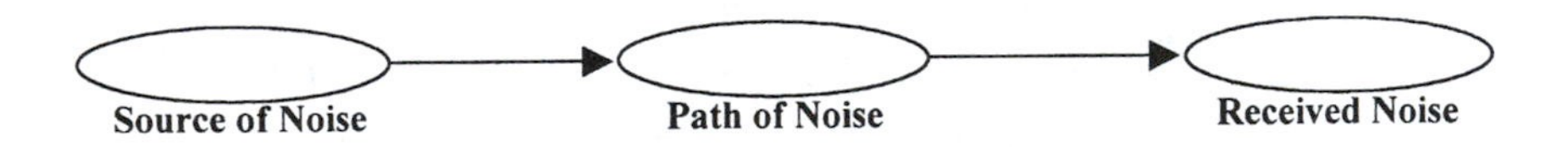

Figure 10-3. Source-Path-Receiver Noise Control Model

Applying noise control involves affecting one of these three elements. For turbine inlet and exhaust noises, the noise is reduced near the *Source* location by installing silencers in the gas *Path*. Controlling the *Path* by use of a turbine enclosure or turbine building reduces the casing noise. The *Receiver* noise or sound level results from controlling either the *Source* or *Path* elements and in some situations, the *Receiver* may be placed inside an enclosure.

As discussed, the turbine has three main sources of noise: inlet, exhaust and casing. In this section we will mainly focus on the inlet and exhaust noise, as those are the most demanding and costly for noise control. The casing noise, as mentioned, is usually not a major concern since the turbine is either inside a building or an enclosure.

Silencer Systems

The principal product of choice for silencing is the parallel baffle (splitter) system where a series of silencer panels are arranged in parallel in a duct to allow air or exhaust gas to flow between the panels as illustrated in Figure 10-4. This passive system typically consists of a sound absorbing material sandwiched between two perforated facing sheets and is the most widely used method for silencing turbine inlets and exhausts and the detailed methods of design and analysis are available in most noise control texts.

Other advanced system designs include reactive or resonator type chambers that are incorporated into the design of the parallel baffle system, which offer the ability to tune to a particular frequency or frequencies. Active noise control has also emerged over the last decade to offer advanced noise reduction capabilities where a disturbing noise tone can be cancelled by introducing a second tone that is 180 degrees out of phase. When the two signals combine the result is a greatly reduced tone or sound level. Early concepts of active noise control involved conventional speakers, which required enormous size and power to effectively cancel a high amplitude tone (and not practical in a hot exhaust stream). In a basic noise cancellation system great care must be exercised in how the system is designed, installed and operated. Performance hinges on the three elements of noise control as shown in Figure 10-3, any slight change to one control feature can negate the effectiveness of active noise cancellation. However, advances in noise cancellation technology and materials are making progress that will greatly reduce or eliminate these restrictions and improve reliability, which is the major concern of the end user.

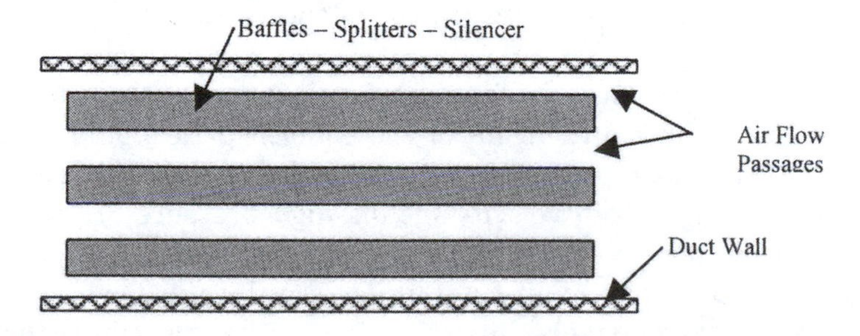

Figure 10-4. Typical Parallel-Baffle Duct Silencer Configuration

Silencer Design and Materials

Silencer design must take into consideration several attributes that constitute the final design of a silencer. They include but are not limited to:

- Required noise reduction in conjunction with materials selection, reliability, and cost
- Allowable size of the silencing system—that is, real estate needed versus availability
- Allowable pressure drop or losses through the silencing system—cost of operation
- Aerodynamics—turbulence and flow distribution—longevity of materials
- Structural and operational design requirements—chemistry of air and exhaust gases
- Mechanical design of the silencers—thermal cycling, vibration, corrosion, seismic

The factors that determine the acoustical performance of a passive duct silencer are based on the following basic design elements:

- Effective (absorptive) length of the silencer
- Thickness of the silencer
- Spacing or air gap between the silencers
- Filler material—acoustical and material properties
- Any covering material used for fiber/fill retention
- Cover sheet perforation—open area ratio, hole diameter, and thickness
- Velocity of air or gas between the silencers

Note that nowhere is listed the height or number of silencer panels. The number and height of the silencer panels only affect the velocity of the gas or air through the silencers. Low velocity means low pressure drop, high velocity means high pressure drop and possibly high self flow noise. Self-noise becomes a limiting factor in very low noise applications. Self-noise is created by the flow of the gas around the silencer panels thus there is nothing that can reduce this secondary noise unless there is a secondary low velocity attenuation system behind the parallel silencers. Typically, the velocity through the panels should not exceed 60 m/s (200 fps) for exhaust applications to limit self noise, aerodynamic turbulence, and fill loss; or, 30 m/s (100 fps) for inlet applications which is mainly for low pressure drop and balanced air flow into the turbine compressor. The reader is referred to reference [13] for an excellent overview on silencer design and pressure drop.

Frequently, the specifying design agency may call for a silencer to achieve a certain dynamic insertion loss (DIL) in decibels. By dynamic, it is meant that the affects of self-noise are to be considered in the design process. On occasion a specification may be written only calling for a DIL but the flow rate is needed in order to calculate the gas velocity between the silencer panels and the sound power level (ahead of the silencer section) is needed to determine the system DIL. Now, here's the surprise. No one can measure the DIL without performing an extensive and very expensive on-site operation that would require the removal and re-installation of the silencers in the duct system. The reason is the definition of (dynamic) insertion loss. It is the measure of the difference in sound level with and without the silencer panels in place. Typically, what can be measured is the relative attenuation of the silencer system as illustrated in the following figure where Lp is the sound level at numbered locations and the bottom duct shows the silencers in the same duct.

In Figure 10-5 the following holds (assuming the noise source is to the left):

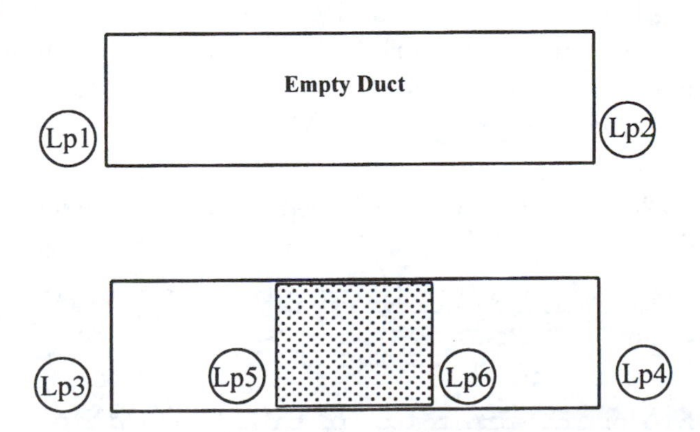

Figure 10-5. Silencer Acoustical Terminology

- The insertion loss (dynamic if under flow conditions): *IL = Lp2 – Lp4*
- The noise reduction is: *NR = Lp3 – Lp4 or Lp5 – Lp6*
- The transmission loss (TL) is: *Same as NR but assumes Lp4 is anechoic (free from echoes and reverberations)*
- Lp1 ≠ Lp3 because of the presence of the silencer

And it is known that IL ≠ NR ≠ TL (but they are usually close). The NR is measured at convenient locations upstream and downstream of the silencers. TL is measured in the same manner as NR but the downstream end must have an anechoic termination; that is, no sound reflection at the end of the duct—the sound is completely absorbed. In measuring the NR and IL, the duct geometry and length play an important role as they affect performance but TL is less affected by the duct geometry since the duct termination must be anechoic. Most all field verification measurements are NR under no flow conditions; while measurements under flow conditions are possible there is variability and uncertainty involved in the process. So, when specifying silencer performance alone be sure to specify the conditions under which the performance is to be evaluated or verified. In certain cases smaller scaled model silencers or full scale can be tested in a laboratory to verify acoustical and aerodynamic performances.[14]

A typical cross section of a silencer panel is shown in the following figure. Shown are the outer perforated sheet, a screen mesh or cloth used to retain the acoustical filler material and the center filled with the acoustical material. The choice of the perforated sheet open area ratio, the screen mesh or cloth, and the acoustical filler material all effect performance.

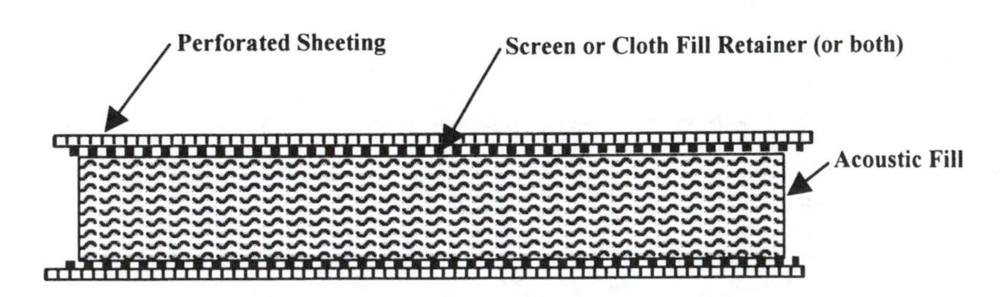

Figure 10-6. Typical Silencer Partial Cross Section

Table 10-4. General Uses and Properties of Acoustical Materials

Silencer Application	Acoustical Material	Retainer Material	Temperature Range	Thermal Cycling	Durability	Resistance to Friability [1]
Inlet	Mineral Wool	Cloth Mesh [2]	N/A	N/A	Good	Good
	Fibrous Glass Blanket/Batt	Cloth Mesh [2]	N/A	N/A	Good	Good
Exhaust	Mineral Wool	Wire Mesh	~ 800°F	Good	Moderate	Moderate
	Ceramic	Wire Mesh	~ 2,000°F	Excellent	Poor [3]	Low [3]
	Basalt	Wire Mesh	~ 1,400°F	Poor [4]	Good [4]	Moderate
	Fibrous Glass Blanket/Batt	Wire Mesh	1,000°F typical limit	Good	Good	Moderate

Notes: [1] Friability—the propensity to breakdown or disintegrate.
[2] A very low flow resistance is required to maintain performance.
[3] Packing of material and low vibration critical for longevity.
[4] Protective thermal shock blanket required for longevity.

The perforated sheet should have an open ratio greater than 35% for inlet applications and greater than 25% for exhaust applications. Table 10-4 provides a general guideline as to choice and applicability of materials. The user should always verify material properties.

Exhaust system temperatures can be in excess of 540° C (1,000°F), which calls for the careful selection of materials. Wire mesh is used in exhaust for material retention because at high temperatures the cloth materials begin to melt and fuse forming a glass sheet that degrades the silencer's ability to attenuate noise. Too heavy or too tightly weaved retainer cloth can degrade performance dramatically and the challenge is to find a cloth suitable for retaining the filler material while still allowing the acoustical sound wave to penetrate the cloth. Mineral wools typically have binders that breakdown at about 425° C (800°F), which causes the material to be blown out the exhaust. Thermal cycling (start-stop of the turbine) also accelerates material breakdown and basalts generally require thermal shock protection usually in the form of a needle-mat layer between the screen mesh and the basalt. Poor aerodynamics can also cause the silencer panels to lose fill by the exhaust flow beating on the panels and duct vibration has the same detrimental effects. Some OEMs require higher operating temperatures in order to have the materials withstand hot starts; thus, fibrous glass batting products may not be suitable, and exhaust gas chemistry must also be considered in the material selection.

The ability of the sound wave to penetrate into the acoustical fill is critical in obtaining the desired performance. The material exhibits an acoustical resistance that attenuates the sound energy that is categorized as airflow resistance and is measured in rayls/meter. The property "rayls/meter" is a measurement of the airflow resistance a material exhibits; it is actually a hydrodynamic measurement[15] that is used as an *indicator* of acoustical performance and the lower the value the better the performance, generally. Most acoustical materials on the market today have flow resistance values less than 40,000 rayls/meter; but, for certain designs, it may be desirable to use materials having higher flow resistances. One must be careful in using airflow resistance values, as they are dependent upon the environmental air or gas temperature and chemistry. Hot exhaust gases typically have viscosity values that are four to six times higher than that of ambient air, (thus, resulting in a higher flow resistance values) and as can be well imagined, material testing is conducted at ambient (room) conditions.

Silencer panels need periodic maintenance and replacement, particularly in exhaust systems. In turbines that operate continuously (thus, free of frequent thermal cycling) the panels may last 10 or more years. However, in turbines that start and stop frequently the thermal cycling causes stress on the silencer construction (welds, etc.) that lead to failures and loss of fill, and may need repair or replacement in just a few years. These expenses need to be considered in the operating system costs and the duct design should allow access to the silencer section.

Duct Wall Design

The duct system is also a key factor in acoustical design as it principally houses the silencer section, controls the flow and direction of the gas flows (air in and exhaust gas out), and must provide a level of noise reduction to avoid breakout and flanking noises from compromising the overall performance. Most have a common grasp that the principal noise from the turbine is by way of its gas path but importantly, as discussed in the introduction, turbines having levels of 140 dB or more can have a significant amount of noise breakout through the duct wall. So, in analysis of the total system noise, the acoustical engineer or designer accounts for both the gas path and the duct wall breakout noises. Also, the duct can act as a *noise short-circuit* around silencers. This is referred to as flanking noise where the silencers may be designed to achieve 60 dB noise reduction but something less is measured because the turbine's in-duct (airborne) noise has coupled to the duct wall structure, travels down the duct liner/walls and re-radiates into the duct downstream of the silencers. This mainly becomes a concern when having to achieve better than 60 dB noise reduction. In these cases a vibration break or isolator must be installed in the duct wall to reduce or eliminate noise from flanking around the silencer section. In cases having extensive duct work exposed in a low noise environment an additional barrier wall or enclosure may be require whereby the main duct is structurally independent of the other acoustical barriers.

Unfortunately, duct walls are not simple to design, especially if having to meet an external shell temperature limit. Most industrial turbines use double wall construction and the internal duct noise is mainly transmitted to the exterior wall by way of the structural connections between the inner liner sheets and the outer duct wall as illustrated in Figure 10-7. Increasing the insulation thickness only plays a minor part in reducing the breakout noise; the transmission loss (TL) of the insulation material alone is fairly low unless using a fairly dense product. In certain cases it may be possible to install an isolator between the structural connector and the liner sheet to minimize noise transmission to the exterior but this is both expensive and labor intensive as each isolator must be installed carefully. Visco-elastic and polymer damping materials and lead septums cannot be used on high temperature exhaust ducts, even if on the exterior shell; the potential for fire, emissions of toxins, and accelerated material degradation will occur.

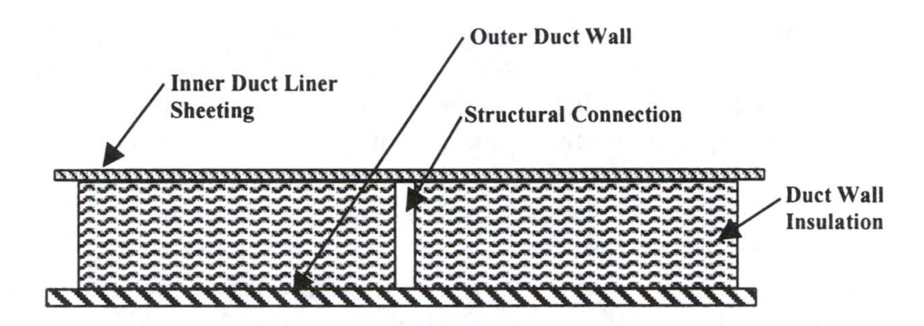

Figure 10-7. Typical Duct Wall Cross Section

There are no published algorithms or simple equations for calculating the noise reduction double wall designs can achieve but there are a limited number of papers[16] that address the design issues and offer limited design tools; reference [17] provides excellent resources as well. In most cases the candidate wall system is built and tested to verify performance[18] and most OEMs and Noise Control Engineers working in the field have extensive databases on duct wall performance. And the orientation of the wall does not affect TL performance; the thinner liner sheet can be on the outside or the inside depending on design latitudes or requirements. In acoustics, this is referred to as the Law of Reciprocity.

Certain mass properties of the shell plate must also be carefully evaluated on the basis of their application. A well-known phenomenon occurs when the stiffness-bending wavelength of the shell plate is coincident with an excitation or forcing frequency such as the inlet bpf. This coincidence frequency (fc) results in a dramatic reduction in the duct wall transmission loss at that frequency because of the matching of the two wavelengths—the airborne to the structure where the wall becomes nearly "transparent" acoustically. This frequency may be estimated by the following equation[17]

$$fc = \frac{c_o^2}{2\pi} \bullet \left(\frac{ph}{B}\right)^{1/2} Hz \tag{10-8}$$

and

$$B = \left(\frac{Eh^3}{12(1-\nu^2)}\right) \tag{10-9}$$

where,

fc = coincident frequency,

c_o = speed of sound in air (or exhaust gas),

B = bending stiffness,

ρ = density of material,

h = thickness of plate,

E = Young's modulus,

ν = poisons ratio.

The critical frequency plays an important parameter when calculating the wall TL for inlet ducts because, as was shown in Figure 10-1, one needs to avoid blade passing frequencies. By examining this figure it is evident that just given an overall sound level of 120.5 dB is not sufficient to adequately design an inlet duct wall, thus it is important to know the principal frequencies of interest in critical applications so always ask for those frequencies.

The duct wall free field sound power level, which is needed to calculate far field noise levels per equation (10-4), is as follows:

$$Lw = Lw(i) - (TL + 6) + 10\ Log\ (S/A)\ dB \qquad (10\text{-}10)$$

where:

$Lw(i)$ = the sound power level inside the duct

TL = the transmission loss of the duct wall

$+ 6$ = accounts for free field acoustical radiation conditions

S = the duct exterior surface area (sq. meters)

A = the duct cross sectional gas path area (sq. meters)

The purpose of this equation is to demonstrate how duct geometry affects the duct wall's sound power level. The goal is to maximize the TL of the wall and the cross area of the duct while at the same time try and minimize the surface area. This is an iterative process and the goal again is to balance acceptable noise emissions while minimizing material cost. Note that the term (TL + 6) is also known as NR, the wall noise reduction.

SUMMARY

The complexity of combustion turbine systems and their ancillary equipment generally require an acoustical engineer well experienced with such systems to fully analyze the acoustical emissions and make a determination whether the expected sound levels meet regulatory requirements, are in balance with the environment or community, or need some level of noise mitigation. Another element of these packages is the ease to upgrade and/or relocate units to new sites. It is strongly advised to investigate the potential environmental noise (as well as regulatory) impact of installing these units as part of the search process before committing to installing units at a particular site. Particular attention must also be paid to regulatory agencies if upgrading or modifying existing combustion turbines particularly the exhaust system as this may impact existing air permitting (EPA exhaust emissions) and/or other regulations that may become applicable in such cases.

References

1. George F. Hessler, Jr. "Controlling noise impact in the community from power plant operations—recommendations for ambient measurements," Noise Control Engineering Journal Vol. 48, No. 5, 141-150 (2000)
2. George F. Hessler, "Beware low-frequency gas-turbine noise," POWER magazine, pp 78-80, July/August 2001
3. Bruce E. Walker, Alan S. Hersh, et al., "Active Control of Low Frequency Turbine Exhaust Noise," *Proc. NOISE CON 2000*, 3-5 Dec. 2000, edited by John Van Houten (The Institute of Noise Control Engineering of the U.S.A. Inc. Washington, DC, 2000),
4. Lisa A. Beeson and George A. Schott, "Low Frequency Noise Considerations for Combustion Turbine Projects," *Proc. International Gas Turbine & Aeroengine Congress & Exhibition*, 2-5 June 1997, (ASME, NY 1997)
5. Acoustics—Attenuation of sound during propagation outdoors—Part 2: General method of calculation, International Standard, ISO 9613-2 (© ISO 1996)
6. *Gas Turbine Installation Sound Emissions*, American National Standards Institute, ANSI B133.8—1977 (R2001) (American Society of Mechanical Engineers, ASME, 2001)
7. *Quantities and Procedures for Description and Measurement of Environmental Sound—Part 5: Sound Level Descriptors for Determination*

of Compatible Land Use, American National Standards Institute, ANSI S12.9-1998/Part 5, (Acoustical Society of America, 1998)

8. "Public Health and Welfare Criteria for Noise," U.S. EPA Report 550/9-73-002, (July 27, 1973)
9. "Information On Levels Of Environmental Noise Requisite To Protect Public Health And Welfare With An Adequate Margin Of Safety," U.S. EPA Report 550/9-74-004, (March, 1974)
10. "Guidelines for Community Noise," World Health Organization, 20, Avenue Appia, CH 1211 Geneva 27 Switzerland (1999)
11. *Pollution Prevention and Abatement Handbook*—Part III, Thermal Power—Guidelines for New Plants, World Bank Group, September 1, 1997
12. General Industry Standards, U.S. code of federal regulations (CFR), Title 29 Part 1910 Section 95, Occupational Noise Exposure
13. Anthony G. Galaitsis and Istan L. Ver, "Passive Silencers and Lined Ducts," Chapt. 10 in *Noise and Vibration Control Engineering-Principals and Applications*, edited by Leo L. Beranek and Istan L. Ver (Wiley, New York, 1992)
14. *Standard Test Method for Measuring Acoustical and Airflow Performance of Duct Liner Materials and Prefabricated Silencers*, American Society For Testing Materials, ASTM E 477—90
15. *Standard Test Method for Airflow Resistance of Acoustical Materials*, American Society For Testing Materials, ASTM C522—1987 (R1993)
16. Ben H. Sharp, "Prediction Methods for Sound Transmission of Building Elements," Noise Control Eng. J. **11**(2), 1978
17. Istan L. Ver, "Interaction of Sound Waves with Solid Structures," Chapt. 9 in *Noise and Vibration Control Engineering-Principals and Applications*, edited by Leo L. Beranek and Istan L. Ver (Wiley, New York, 1992)
18. *Standard Test Method for Laboratory Measurement of Airborne Sound Transmission Loss of Building Partitions*, American Society For Testing Materials, ASTM E 90—1987

Chapter 11

Microturbines

MICROTURBINES

The size of a gas turbine, or its name (heavy industrial, aero-derivative, mini-turbine, microturbine) does not change the fact that all gas turbines are mass flow machines and are thermodynamically the same. To maximize the output power from a gas turbine the tips of its rotors have to turn close to the speed of sound. Therefore, the smaller the diameter of the turbine the higher the speed required to maximize the power output. Aero-derivative and heavy industrial gas turbines generally fall in the 1 megawatt and up range.

Table 11-1.

Type$^{\alpha}$	*Power Range*
Heavy Industrial & Aero-derivative	1000 kW and up
Mini-Turbine	500 kW – 1000 kW
Microturbine	20 kW – 500 kW

$^{\alpha}$See Appendix A-1 Gas Turbine Manufacturers and A-2 Microturbine Manufacturers

The capability of the materials used in a gas turbine is the only limitation it faces. The aero-derivative produces almost as much power as the largest heavy industrial machine. It accomplishes this by running at high rotor speeds (up to 20,000 rpm) to provide the required mass flow.

Microturbines are high-speed (up to approximately 100,000 rpm) combustion gas turbines with outputs from a low of 20 kW to a high of 500 kW. These units evolved from automotive and truck turbocharger components, small jet engines (in turboprop applications), and auxiliary power units commonly used for ground power for aircraft. Microturbine components typically consist of a centrifugal compressor and radial turbine components mounted on the same shaft, a combustor, and a recuperator. In some designs a separate turbine wheel, the power turbine wheel, is also provided. The recuperator is used to capture exhaust waste heat to heat up compressor discharge air. Heating the compressor discharge air reduces fuel consumption, which reduces NO_x formation (NO_x formation is discussed in detail in Chapter 9) and increases the overall efficiency. Microturbines are coupled to alternators or generators (inductive or synchronous generators) for the production of electricity. In mechanical drive applications the microturbine may be coupled to a compressor in refrigeration service or a pump in pumping service. In single and split-shaft microturbine designs the electric generator may be mounted on the same shaft as the compressor and turbine components or a speed-reducing gearbox may be employed between the gas turbine output shaft and the electric generator (or compressor or pump depending on the application).

Due to the high speeds of the microturbines special attention must be given to the electric generators. These generators are either specifically designed to produce a 60 Hz output or have specialized electronics (rectifiers and converters) to convert the power generated to facilitate utility grid connection.

APPLICATIONS

Microturbines are used in distributed power and combined heat and power applications. With recent advances in electronic, microprocessor based, control systems these units can interface with the commercial power grid and can operate "unattended."

Table 11-2 list typical applications for microturbines and the power range utilized for each application.[1]

Table 11-3 demonstrates the relative cost of reciprocating engines, microturbines, and fuel cells. Since microturbines and fuel cells are relatively new to the marketplace their price & performance are drawn from comparatively few samples.

Table 11-2.

Convenience/Fast Food Store	40-50 kW
Restaurant Chain/Gas Filling Station	50-70 kW
Supermarkets (Old)	150-300 kW
Supermarkets (New)	300-2,000 kW
Hospitals	100-6,000 kW
Large Office Building	400-3,000 kW
Factories	500 kW and up

Table 11-3.
Cost and Performance of Micropower Technologies[2]

Technology	*Engine*	*Microturbine*	*Fuel Cell*
Size	30 kW–60MW	30-200 kW	100-3000 kW
Installed Cost ($/kW)[α]	200-800	350-900	900-3300
Elec. Efficiency (LHV)	27-38%	15-32%	40-57%
Overall Efficiency[β]	~ 80-85%	~ 80-85%	~ 80-85%
Variable O&M ($/kWh)	0.0075-0.02	0.004-0.01	0.0019-0.0153
Footprint (sq_ft/kW)	0.22-0.31	0.15-0.35	0.9
Emissions (lb/kWh unless otherwise noted)	Diesel: NO_x: 0.022-0.025 CO: 0.001-0.002 Natural Gas (NG): NO_x: 0.0015-0.037 CO: 0.004-0.006	NO_x: 3-50ppm CO: 3-50ppm	NO_x:<0.00005 CO: <0.00002
Fuels	Diesel, NG, Gasoline, larger units can use dual fuel (NG/Diesel) or heavy fuels	Diesel, NG, kerosene, naphtha, methanol, ethanol, alcohol, flare gas	NG, propane, digester gas, biomass and landfill gas

[α]Cost varies significantly based on siting and interconnection requirements, as well as unit size and configuration.
[β]Assuming CHP

HARDWARE

The difference between a microturbine and the larger, heavy industrial and aero-derivative, gas turbines is size. While microturbine components are smaller than the multi-megawatt gas turbines, they perform identical functions. That is they follow the laws of thermodynamics and are represented by the Brayton Cycle. In most cases microturbine components are made of the same materials used in larger gas turbines (see Figure 11-1). The primary areas of new development are the combustor and the recuperator.

Figure 11-1. Courtesy of Capstone Turbine Corporation. The C30 & C60 microturbine systems are ultra-low-emission combined heat and power generator sets providing up to 30 kW and 60 kW of power, 85 kW and 150 kW of heat, at heat rates of 13,100 Btu/kW-Hr and 12,200 Btu/kW-Hr, respectively.

Compressor

Microturbine compressor components are almost exclusively centrifugal designs. This design type has several advantages: burst speeds are well above even the highest operating speeds, materials are readily available, this component can be precision cast with minimal machining, and some components (derived from small turboprop engines, turbochargers, APUs, etc.) are already available.

For example, United Technology Corporation, in partnership with DTE Technologies, Kyocera and others has developed a microturbine derived from the Pratt & Whitney ST5 helicopter engine. The ST5 engine was derived from the PWA200 (Figure 11-2), which in turn was derived from the PT6 (Figure 2-3) turboprop engine. These engines use a combination of centrifugal and axial components. Conversely, the Ingersoll-Rand's PowerWorks 250 was scaled down from the Ingersoll-Rand KG2, a 1.7-megawatt gas turbine, which employs centrifugal designs for both the compressor and turbine components (Figure 11-3).

Figure 11-2. Courtesy of United Technologies Corporation, Pratt & Whitney Aircraft. The PW206 turboprop engine is a 700 shaft horsepower engine used primarily in helicopter applications. Current applications include the Agusta A109E *Power*, Bell M427, Eurocopter EC135P1/P2, and the MDHI MD*900/902*.

The compressor component, turbine component and in some designs the electric generator rotor are fixed to the main shaft (Elliot Energy Systems TA 100 inertia weld these components onto the main shaft). The compressor and main shaft (see Figure 11-4) are typically made of stainless steel. Specific materials used in microturbine air inlets and compressors are similar to materials used in the larger gas turbines and are listed in Table 3-2.

Turbine

The majority of microturbines employ the radial inflow turbine wheel design. The design looks much like the compressor component (except of course the geometry is reversed). There are some designs that combine an axial stage (as in multi-megawatt

Figure 11-3. Courtesy of Ingersoll-Rand Energy Systems. The PowerWorks 250 microturbine, a 250 kW single shaft gas turbine operating at 45,000 rpm with a 4:1 compression ratio, was derived from the KG-2 1.7 MW gas turbine also manufactured by Ingersoll-Rand.

Figure 11-4. Courtesy of Capstone Turbine Corporation. The C30 & C60 microturbine systems are ultra-low-emission combined heat and power generator sets providing up to 30 kW and 60 kW of power, 85 kW and 150 kW of heat, at heat rates of 13,100 Btu/kW-Hr and 12,200 Btu/kW-Hr, respectively.

gas turbines) with a radial stage. The selection of turbine wheel design types is primarily a function of parentage (that is, the origin of the microturbine design).

While current turbine wheels are made from the same materials used in the larger gas turbines, development of turbine wheels made from monolithic silicon nitride material (a ceramic) is in progress. These new materials will provide improved high temperature creep rate and oxidation resistance.

When developed these designs and manufacturing techniques will undoubtedly be incorporated into the multi-megawatt gas turbines—just as advances made in jet engine technology was, and continues to be, transferred to the large industrial gas turbine engine designs.

Recuperator

Recuperators transfer heat from the turbine exhaust to the combustor inlet (or compressor discharge). In a recuperator the heat transfer occurs through the passage walls (hot exhaust gases on one side and cool compressor air on the other—Figure 11-5). Recuperators are essential to achieving increased efficiency from microturbines. "With recuperation electric efficiencies are typically 26-32% versus 15-22% for non-recuperated units."[3] Another advantage of recuperators is the reduction in NO_x. This reduction is due to the decrease in the amount of fuel (evidenced as an increase in efficiency) and the subsequent decrease in fuel-bound nitrogen (See oxides of nitrogen in Chapter 9 for a detailed discussion of exhaust emissions).

The renewed interest in recuperators is extending into the multi-megawatt sizes. Recuperated gas turbines can take advantage of their lower compressor pressure ratios and combustion temperatures to reduce NO_x formation. "Recuperated turbine DLE combustors show potential to meet ultra-low emissions levels without any after-treatment."[4]

High temperature recuperator materials are being developed in order to achieve greater than 40% efficiency at a cost less than $500.00 per kilowatt. These materials are expected to operate between 1290°F (700°C) and 1830°F (1000°C). Among the materials being studied are 347SS, Alloy 230, Modified Alloy 803, Alloy 120, Thermie Alloy, Alloy 625 and Alloy 214.[5]

Figure 11-5. Courtesy of Ingersoll-Rand Energy Systems. The recuperator is integral with the combustor (not shown) thereby eliminating transition parts. The combustor fits into the large circular chamber on the right side of the photograph and is mounted directly above the compressor-turbine components.

Combustor

There is no significant difference between microturbine combustors and multi-megawatt gas turbine combustors. Combustors for microturbines, like combustors for larger gas turbines, are either annular or single combustor designs. They are fabricated from martensitic and ferritic iron base alloys or nickel base alloy (see Table 3-4). The single combustor, which lends itself to the reverse flow design concept, has become the design of choice.

Under development are catalytic combustors. The catalytic combustor promises to achieve emissions below 3 ppm (on natural gas) without the need for exhaust gas treatment.

Controller

Controllers used on microturbines are generally the same as the controllers used on the multi-megawatt gas turbines. In electric generation applications there are some added control tasks. The advent of microturbines and their high-speed generators has necessitated the marriage within the controller of power output conditioning with gas turbine engine control. This controller must not only manage the fuel valve during Start, Stop and Governing sequences, provide machine protection against high vibration, over-speed and over-temperature, but must also condition the high frequency, high voltage power produced by the generator to grid quality.

Also, the controller must be able to communicate with other systems {such as the plant Distributed Control System (DCS), and the Human Machine Interface (HMI)}. Communication protocols such as Modbus®, Ethernet TCP/IP, Ethernet UDP, OPC (Ethernet), DDE (Dynamic Data Exchange), EGD (Ethernet) must be available so that the user/operator can easily interface the control to existing or new plant systems and to maintenance systems.

To achieve suitably fast response the fuel valve must be sized for the microturbine. It is also helpful if the fuel valve can handle two flow paths—one for starting flow (pilot flow) and another for running flow (primary flow).

For power generating applications a power conditioning control must manage load (KW) control, frequency control, synchronizing, load sharing and KW droop.

For mechanical drive applications (compressor or pump drives) the controller must manage suction and discharge pressure & temperature control, bypass or recirculation flow control & cooling, surge control (for centrifugal compressors) and miscellaneous plant valves.

Generator

Microturbine manufacturers are currently employing two design variations: high speed single shaft and split shaft. The single shaft type drives the high-speed synchronous (external field coil excitation or permanent magnet) or asynchronous (induction) generator at the same speed as the gas turbine. This generator produces very high frequency AC power that must be converted first to DC power and than to 60 Hz AC power. This is accomplished using a rectifier and an inverter. The inverter rectifies the high frequency AC voltage produced by the alternator into unregulated DC voltage. It then converts the DC voltage into 50 Hz or 60 Hz frequency and 480 volts AC.

With the split shaft design the electric generator is driven by the free power turbine, usually through a speed reduction gearbox, at 3600 rpm. In this design the generator produces 60 Hz power and does not require a rectifier and an inverter.

There are three general types of converters:

- DC Link Converter
- High Frequency Link Converter
- Cycloconverter

A list of microturbine and power converter manufacturers is provided in Appendix C-15. Power electronic converters operate either in voltage mode or current mode. Current mode is used when the microturbine is connected to the grid. Voltage mode is used when the microturbine is in the "island" or stand alone mode.

Bearings

Because operating conditions are very demanding at the high rotational speed (up to 100,000 rpm), bearings are a critical component of the microturbine. Conventional hydrodynamic and anti-friction bearings, employing a pressurized lube oil system (including a pump and an oil cooler) are still employed by some manufacturers. However, advances have been made into the use of "air," "gas" of "film" bearings.

This type of bearing uses a thin film of pressurized air to support the shaft. Fluid film bearings utilize a thin film of air between the rotating shaft and the stationary housing and this film transfers the forces from one to the other. The fluid film is accomplished by delivering airflow through the bearing compartment. One design technique creates the air film with an orifice. Another design technique delivers the air through a porous medium thereby ensuring uniform pressure across the entire bearing area.

The result is that air bearings provide a frictionless load-bearing interface between surfaces that would otherwise be in contact with each other. Since air bearings are non-contact, they avoid the traditional bearing-related problems of friction, wear, and the need for a lubricant.

Air bearings are typically about 60% efficient. Therefore the lifting capability can be determined by the following equations:

Maximum Theoretical Lift $\equiv L_{MT} = P_B \times B_A$

where P_B = Air pressure in psia

B_A = Bearing area in sq inches

and

Actual Lift $\equiv L_A = 60\% \times L_{MT}$

While this is a reliable design guideline for air bearings, lower air pressures, increased damping, increased stability and increased stiffness are all improved with increased surface area.

A refinement of the air bearing design is the "foil" bearing. In a "foil" bearing design, when the shaft is stationary there is a small preload between the shaft and the bearing caused by the foils. When the shaft starts to turn, hydrodynamic pressure builds up that pushes the foil away from the shaft causing the shaft to become airborne

Another particular bearing design can be described as follows: Water or water vapor discharged through ultra-fine porous medium is used as the working fluid. The bearing surface is virtually hydraulically smooth. The bearing is enclosed by the water storage, and the water pressure is controlled so as to be higher than the atmospheric pressure. At rest the clearance is filled with liquid water. As the speed of rotation increases a phase change occurs through viscous dissipation and heat transfer from the high-temperature journal to the bearing surface converting the "liquid water" to "water vapor." Water vapor evaporation from the bearing surface elevates local pressure in the bearing clearance area; this stabilizing effect due to "hydrostatic" pressure is than superimposed on the conventional stabilization due to "hydrodynamic" pressure.[6]

The Elliott Energy Systems TA 100 microturbine (Figure 11-6) utilizes two primary bearings. The inboard bearing is a conventional hydrodynamic design and the outboard bearing is a rolling element bearing utilizing ceramic balls. Both bearings are lubricated and cooled by synthetic oil.

References

1 "Microturbines: installation and Operation" presented by Jim Watts, Ingersoll-Rand, Distributed Energy Road Show—Sturbridge, Mass, May 2003

2 "Industrial Applications for Micropower: A Market Assessment" Prepared for Office of Industrial Technologies, DOE and Oak ridge National Laboratory, Prepared by Resource Dynamics Corp., November 1999.

3 "Industrial Applications for Micropower: A Market Assessment" Prepared for Office of Industrial Technologies, DOE and Oak ridge National Laboratory, Prepared by Resource Dynamics Corp., November 1999

4 "Clean Distributed Generation Performance and Cost Analysis" Subcontract # 400002615, Prepared for Oak Ridge National Laboratory

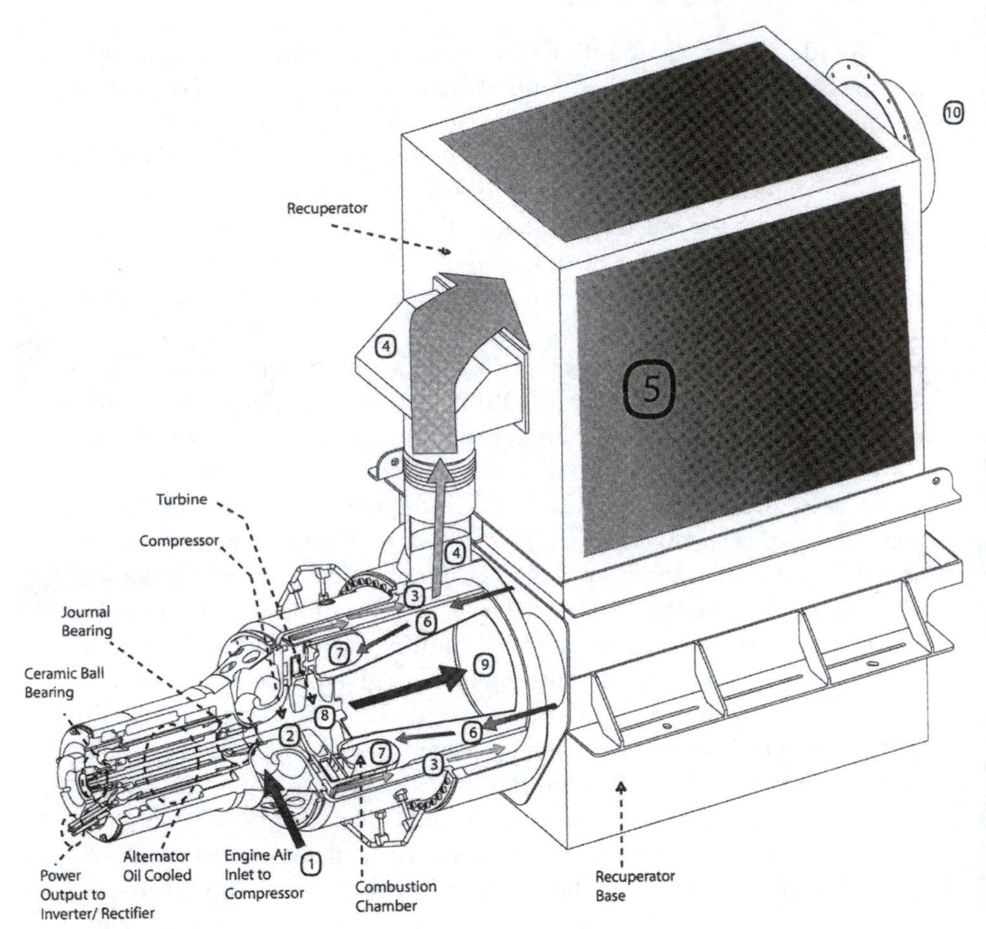

1. Engine inlet air to Compressor
2. Air is compressed by Impeller.
3. Compressed air is sent to Recuperator.
4. Duct directs compressed air into Recuperator
5. Recuperator heats the compressed air using exhaust energy.
6. Compressed and now heated air is directed towards Combustion Chamber
7. Heated Air is mixed in the Combustion Chamber with natural gas and ignited.
8. Exhaust from combustion process is directed through turbine wheel causing alternator to rotate, which provides an output of 105 W
9. Ehaust is passed to the Recuperator to heat the Compressor delivery air.
10. Exhaust exits the Recuperator.

Figure 11-6. Courtesy of Elliott Energy Systems, Inc. The TA100 CHP microturbine is packaged as a combined heat and power generator set capable of producing 100 kW of electrical energy and 172 kW of thermal power per hour. Cogeneration usage can consist of hot water, absorption chiller or drying system applications. Depending upon the user application, overall total thermal efficiency could be greater than 75%. The microturbine operates at a 4:1 compressor ratio and a simple cycle heat rate of 12,355 Btu/kW-Hr.

and U.S. Department of Energy by DE Solutions, Inc. April 2004
5 "DER Materials Quarterly Progress Report—January 1, 2004 to March 31, 2004" prepared by David P. Stinton, Roxanne A. Raschke DER Materials Research ORNL for DOE
6 "Clarification of Thermal Phenomena and Development of Novel Thermal System" Hideo Yoshida, Hiroshi Iwai, Motohiro Saito; Kyoto University, Graduate School of Engineering

Chapter 12

Waste Heat Recovery

Introduction

Waste heat recovery is the recovery of heat from waste gas sources (such as, gas turbines, gas and diesel engines, process plants, engine cooling water, etc.). Of all prime movers the gas turbine produces the highest exhaust temperature and mass flow per unit of simple cycle power. Waste heat recovery will be defined first in the broadest sense and then with specific definitions for our purposes it will be discussed relative to gas turbines and more specifically to cogeneration and combined cycle applications as part of a gas turbine plant.

The thermodynamic principles as applied to waste heat recovery and the linked technologies are discussed along with the physical realities of the equipment that make up a waste heat recovery (WHR) venture. Various applications of WHR are addressed including the influences of the local climate, the advantages and disadvantages of WHR and the barriers to WHR including environmental constraints. While environmental constraints are primarily political in nature they can be insurmountable.

Waste heat recovery utilizes the hot exhaust waste gases to improve plant efficiency and power output, usually without increase in fuel consumption or exhaust emissions. It is a source of heat from one process to generate steam, hot water, hot oil, etc. to be used in another process. Depending on the application a waste heat recovery process may be referred to as a combined cycle process or a cogeneration process.

An example, with calculations is included in Appendix C-16 to enable the reader to perform rough sizing and costing estimates for new projects, and performance calculations for existing plants.

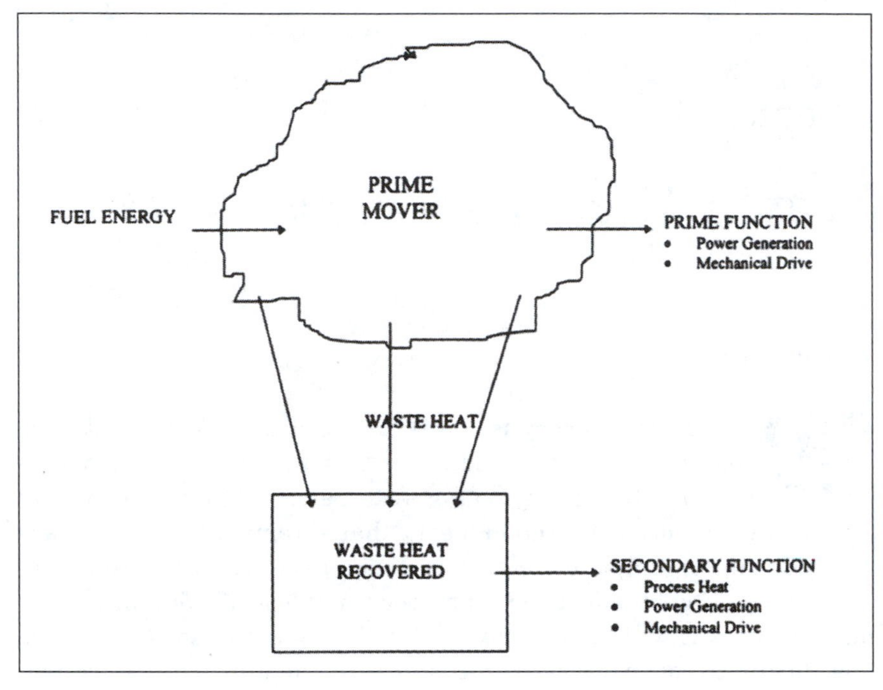

Figure 12-1A. Generic Waste Heat Recovery System

> **The heat recovery process itself does not reduce air or water pollution (nor in most cases will it increase pollution)!**

Definitions

The more obvious sources for energy recovery are gas turbine exhaust, reciprocating engine exhaust, incinerator exhaust, fluidized bed exhaust, and low efficiency boilers. These are often referred to as the topping units or topping cycle. The sensible heat in the waste exhaust may be recovered in a bottoming cycle. The bottoming cycle may be utilized directly as process heat or in a steam generator.

Waste Heat Recovery

Waste heat recovery is a synergistic combination of cycles operating at different temperatures. The heat rejected by the higher

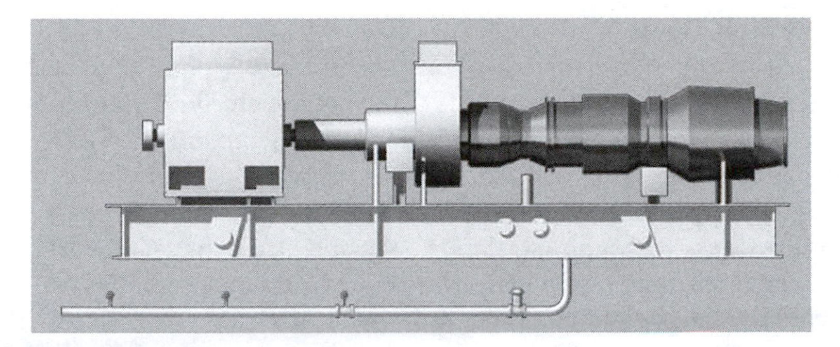

Figure 12-1B. Typical Gas Turbine. ***Courtesy of Freeport-McMoran Oil & Gas LLC (formerly Plains Exploration & Production Company.)*** **This graphic is based on the Gas Turbine Cogeneration Plant Operator Interface Screen.**

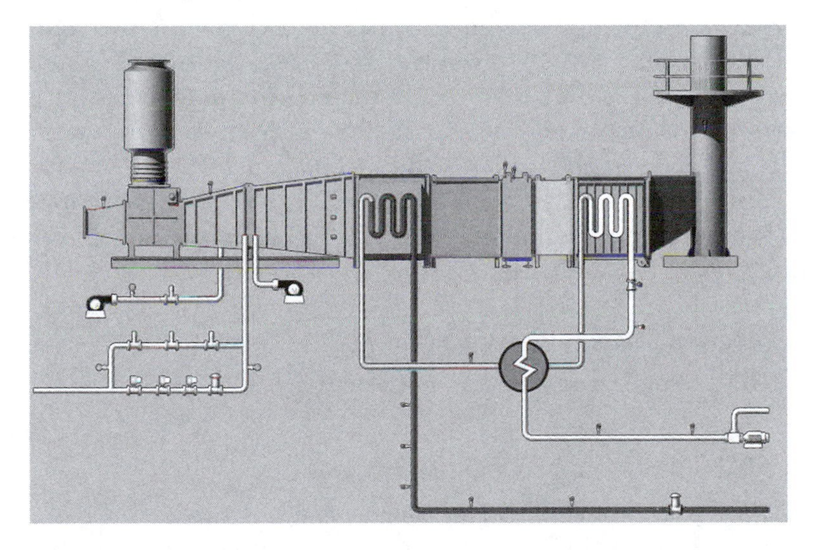

Figure 12-1C. Typical Waste Heat Recovery Boiler. ***Courtesy of Freeport-McMoran Oil & Gas LLC (formerly Plains Exploration & Production Company.)*** **This graphic is based on the Gas Turbine Cogeneration Plant Operator Interface Screen.**

temperature cycle (topping cycle) is recovered and used by the lower temperature cycle (bottoming cycle) to produce additional power to realize an improved overall efficiency. Two types of bottoming cycles will be explored: steam bottoming cycles and organic bottoming cycles.

Combined Cycle

A combined cycle process consists of two (or more) systems operating at different temperatures—each of which could operate independently given the right conditions. Heat rejected in the exhaust of the prime mover (gas turbine) is used to drive the lower temperature cycle to generate steam which, in turn, is used to generate shaft power via a steam turbine. The shaft power may then be used to drive an electric generator, pump or compressor. Another example would be to vaporize an organic fluid in a fluid boiler and then expand the high pressure-high temperature gas through a turbo-expander to generate shaft power. A combined cycle system is a specific form of WHR. To be a combined cycle the different cycles must operate on separate fluids.

A cogeneration cycle process is a specific type of combined cycle process where the exhaust heat from a prime mover is the source of heat to generate steam, hot water, hot oil, etc. to be used in another process. An example would be a gas turbine-compressor power plant where the exhaust from the gas turbine is used to generate steam which in turn is applied through a heat exchanger to provide heat to a separate process. That separate process could be a cheese processing plant, a potato chip plant or a distillation column in a refinery. The following list the various forms of heat energy recovery:

- Medium temperature water heater
- High temperature organic fluid heater
- Low pressure steam generator
- Medium pressure steam generator
- High pressure steam generator with economizer
- High pressure steam generator with economizer & superheater
- High pressure steam generator with non-condensing steam turbine
- High pressure combined cycle with condensing turbine, a low pressure steam section on a boiler for low pressure process steam generation.

Cogeneration systems are most adaptable to industries having moderately large and stable energy needs!

Steam Bottoming Cycles

A bottoming cycle utilizes the rejected heat from the topping cycle process to generate additional shaft power. The typical bottoming cycle collects waste heat from the topping cycle process and sends it through a waste-heat-recovery boiler or HRSG to convert this thermal energy to steam which is supplied to a steam turbine, extracting steam to the process and also generating mechanical shaft power.

Organic Bottoming Cycles

Similar to the steam bottoming cycle, the organic bottoming cycle also uses the "Bottom" portion or the rejected heat as the heat source for the waste heat recovery system.

CLIMATIC INFLUENCES

The effect of changes in ambient conditions on prime movers, especially gas turbines, is an important consideration. The gas turbine's power output decreases with increasing ambient temperature, and the exhaust gas temperature drops with decreasing ambient temperature as shown in Figure 12-2.

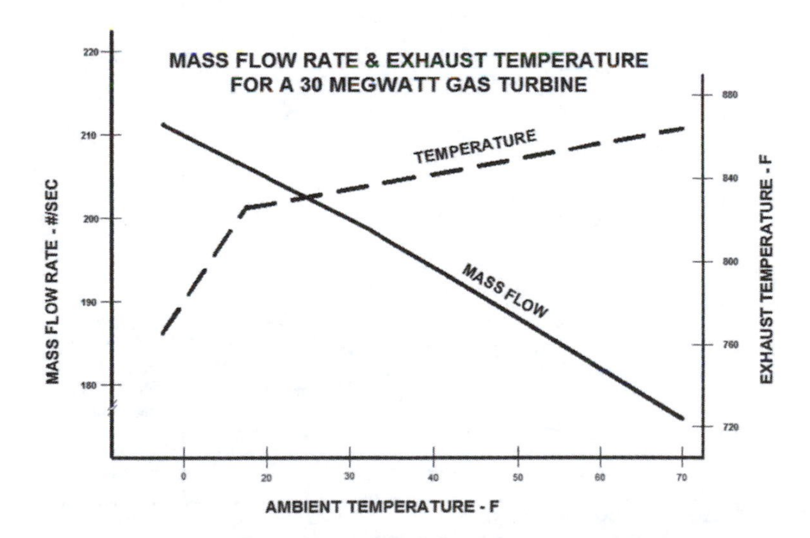

Figure 12-2. Curve of Mass Flow & Exhaust Gas Temperature vs. Ambient Temperature

Therefore, the heat recovery system output varies throughout the ambient temperature range. The effects on the gas turbine also impact on the boiler as shown in the variations in efficiency and power (Figure 12-3).

Climatic conditions can also have an ill effect on steam systems after shutdown and prior to start-up. To deal with this in cold climates, a steam system must be designed with rapid drains and heaters in the condensate tank. This requirement is not necessary for the organic system.

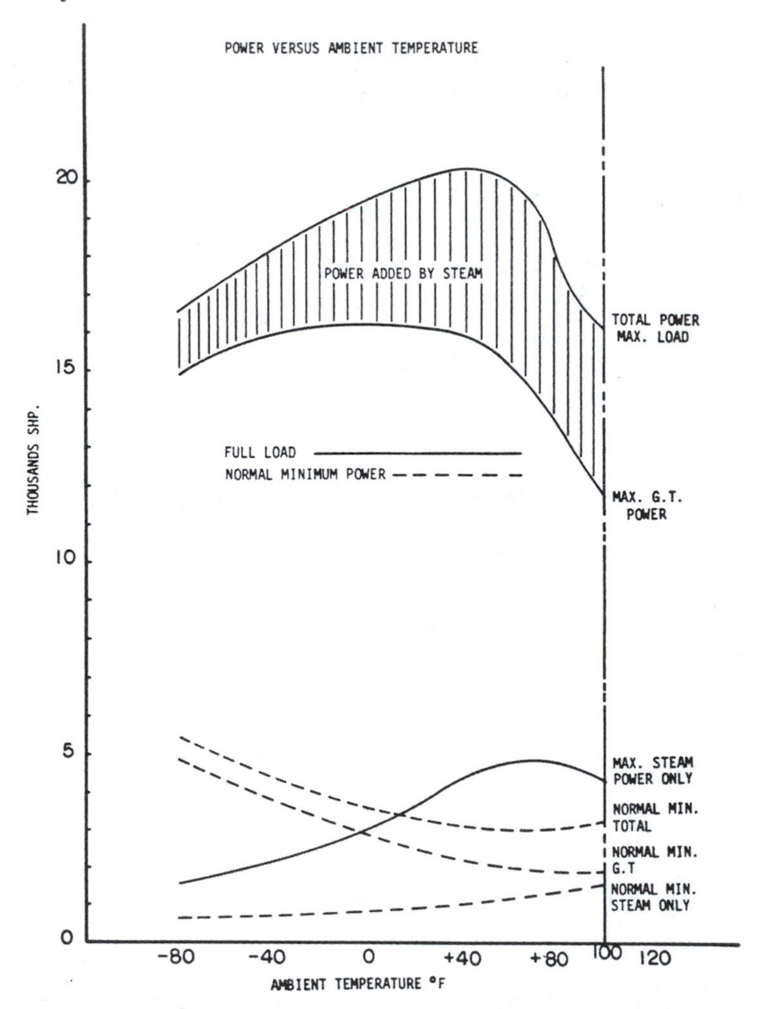

Figure 12-3. Power vs. Ambient Temperature

APPLICATIONS

The requirement for waste heat recovery systems is based primarily on a desire to obtain the most efficient plant heat rate possible. The actual economic advantage from any form of waste heat recovery depends on the availability and cost of the fuel being considered regardless of whether it is a fossil fuel or some form of waste material suitable for use as a fuel.

The large number of combinations of single and combined cycle units makes the selection of the best package for a given project complicated. For example, a 15,500 brake horse power (first generation[1]) unit without WHR is only 27% efficient, but with waste heat recovery its efficiency can reach 38%. Which is more efficient than a 16,000 Brake Horse Power (second generation[2]) unit with a simple cycle efficiency of 33%. This is an important consideration when considering plant upgrades. To trim the selection task to a workable level, other criteria must be considered. Specifically,

- Plant location,
- Effects of changes in ambient conditions,
- Laydown and maintenance space,
- The amount of waste heat recovery desired relative to availability, and
- The waste heat recovery application (either direct drive into the same unit driven by the main gas turbine, or driving another unit in parallel with the primary unit or a supplementary or auxiliary use (such as, driving an electric generator, or generating heat).

Examples of industries that utilize WHR are shown in Table 12-A. In electric generation alone there are several types of generation plants to be considered. These are synchronous generation, isolated generation and self-generation.

- In synchronous generation the power producer is tied into a grid and therefore must operate at the same frequency as the grid (in this example the operator is said to be "in sync"). This plant includes "synchronizing" equipment "hardware and software" which will synchronize his unit with the grid before he is able to connect to the grid.

Table 12-A

INDUSTRY	SOURCE	PRODUCTS
Oil Recovery	Gas turbines Gas engines	Steam
Refineries	Gas turbines Gas engines	Steam
Sewer Treatment	Gas turbines Gas engines	Steam
Breweries	Gas turbines Gas engines	Steam
College Campuses Business Complex Hotels	Gas turbines Gas engines Fuel Cells	Steam Hot Water
Cement Kilns	Gas turbines Gas engines	Hot Air
Glass Mfg.	Glass Melting Furnaces	Steam
Electricity	Municipal Incinerator	Steam

- Isolated generation, also known as "Within the Fence Generation." This is where the user-operator depends solely on his ability to generate the power and heat that his operation requires. He does not have an outside source to draw power from in the event he cannot generate the power he requires.

- Self-generation can fall into either of the above categories. While the operator may generate his power for his own needs, he maintains a connection to the grid so that he can draw power from the grid, if and when, he needs it. Generally through an agreement with the utility or power pool he synchronizes with the grid and sends his power to the grid while simultaneously drawing power from the grid for his operation. In this case he pays a surcharge for the connection and for the difference in the power he draws from the grid less the power he sends to the grid. In some cases he may be paid for the excess power he sends to the grid.

ADVANTAGES OF WHR

The advantages of waste heat recovery systems specifically as compared to utility sites are very impressive. First there is the high degree of flexibility relative to the scale of the project for site critical applications. In existing plants the heat recovery system can be fitted into existing systems. This provides an excellent opportunity to select variations in the physical size of the waste heat boiler (economizer, the evaporator and the superheater). Another advantage is the improvements in efficiency without modifications to the prime mover (gas turbine). The recovered waste heat can be utilized for shaft horsepower (generation, compression, pumping), process heat, or a combination of these. By converting the waste heat to shaft power up to 35% more power may be generated without additional fuel usage or pollution. This potential increase in energy conservation is not often recognized nor given significant value.

For new plants the gas turbine can be selected as a function of a balance between the power and the heat recovery required. Furthermore, proper location of the plant will result in a reduction in energy transmission distances thereby decreasing losses by placing the equipment where its output will be utilized. Finally the burning of non-standard fuels such as coal, solid waste, sludge, etc. may be considered because efficiency of gas turbines alone is not the main consideration.

Payback

The cost of every project consists of the equipment capital cost, installation cost and operating cost (fuel, manpower, etc.). To be truly effective a project should pay for itself in a reasonable amount of time. Considering the equipment capital cost, installation cost, simple & combined cycle horsepower and heat rates and the price of fuel the payback time may be calculated. In the following examples the price of fuel is taken at $5.00/million Btu.[3]

$$\text{Payback} = \frac{\text{Cost}_{\text{Equipment+Installation}}}{\text{HR}_{\text{simple cycle–combined cycle}} * \text{BHP}_{\text{combined cycle–simple cycle}} * \dfrac{\$5.00}{10^6\ \text{Btu}} * 8760\ ^{\text{Hours/}}{}_{\text{Year}}}$$

This information is shown on Table 12-B, along with simple cycle and combined cycle horse powers and heat rates.

Table 12-B. Waste Heat Recovery (Iso Conditions) Steam

SIMPLE CYCLE			COMBINED CYCLE			
BHP	HEAT RATE	OVERALL THERMAL EFFICIENCY	BHP	HEAT RATE	OVERALL THERMAL EFFICIENCY	PAYBACK TIME YEARS
28,923	7365	34.6	37420	5693	44.7	3.9
16,330	7618	33.4	21317	5836	43.6	3.9
15,461	9292	26.3	21593	6882	37.0	2.6

Based on a fuel cost of $5.00/10^6 Btu payback time for the 28,923 and the 16,330 horsepower units is 3.9 years while payback time for the 15,461 horsepower unit was only 2.6 years. However, what may be justified at $5.00 dollar fuel may not be justified at $2.50 fuel (as payback time would double to 7.8 and 5.2 years, respectively).

Advantages of Steam Bottoming Cycle

The main advantage of the steam system is that it has been in use both in steam plants and in combined cycle applications for a considerable number of years. As a result there is considerable experience in the design, construction, and operation of the steam cycle.

Disadvantages of the Steam Bottoming Cycle

The disadvantages of the steam cycle also evolve around the fluid itself; steam has a low molecular weight, a high latent heat, and a high critical temperature and pressure. These characteristics of steam result in its being described as a wet fluid (that is, the fluid must be adequately superheated before being expanded in the turbine in order to avoid turbine blade erosion).

Furthermore, steam is a corrosive fluid, especially when air is allowed to become entrapped in the fluid circulated throughout the system. Inadequate superheating results in "water slugging" the turbine, which can, in instances, causes considerable turbine blade damage. Water droplets in the boiler straight tube runs and the inner bends of the tubes can cause severe erosion and excessive vibration; resulting in leaks in areas where accessibility may be difficult. Whether due to equipment design or the steam fluid itself, high pressure (600 psi) boilers have been known to experience severe vibration. This vibration has, in many instances, resulted in failure of the heat transfer tubes within the boiler. Also, operation in arctic regions can cause operating problems duc to freeze-ups and the need to drain the system during even brief outages.

Advantages of the Organic Bottoming Cycle

The organic bottoming cycle derives its name from the use of organic fluids. Organic cycle systems are used primarily for process heat as opposed to electrical power generation. Examples of organic fluids are propane, ammonia, pyridine and toluene. Advantages of the organic fluids is their high temperature-low pressure characteristics, versatility in operating temperature, chemical stability and lower capital cost of unpressurized or low pressure systems. Tables 12-C and 12-D, on pages 215 and 216, detail the properties of these fluids.

Disadvantages of the Organic Bottoming Cycle

The two major drawbacks of organic fluids are flammability and toxicity and to a lesser degree thermal instability for some fluids.

Barriers to Cogeneration

Institutional and bureaucratic barriers pose a threat to the development of cogeneration facilities in such areas as the sale of excess power back to the utility, proper ownership, the effect of cogeneration proliferation on the public utilities baseload demand and its effect on rate structures, the guarantee of the utility to provide standby power and peaking power and plant siting.

Legal difficulties include permitting jurisdiction and permitting requirements. This is primarily based on environmental concerns. However, converting a simple cycle plant into a cogeneration or combined cycle plant does not add to the emissions level while it does increase power output. This in effect results in fewer emissions per unit of power produced. Yet permitting requirements can be just as complex as if a new, emissions producing, power plant was being considered. Environmental protection laws impose additional cost, delays, and risk to a facility planning a cogeneration plant—even though the overall effects of environmental pollution for the region would be reduced as a result of the fuel saving benefits of cogeneration.

If the cogeneration plant is owned and operated by the utility, the sale of the electrical component is simplified. However, it becomes more complex if the plant is generating electricity for sale to the utility and simultaneously providing thermal energy for sale to a separate customer or power pool.

Environmental Restraints

The environmental concerns involve noise emanating from the plant and air & ground pollution. Compliance with the limitations on noise levels can be accomplished without much difficulty using readily available acoustic barrier equipment. Oftentimes the expense of acoustic barriers is justified on an emotional basis rather than a technical basis. For example, on a recent project it was mandated that a sound enclosure be installed around new equipment. The justification was that the plant was required to limit a noise level at their perimeter fence to 85dba[4]. What made this requirement unreasonable was a) the acoustic wall was required and designed based on the assumption that the equipment noise level would be 98dba three feet from the equipment base, and b) prior to the equipment installation the ambient noise level at the fence perimeter was actually in excess of 95dba. After commissioning the equipment the noise level tested at 83dba. Today that unit operates 24/7 with most of the sound barrier components stored off site.

Air pollution restraints are more difficult to meet. Most gas turbines require one or more of the following to meet today's environmental restraints:

1. Water (or steam) injection into the combustor for NO_x control.
2. Exhaust gas treatment, such as
   a. Selected catalytic reduction for NO_x control.
   b. CO catalyst for carbon monoxide control.
   c. SO catalyst for sulfur oxide control.

These modifications do not come without cost (either first cost or maintenance cost). The water or steam injected into the combustor must be high quality demineralized water. The exhaust catalyst must be sized for the operating temperature profile of the gas turbine exhaust. Operation out of this range will reduce the effectiveness and shorten the life of the catalyst.

Gas turbine manufacturers are currently developing dry, low NO_x combustors that will achieve 25 ppm NO_x. However, environmentalists are already looking to reduce the current 9 ppm limit to less than 2 ppm (primarily in California in the South Coast Air Quality Management District—SCAQMD).

It is agreed that air quality is important and must be protected. However, there should be a trade off that allows advanced technology to build acceptably clean plants while constantly improving the technology as experience is gained, leading to the day when more acceptably clean and efficient power sources are available.

REVIEW OF BASIC THERMODYNAMICS

As with all heat engines and waste heat recovery systems the laws of thermodynamics apply. Most text books claim that there are two laws of thermodynamics. However, some text books include a basic law which is often referred to as the Zeroth Law. The *Zeroth Law of Thermodynamics* states that when two bodies have equality of temperature with a third body, they in turn have equality of temperature with each other[5].

The *1st Law of Thermodynamics* (often referred to as the conservation of energy) states that the sum of the work delivered to the surroundings is equal to the sum of the heat received from the surroundings. Therefore heat & work are different forms of a single entity (energy) which is conserved.

The *2nd Law of Thermodynamics* states that it is impossible to achieve 100% efficiency. That is the entropy of an isolated macroscopic system never decreases, or, equivalently, that perpetual motion machines are impossible. These are the laws that govern the transfer of waste heat energy. Two thermodynamic cycles are prominent in this process a) the Brayton Cycle shown in Figure 12-5 and b) the Rankine Cycle shown in Figure 12-6. To better understand these thermodynamic cycles lets first look at the Carnot Cycle. The Carnot cycle is an idealized cycle as it does not actually exist in reality. However, it leads the way to understanding the Brayton and Rankine Cycles.

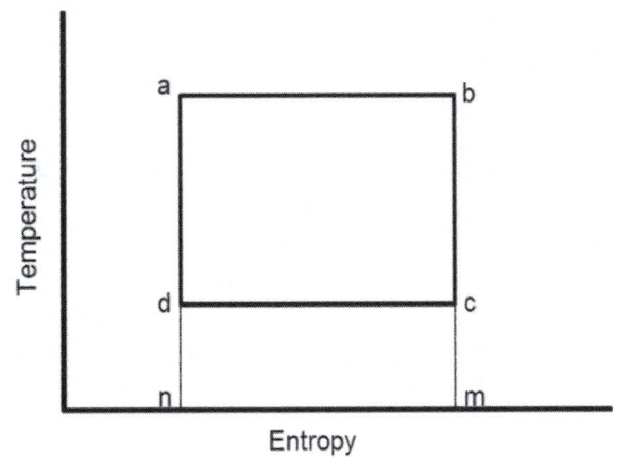

Figure 12-4. Carnot Cycle

Note that the area a-b-c-d represents the net heat transferred which equals the net work of the cycle, the area a-b-m-n represents the heat supplied to the cycle and the area c-m-n-d represents the heat rejected

Nicolas Carnot (June 1, 1796-August 24, 1832) first described this cycle in 1824 in a paper titled "Reflections on a Motive Power of Fire." This described the first theoretical explanation of heat engines and laid the foundations for the 2nd Law of Thermodynamics.

Although some crude internal combustion engines were already in existence the scientific basis of their operation was not fully understood. Carnot sought to explain a) whether or not the work available from a heat source was potentially limitless and b) whether or not heat engines could, in principle, be improved by replacing steam with some other working fluid or gas.

Carnot's approach was to consider an idealized engine in order to understand the fundamental principles applicable to all heat engines, independent of the particular design choices made. In so doing he showed that the efficiency of his idealized engine was a function of the temperatures of the two reservoirs between which they operated. The exact form of the function is

$$\frac{T_1 - T_2}{T_1}$$

where T_1 is the absolute temperature of the hotter reservoir and T_2 is the absolute temperature of the cooler reservoir.

The second point Carnot sought was that the maximum efficiency attainable did not depend upon the exact nature of the working fluid (be it steam or some other gas).

THERMODYNAMICS OF WASTE HEAT RECOVERY

Waste heat recovery is visually displayed via the Brayton and Rankine Thermodynamic Cycles. Brayton or open-cycle gas turbines are commercially proven power systems that permit the easy recovery of large quantities of waste heat. Currently most of the larger industrial cogeneration systems utilize the gas turbine in one form or another (heavy frame gas turbine or aircraft derivative). The gas turbine converts approximately 25% to 35% of the fuel energy input to drive the compressor and 20% to 40% (depending on gas turbine generation[1,2]) into useful shaft output. The remaining fuel energy—25% to 55%—is rejected in exhaust flow at high temperature and ambient pressure with a small percentage consumed in mechanical losses, oil losses, and unit heat radiation. The sensible heat in the gas turbine exhaust may be recovered as previously stated. Most gas turbines in cogeneration applications are fueled with natural gas, although there are some that are fueled with distillate fuel oil. With the development of gasifiers, pressurized fluidized bed and atmospheric fluidized bed combustion systems gas turbines can fire heavy oils and even solid fuels (coal).

Retrofitting existing gas turbines into cogeneration or combined cycle systems is not difficult if adequate space is available for the installation of the heat recovery boiler and auxiliary equipment. In this application the gas turbine is often referred to as the topping unit or topping cycle. The sensible heat in the waste exhaust may be recovered in a bottoming cycle. The bottoming cycle may be utilized directly as process heat or in a steam generator. If utilized in a steam generator that steam may then be utilized in a steam turbine to generate shaft power.

Topping Cycle—The high temperature cycle is commonly called the topping cycle and is Otto, Brayton or Rankine cycles.

Bottoming Cycle—The lower temperature cycle is called the bottoming cycle. All bottoming cycles are Rankine Cycles.

The Rankine Cycle (Figure 12-6) is often referred to as the bottoming cycle. This becomes obvious when viewing Figure 12-7 which depicts both the Brayton Cycle and the Rankine Cycle. The Brayton Cycle is at the top of the graph and the Rankine Cycle is at the bottom of the graph.

Brayton and Rankine Cycles

Combining the Brayton and the Rankine Cycles demonstrates how the waste heat out of the gas turbine exhaust can convert water to steam. The steam can then be expanded through a condensing turbine and the steam turbine exhaust recovered in the condenser and converted back to liquid water. These processes are depicted in Figure 12-8.

The Rankine cycle is easily visualized in the Temperature-Entropy Chart for water and steam shown in Figure 12-8 within the heavy lines.

Charts shown later in this chapter will detail the various points as water is first converted to steam, than converted to superheated steam and then expanded through a condensing steam turbine and a condenser back to water.

It should be noted that the Rankine steam cycle is also frequently depicted on the Mollier diagram. The Mollier diagram is a graphical representation of the thermodynamic properties and

BRAYTON CYCLE

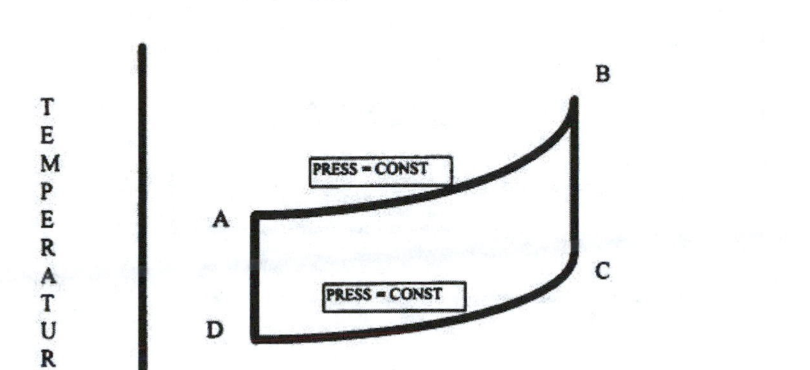

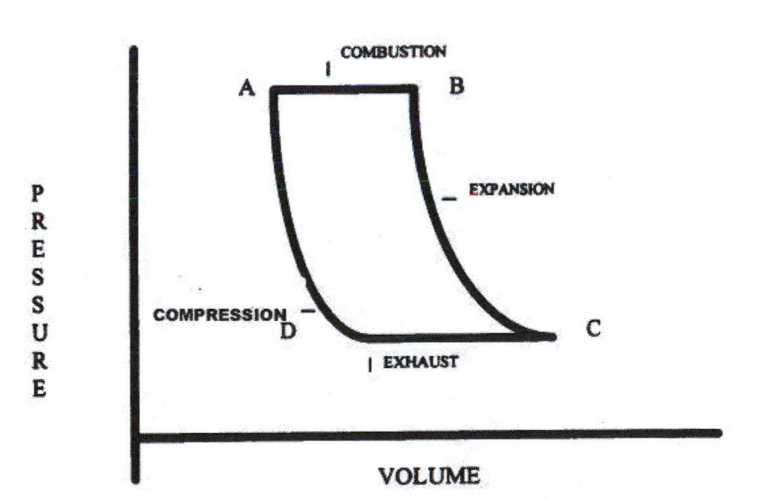

Figure 12-5. Brayton Cycle for Gas Turbines

states of materials involving enthalpy, entropy, temperature, pressure and volume (see Figure 12-9). The Mollier diagrams are named after Richard Mollier (1863-1935), a professor at Dresden University who pioneered this graphical display. Mollier's Enthalpy vs.

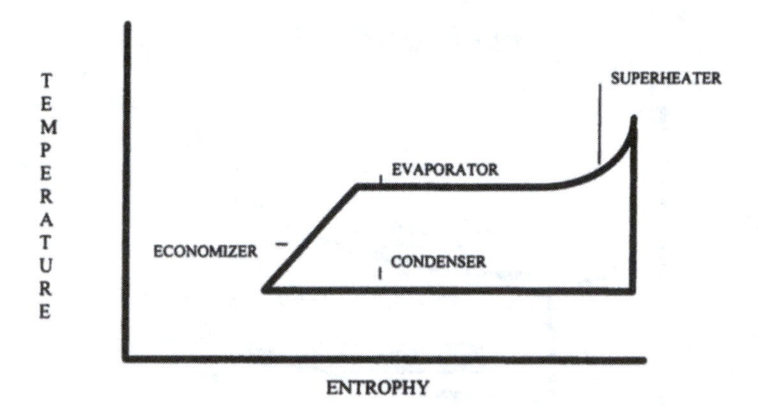

Figure 12-6. Rankine Cycle for Steam Turbines

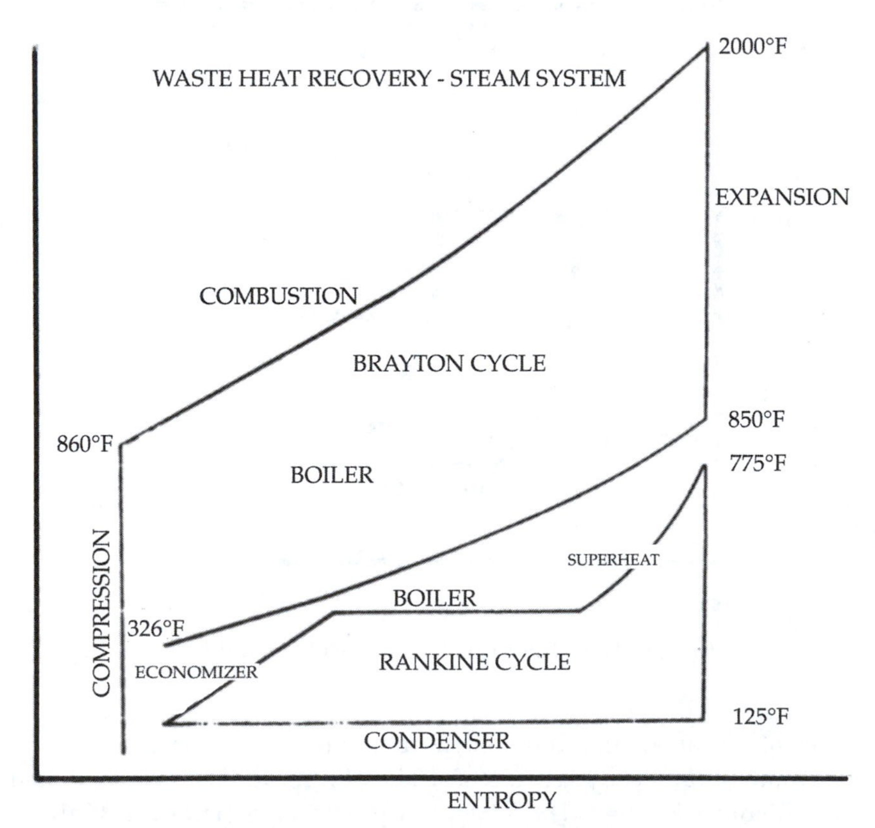

Figure 12-7. Combined Brayton and Rankine Cycles

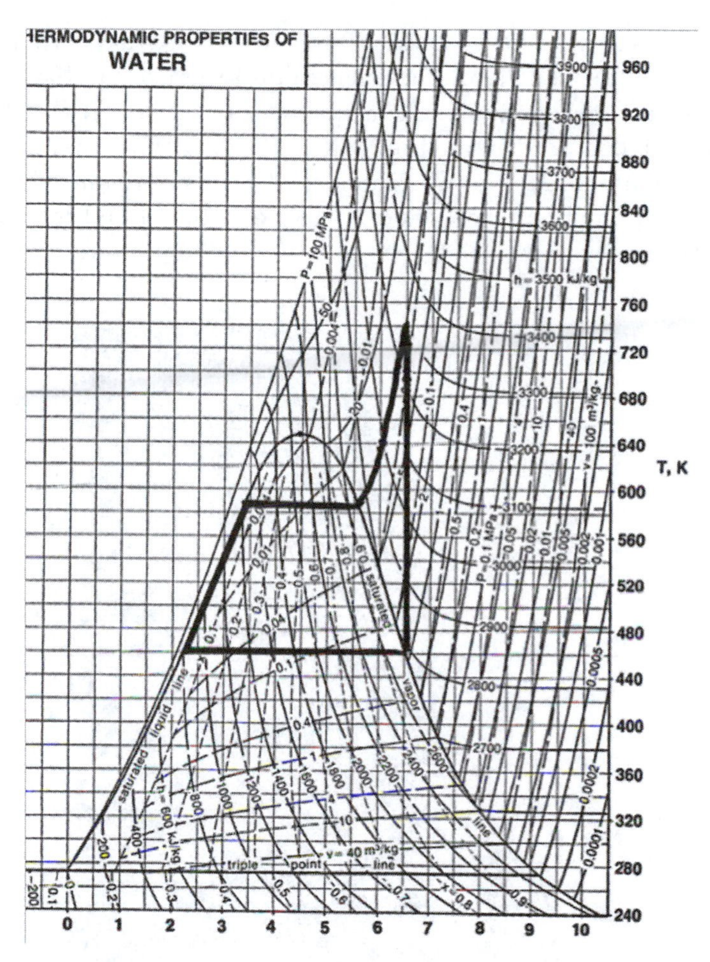

Figure 12-8. Temperature-Entropy Chart for Water/Steam Courtesy of Gas Processors Suppliers Association

Entropy (H-S) diagram is a graphical representation of the Steam Tables[6] and a logical extension of the Temperature vs. Entropy (T-S) diagram. The advantages of the Mollier diagram are that vertical lines (entropy) represent reversible processes and horizontal lines (enthalpy) represent lines of constant energy. Mollier diagrams are routinely used in the design of power plants, heating and refrigeration systems and even wet steam production for tertiary oil recovery. These cycles are easily represented on H-S diagrams where work is calculated directly from vertical distances as opposed to areas on Temperature-Entropy (T-S) and Pressure-Volume (P-V) diagrams. Additionally, inefficiencies due to irreversibility in real processes are easily depicted on an overlay on the H-S diagram shown in Figure 12-9. While the Mollier Diagram for water/steam is the most common, Mollier diagrams exist for a multitude of materials (such as, ammonia, propane, pyridine, toluene and isopentane to mention just a few).

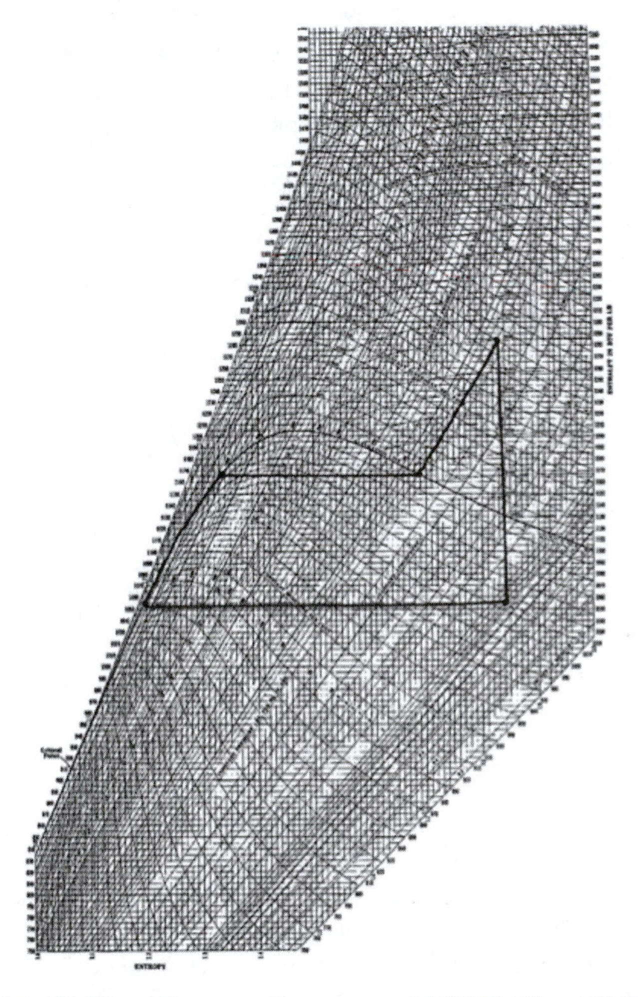

Figure 12-9. Mollier Diagram Courtesy of ASME. The ASME Mollier Diagram was originally published in "Transactions of ASME" February 1941

Rankine cycles, such as that in a combined-cycle system utilizing a condensing steam turbine, generally are not considered true cogeneration systems because they do not produce a source of process heat. However, the open-cycle, or non-condensing steam turbine can be a candidate for cogeneration. The steam turbine exhaust at some pressure above atmospheric for low pressure process, heating, or air conditioning systems. Also steam can be generated by a gas turbine waste heat recovery boiler or fired-power boiler at higher pressure or temperature, such as 1200 psig, 900°F, and expanded through the steam turbine down to 124 psig for process, thereby producing substantial power plus process heat. Consideration should be given to the fact that non-condensing steam turbines are relatively inefficient—with efficiencies varying from 45% to 65% depending on speed and steam conditions. By contrast, condensing steam turbines have efficiencies in the 75% to 80% range.

The components used for all prime mover packages in combined cycle applications are similar and will vary only in size, capacity, flow rates, and heat transfer rates. These components include the waste heat boiler, the condenser, the steam (or expansion) turbine, and the condensate pump.

Fundamentals

To provide the most economical design several parameters must be considered. These are the

- gas turbine exhaust gas flow and temperature;
- steam pressure & temperature;
- pinch point (the distance, measured in temperature, between the exhaust gas temperature and the evaporator inlet temperature);
- approach points of the superheater and the economizer (the distance, measured in temperature between the exhaust gas temperatures and the superheater outlet temperature and the economizer inlet temperature); and the
- exhaust stack outlet temperature.

The diagram in Figure 12-10 shows gas turbine exhaust temperature decreasing from 808°F to 288°F and the water/steam increasing from 125°F feedwater to 600°F superheated steam. In this example the pinch point is 20°F, the economizer approach point is 163°F and the superheater approach point is 208°F.

Fluid Properties

Water / Steam

The steam fluid itself has several advantages: it is chemically stable, it is inert, and it has a high specific heat. Table 12-C list the properties of water/steam.

Fluid (Water/Steam) Properties

See Table 12-C.

Organic Fluids

Table 12-D lists the properties of several organic fluids.

Hot Oils

Hot oils are synthetic heat transfer fluids designed to provide reliable, consistent heat transfer performance over a long life. Synthetic

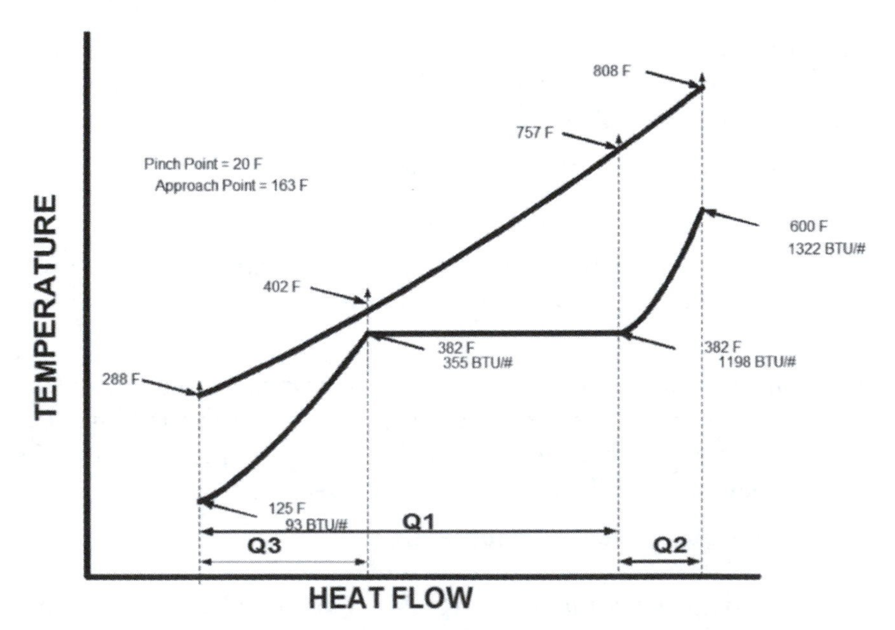

Figure 12-10. WHR Boiler Energy Balance

Table 12-C

STEAM/WATER PROPERTIES	
Chemical Formula	H_20
Molecular Weight	18
Flammability	Non-Flammable
Freezing Point - ^{0}F	32
Boiling Point - ^{0}F	212
Critical Temperature - ^{0}F	705
Critical Pressure - PSIA	3206
Latent Heat Vaporization- BTU\LB	970
Specific Heat - BTU\LB ^{0}F	1
Toxicity	Non-Toxic
Materials Compatibility	Corrosive
Thermal Stability	Stable
Leak Detection	Difficult < Atmos. Press

heat transfer fluids are superior cost-performance alternatives to common mineral oil-based heat transfer fluids.

Synthetic heat transfer fluids have operating ranges, depending on fluid selection, from a low of -120°F to +550°F to a high of +54°F to +750°F. Compare this to water at an operating range of +32°F to +700°F. These fluids are designed for use in non-pressurized or low pressure, indirect heating systems. They deliver efficient, dependable, uniform process heat with no need for high pressures.

Synthetic heat transfer oils are non-corrosive to metals commonly used in the design of heat transfer systems, and they are

Table 12-D

ORGANIC FLUID PROPERTIES				
FLUID	**PROPANE**	**AMMONIA**	**PYRIDINE**	**TOLUENE**
CHEMICAL FORMULA	C_3H_8	NH_3	C_5H_5N	$C_6H_5CH_3$
MOLECULAR WEIGHT	44	17	79	92
FLAMMABILITY	FLAMMABLE			
IGNITION TEMP °F	840	1200	135	997
FREEZING POINT TEMP °F	-310	-110	10	-140
MAXIMUM HISTORICAL TEMP USE °F	356 WHR Geothermal	660 Laboratory	750 Laboratory	750 Laboratory
TOXICITY	Non Toxic	Toxic	Toxic	Mildly Toxic
THERMAL STABILITY	Stable	Unstable at Temps > 660 °F	Stable	Stable
VAPOR PRESSURE AT °F psia	110@59F	108 @ 0 °F;	1@50 °F	1@59 °F
LEAK DETECTION	With Difficulty	Easily Detected	With Difficulty	With Difficulty
Latent Heat of Vaporization-BTU/Lb	184	590	193	155
Critical Temp °F	205	270	701	605
Critical Press	618	1636 PSI	649.76 psi	596 psi
Boiling Point - °F	-46.6	-28	239	232
Flash Point – °F	-156	52	70	43
Odor	No	Yes-Pungent	Yes-Weak	Yes
Appearance	Colorless			

significantly less sensitive than mineral oils to the negative consequences of thermal oxidation (sludging, fouling, etc.).

HEAT RECOVERY BOILERS

A waste heat boiler consist of a series of heat exchangers in which heat is transferred from the gas turbine's exhaust gas flow to raise the temperature of feedwater (in the economizer) up to the saturation temperature; to evaporate water to steam at constant saturation temperature (in the evaporator) and to superheat the steam (in the superheater).

Boilers are classified as follows:

- Natural Circulation,
- Forced Circulation,
  - Once Through Cycle
  - Spillover Cycle

- Forced Recirculation
- Supplementary Fired

The method of producing flow in boiler circuits, whether natural or forced, has practically no bearing on the effectiveness of the heat absorption surfaces as long as the inside surface is wetted at all times by the water in the steam-water mixture of suitable quality to maintain nucleate boiling. If this requirement is met, the water-film resistance to heat flow is negligibly small and the overall heat conductance depends on the gas side conditions.

With most boiler configurations (with the exception of the once-through-boiler) water flows into the economizer and then into the evaporator where most of the liquid is vaporized to steam. The flow then continues into the boiler drum which acts as a demister to remove water droplets from the steam. Steam is taken off the top of the drum and through the superheater (to further insure that any remaining water droplets are removed). The steam proceeds to a steam turbine or other process.

NATURAL CIRCULATION BOILER

The natural circulation boiler is the largest of all boiler types (see Figure 12-11). Natural circulation is best employed when large changes in density occur as a result of heat absorption. Therefore, natural circulation is restricted to subcritical applications where there is a substantial difference in densities between the downcomer fluid and the riser fluid. In natural circulation boilers the flow through the heat exchangers depends upon the difference in densities of the fluids in the downcomers and risers. The downcomers carry the water from the steam drum (at the top of the boiler) to the mud drum (at the bottom of the boiler). The risers return the water from the mud drum to the steam drum. Water circulation increases with increased heat input until a point of maximum fluid flow is reached. Beyond this point any further increase in heat absorption results in a decrease in flow. Larger heat recovery applications usually employ the natural recirculation system, mostly the two-drum variety.

FORCED CIRCULATION

A primary difference between natural circulation boilers and forced circulation boilers (see Figure 12-12) is the addition of the

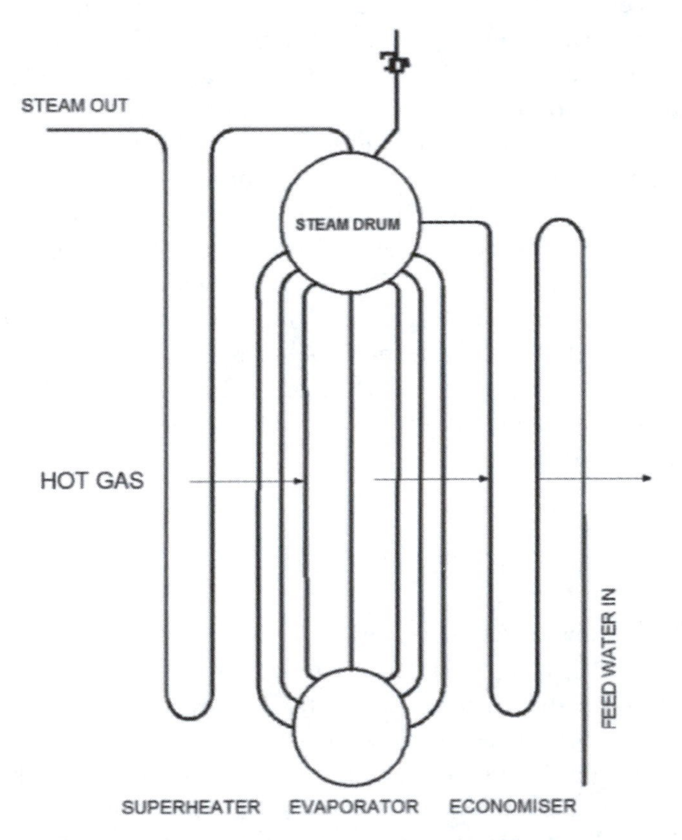

Figure 12-11. Natural Circulation Boiler

circulation pump. Water flowing from the condensate pump first enters the economizer, than flows into the evaporator where all the liquid is vaporized to steam. The flow then continues into the boiler drum which acts as a demister to remove water droplets from the steam. Steam is taken off the top of the drum, through the superheater, to further insure that any remaining water droplets are removed. The flow then goes to the steam turbine.

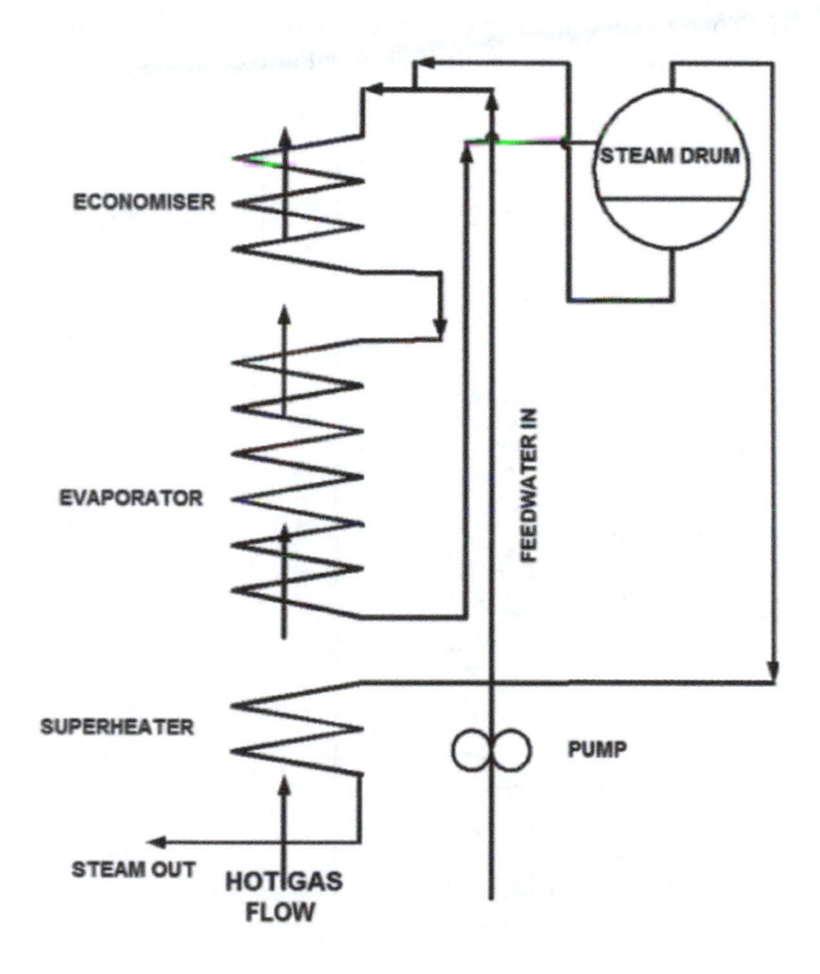

Figure 12-12. Forced Circulation Boiler

The forced circulation once-through cycle uses an essentially monotube boiler, with no intervening drums or separators (see Figure 12-13). All the water entering the inlet is heated, evaporated, and superheated within the tube. Most supercritical utility boilers are of this type. This cycle depends on having steam in superheat for control. A once-through boiler for production of saturated steam only cannot be controlled accurately because the degree of wetness cannot be precisely controlled. Furthermore, water input must be accurately proportioned to heat input or the system will cycle from wet to superheat. Control of multiple circuits is critical and complex.

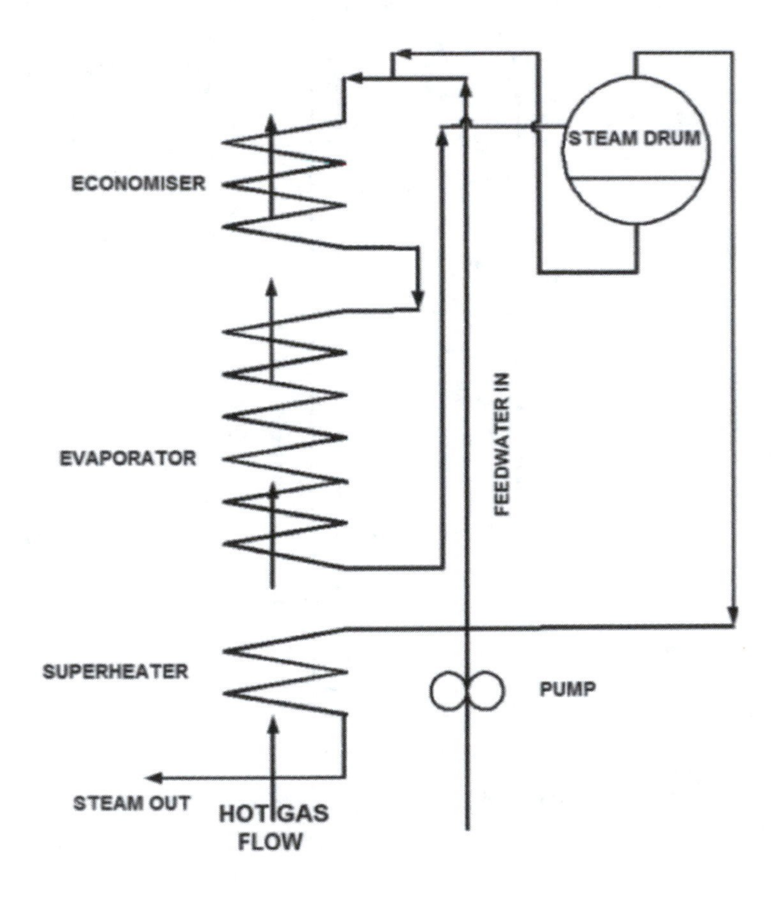

Figure 12-13. Once-Through Cycle Boiler

During periods of load change or instability water slugs might carry over into the steam turbine. Impurities in the water, whether soluble or insoluble, deposit out in the superheat region. While solubles can be flushed out periodically, insolubles must be removed by acid cleaning. Water should be treated to "zero" hardness, and a pH value of 8 to 9.5.

A modification of the once-through monotube cycle is the once-through spillover cycle (Figure 12-14). In the spillover cycle controls are set to produce wet steam (approximately 90%) at the exit end of the evaporator zone. At this point a cyclone separator is used

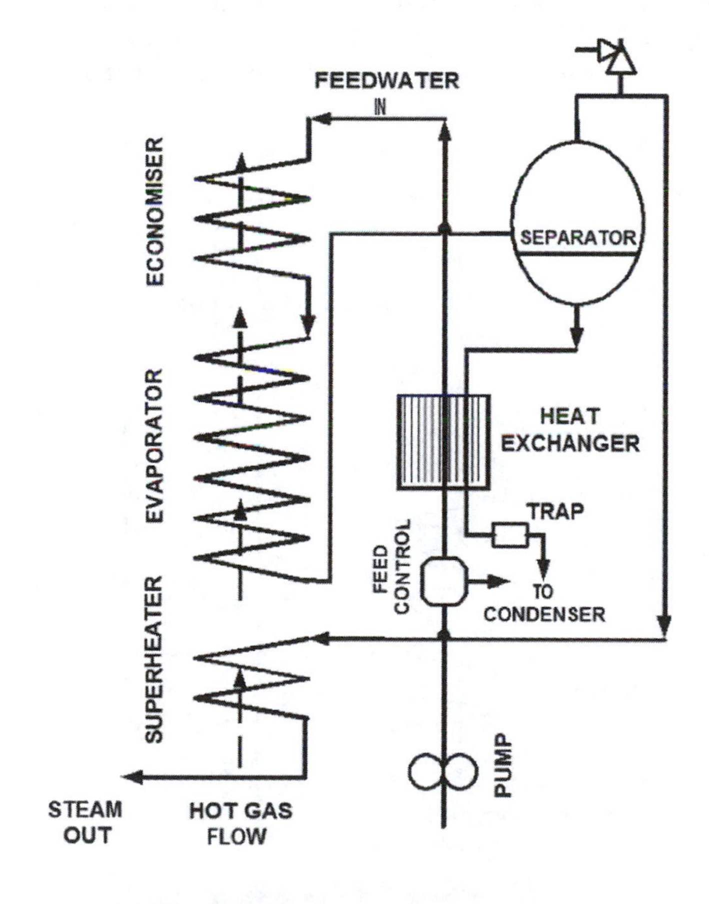

Figure 12-14. Spillover Cycle Boiler

to separate the 10% water from the wet steam. Saturated steam from the separator then goes to the superheater section for further heating. The spillover cycle is not as sensitive to water quality as the monotube cycle because solubles come out with the 10% water.

FORCED RECIRCULATION BOILER

A forced recirculation boiler utilizes a pump to recirculate water from the steam drum, which contains approximately 10% steam and 90% water, back through the evaporator section of the boiler (see Figure 12-15). This assures that there is always forced water circulation in all the boiler tubes. In so doing, vapor separation

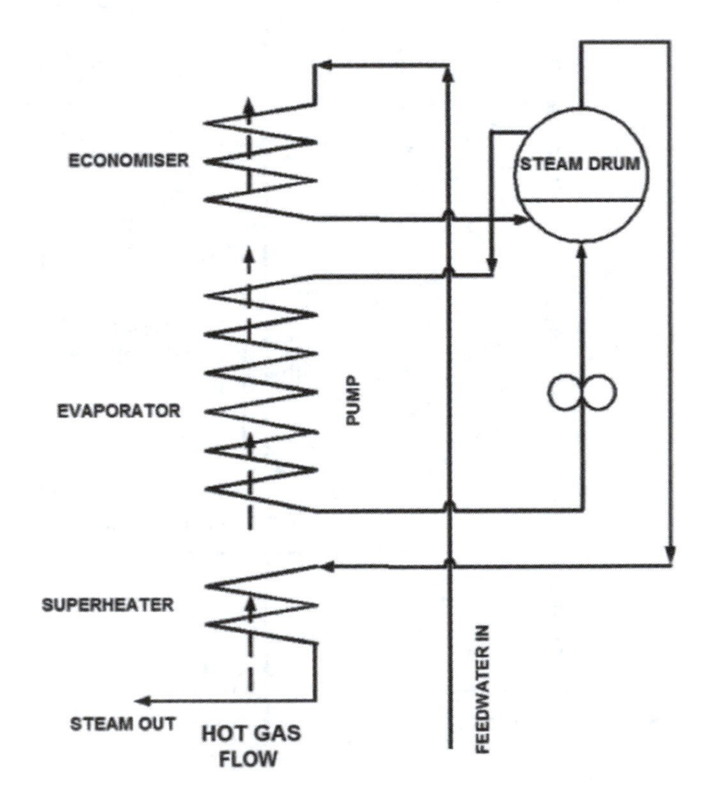

Figure 12-15. Forced Recirculation Boiler

within the steam generating tubes is prevented. The recirculation rate is usually 5 to 10 times the maximum steaming rate of the boiler.

Forced recirculation boilers use smaller sized tubes, which reduce the size and weight of the boiler. Also reduced is the time required for steam up. Furthermore, heating surface design can be quite flexible: using vertical, horizontal, or circular configurations. The circular coil design provides flexibility in handling different stresses which are likely to occur in most heat recovery projects.

Water and steam velocities in the tubes of a forced recirculation boiler vary with load and pressure. As a general rule, the pump is sized to produce a water velocity of 2 to 4 feet per second at the entrance to the tubes where the fluid is all water. The most critical component in this system is the pump.

SUPPLEMENTARY FIRED BOILER

Boilers may be either unfired or fired. Unfired boilers use only the waste heat from the gas turbine exhaust, while fired boilers add additional fuel via afterburners into the hot gas path upstream of the boiler inlet. Afterburners receive their oxygen from the gas turbine exhaust gas, which is sufficiently rich in oxygen to support combustion. The afterburners are activated when the gas & steam turbines are loaded and additional power is required. The power realized from this system can be substantially increased. The limit on supplementary firing, for conventional heat recovery boiler construction, is usually 1400°F but, with installation of higher quality materials in the superheater section and the final stack, 1700°F can be achieved.

The laydown space required by the prime mover, steam boiler or waste heat recovery generator, and steam (or expansion) turbine and condenser is a major variable. While the gas turbine and steam turbine laydown space is well defined for ranges of power, the boiler and condenser will vary as a function of the boiler configuration and the amount of heat recovered. The Figure 12-16 shows a typical arrangement for a gas turbine-compressor plant with a space saving forced circulation boiler, steam/air condenser and a separate steam turbine-compressor.

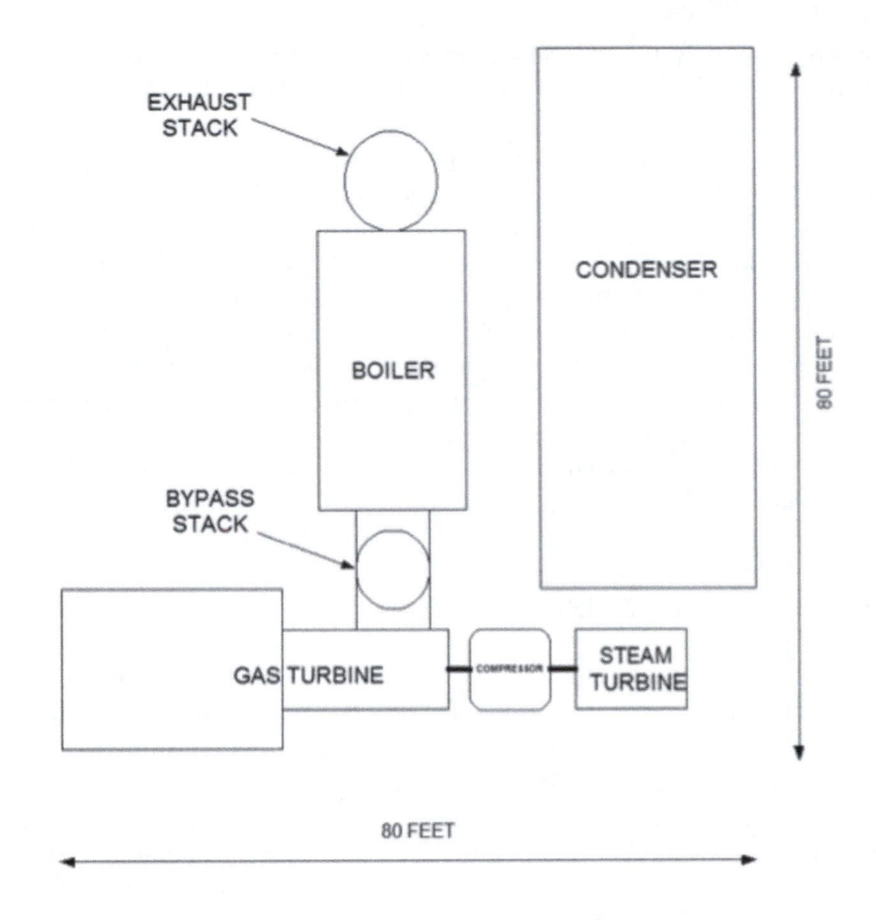

Figure 12-16. Combined Cycle Plant Laydown Area

BOILER HARDWARE

Single Pressure Boiler

Figure 12-17 depicts a single pressure boiler system. Note that the expansion from 3 to 4' instead of 4 is indicative of the efficiency of the steam turbine. The 1-2-3-4 area is the work being done and is a function of the heat energy into the system 1-2-3 and the heat energy

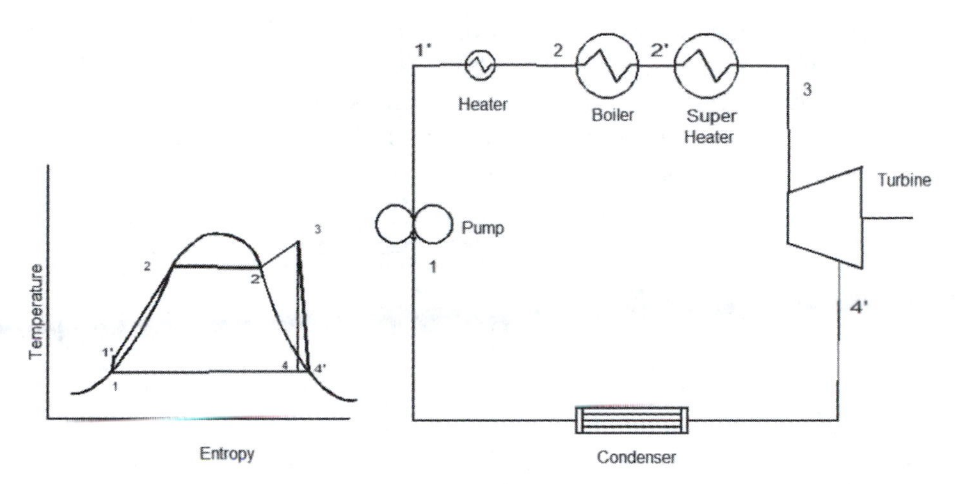

Figure 12-17. Single Pressure Boiler Schematic

extracted from the system as turbine or shaft work and condenser capabilities (size and cooling media).

This graphic takes the reader on a path through the Rankine Cycle for a heat recovery steam boiler, a condensing steam turbine and a condenser. See Table 12-E.

Table 12-E

PATH	COMPONENTS
1 – 1'	Water is pumped into the economizer
1' - 2	Water travels through the economizer
2 – 2'	Water enters the boiler and leaves as wet steam
2' - 3	Wet steam is superheated
3 – 4'	Superheated steam is expanded through the steam turbine at near constant entropy exiting below atmospheric pressure
4' - 1	Low pressure steam travels through the condenser where it is returned to liquid water.

The path of constant entropy would be from 3 to 4 however, inefficiency of the steam turbine results in a path from 3 to 4'. The increase in entropy represents the efficiency (or inefficiency) of the steam turbine.

Dual Pressure Boiler with Reheat

Dual Rankine cycle with reheat is pictured in Figure 12-18. Note that the cycle path from 1 to 3 is the same as the single pressure boiler. However, work is performed from 3-4 and from 5-6 with a re-heater between 4 & 5. See Table 12-F.

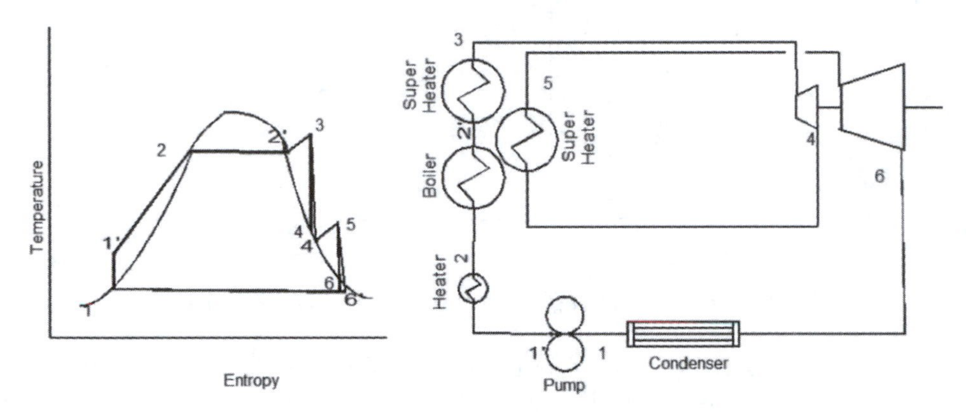

Figure 12-18. Dual Pressure Boiler Schematic

CONDENSERS

For optimum efficiency, the condenser should operate at as low an absolute pressure as possible. Initially a one (1) psia pressure might be considered, however, this low pressure will result in a high cost, large area condenser. A more practical approach is to consider a higher total pressure, for example, two (2) to four (4) psia. The condensers may be water cooled or air cooled, utilizing force draft fans, louvers, and some recirculation for extremely cold applications. Discharge from the condenser enters the condensate pump for recycling back through the waste heat boiler. At the point between the condenser and the condensate pump, make-up water is provided for the water lost in the condenser, steam turbine seals, etc.

BOILER TUBING

The selection of boiler tubing material is based on the design temperatures in the various heat exchangers. Table G lists several materials for use in a boiler and the maximum allowable temperatures.

Table 12-F

PATH	COMPONENTS
1 – 1’	Water is pumped into the economizer
1’ - 2	Water travels through the economizer
2 – 2’	Water enters the boiler and leaves as wet steam
2’ - 3	Wet steam is superheated
3 – 4’	Superheated steam is expanded through a steam turbine section at near constant entropy exiting as wet steam
4’ - 5	Wet steam is superheated in a re-heater
5 – 6’	Superheated steam is expanded through additional steam turbine sections exiting below atmospheric pressure at near constant entropy
6’ - 1	Low pressure wet steam travels through the condenser where it is returned to liquid water.

Table 12-G
Tube Material Specification & Maximum Allowable Tube Wall Temperature

ASME Material Specification	Maximum Allowable Tube Wall Temperature ^{0}F
SA 178 GRADE A	700
SA 178 GRADE C	800
SA 250 GRADE T1b	850
SA 209 GRADE T1/GR T1b	950
SA 213 GRADE T1/GR T1b	1000
SA 213 GRADE T22	1050

The size (diameter and wall thickness) of boiler tubes is based on the average internal flow velocities through the various heat exchangers.

Table 12-H
Recommended Internal Flow Velocities

Heat Exchanger	Velocity Range FT/Min
Feedwater Heaters	150 - 300
Economizers	150 - 300
Reheaters	2000 - 5000
Evaporators	70 - 700
Superheaters	2000 - 3000

Bare tubing is the least expensive while finned tubing provides greater heat transfer. These concerns need to be addressed in the detail design of a heat recovery boiler. However, these concerns are not necessary to make preliminary estimates.

STEAM TURBINES

A steam turbine converts the energy of high pressure-high temperature steam through extraction to produce shaft power. Steam turbines are available as condensing turbines, non-condensing (or back pressure) turbines, reheat-condensing turbines and extraction & induction turbines. A condensing steam turbine discharges into a vacuum created by the cooling of the steam fluid in the condenser. Non-condensing turbines are primarily used in process plants where the high pressure steam is used to generate some shaft power and the lower pressure steam is used in a plant process (i.e., heating). A reheat-condensing turbine is a variation of the condensing turbine. Extraction and induction turbines are specific configurations of turbines for specific applications.

OFF DESIGN CONSIDERATIONS

The choice of the gas turbine must be made on a site-specific basis, where the ratio of thermal to electrical loads, fuel type and availability, and environmental constraints are known. The generators, heat exchangers and condensers can be specified and designed using standard techniques and components. Development of systems capable of burning alternate fuels (other than distillates & natural gas) has become more important as the cost of fossil fuels increases, and decreases in emissions from fossil fuel facilities is mandated. Fluidized bed technology is one means of burning low grade fuels. The use of this technology will increase as fuel cost increase. The use of landfill and sewer gases is another source of low Btu fuel.

A bypass stack allows the gas turbine to operate in the simple cycle mode—usually upon start up and shutdown. It can be used to instantly switch from a combined cycle mode to a simple cycle mode in the event of a feedwater pump failure, thereby eliminating the possibility of over-heating the boiler tubes.

A NO_x catalyst (selective catalytic reduction—SCR) can be installed before, after or within the WHR boiler depending on the optimum temperature range of the SCR. The same is true for a CO catalyst.

References

1. First Generation is a designation assigned to the initial design & manufacturer of a gas turbine engine model.
2. Second Generation is a designation assigned to follow-on designs employing significant improvements.
3. U.S. Energy Information Administration, Natural Gas Average July 2012
4. A decibel is a sound measurement. The subscript "a" refers to the A weighted scale. See Chapter 10 "Gas Turbine Acoustics And Noise Control" for details on sound attenuation.
5. *Introduction to Thermodynamics Classical and Statistical*, 2nd Edition, Richard Sonntag and Gordon J. Van Wylen
6. Steam Tables: Properties of Saturated and Superheated Steam, Combustion Engineering, Inc. and ASME Compact Edition

Chapter 13

Detectable Problems

As with any piece of mechanical equipment, the gas turbine is susceptible to a wide variety of physical problems. These problems include such things as dirt build-up, fouling, erosion, oxidation, corrosion, foreign object damage, worn bearings, worn seals, excessive blade tip clearances, burned or warped turbine vanes or blades, partially or wholly missing blades or vanes, plugged fuel nozzles, cracked and warped combustors, or a cracked rotor disc or blade. Gas turbine problems may be detected by concentrating on four general areas: 1) the thermodynamic gas path, 2) vibration of bearings, rotors, etc., 3) lubrication, and 4) controls. The information gathered will often be the result of an interdependent parameter relationship and will assist in verifying a diagnosis and more precisely isolating a fault.

Some of these problems will become evident as vibration increases, others will be detected by a change (increase or decrease) in lubrication oil temperature. However, some problems (often the most serious problems) can be detected only through gas path analysis. The gas path, in its simplest form, consists of the compressor(s), combustor, and turbine(s). The operation of each component follows predictable thermodynamic laws. Therefore, each component will behave in a predictable manner when operating under a given set of conditions.

Detectable faults make themselves evident through changes in observed engine parameters. Actually these faults are the result of changes in component geometry. Therefore, it is necessary to identify the geometry changes in order to evaluate the severity of the problem. Lou Urban, in his book "Gas Turbine Parameter Interrelationships"[1], defined the measurable engine parameters (rotor speed, temperature, pressure, fuel flow, and power output) as dependent variables and pumping capacity, efficiency, and effective nozzle areas as independent variables. To further simplify matters it is usually not necessary to actually calculate pumping capacity, efficiency, or effective nozzle areas. It is sufficient to consider only the changes in the dependent variable parameters that define the independent variables. It should also be noted that a change in any one dependent parameter does not necessarily indicate a particular independent parameter fault. For example, a change in compressor discharge pressure (CDP) does not necessarily indicate a dirty compressor. The change could also be due to a combined compressor and turbine fault, or to a turbine fault alone.

The objective of the gas path analysis is to detect as many problems as is sensibly and economically feasible through the observation of suitably chosen parameters. To be detectable, the problems must be of a nature and magnitude to produce an observable change. Cracks in combustors, stators, nozzles, blades, or discs are not detectable through gas path analysis. They are, however, detectable through boroscope inspection. Corrosive attacks on airfoils may also be detected by boroscope inspection. Severe corrosion, as it changes turbine airfoil geometry, is detectable through gas path analysis. Physical problems that are the result of changes in component geometry will degrade component performance, which will produce changes in certain measurable parameters. The changes in these measurable parameters will lead to the identification of the degraded component and subsequent correction of the physical problem.

While input data accuracy is a concern, what is of major importance is data repeatability. Measurement inaccuracies can be eliminated (or at least significantly reduced) as an error source by periodically comparing data to an initial baseline and trending that data. Care must be taken when replacing sensors to re-establish the baseline.

GAS TURBINE—GAS PATH ANALYSIS

In determining whether gas turbine engines are operating normally, data taken from the engine instruments must be corrected to a standard condition and compared to baseline data or data known to be correct for that specific engine model. The baseline data usually consists of a sufficient number of data points to generate a series of curves representing the engine's thermodynamic cycle signature.

These curves are drawn relative to either corrected gas turbine power output (or load), corrected turbine inlet temperature, or corrected rotor speed. Power and turbine temperatures are the most used "reference" parameters. For our discussion we will use corrected power as the "reference" parameter. The first step in analyzing engine data is converting it to standard conditions. Table 13-1 lists the correction factors and their use:

Table 13-1. Performance Correction Factors.

Pressure Correction Factor $\equiv \delta \equiv P_x/29.92$

Temperature Correction Factor $\equiv \theta \equiv T_x/520$

Pressure Corrected $\equiv P_{X\ corr} \equiv P_{X\ obs}/\delta$

Temperature Corrected $\equiv T_{X\ corr} \equiv T_{X\ obs}/\theta$

Speed Corrected $\equiv N_{X\ corr} \equiv N_{X\ obs}/\sqrt{\theta}$

Fuel Flow Corrected $\equiv W_{f\ corr} \equiv W_{f\ obs}/\delta\sqrt{\theta}$

Specific Fuel Consumption Corrected $\equiv SFC_{corr} \equiv SFC_{X\ obs}/\sqrt{\theta}$

Airflow Parameter Corrected $\equiv W_{a\ corr} \equiv W_{a\ obs}\sqrt{\theta}/\delta$

BHP Corrected $\equiv BHP_{corr} \equiv BHP_{obs}/\delta\sqrt{\theta}$

where the subscript 'x' represents the particular stage or location of the measurement and the subscript "obs" represents the observed data. For example, 'x = Amb' designates ambient pressure and temperature and $\mathbf{P_{Amb}/29.92}$ and $\mathbf{T_{Amb}/520}$ indicates that all corrections are referenced back to ambient conditions. The standard day has been defined as **59°F, 29.92"HGA** by the International Organization of Standardization (ISO). Hence this is referred to as the ISO Standard. When inlet air treatment (heating or cooling) is used, it is sometimes more convenient to correct to the 'standard' temperature and pressure conditions at the inlet to the compressor. This value must be decided upon by the owners, operators, manufacturers, etc. For the purposes of this discussion we will use ISO Standard.

Compressor Discharge Pressure Corrected = CDP/δ = $\mathbf{P_3/\delta}$

where 'x=3' is the axial location for the compressor discharge on single spool gas turbines.

Exhaust Gas Temperature Corrected = EGT/Θ = T_7/Θ

where 'x=7' is the axial location for the turbine exhaust on dual spool gas turbines.

In the following discussions parameter changes are all measured at constant power. Correcting all parameters to the ISO Standard eliminates the variables due to daily and seasonal fluctuations in ambient temperature and pressure.

In electric power generation applications electric power is easily determined as follows[2]:

$$P_e = 3.6\left(\frac{K_h}{t_w}\right) \qquad (13\text{-}1)$$

where

K_h = watt-hours/meter disk revolution
t_w = seconds/meter disk revolution

In applications where a process compressor is driven by a single shaft gas turbine, power is determined by the heat balance on a driven process compressor (the load).

$$P = W_g\,(h_i - h_o) - Q_r - Q_m \qquad (13\text{-}2)$$

where

P = power input to the load device, Btu/sec
W_g = weight flow of gas entering the process compressor, lb/sec
h_i = enthalpy of gas entering the process compressor, Btu/lb
h_o = enthalpy of gas exiting the process compressor, Btu/lb
Q_r = radiation and convection heat loss from the process compressor casing, Btu/sec
Q_m = mechanical bearing losses for the bearings of the process compressor, Btu/sec

For free power turbine configurations, regardless of the application, the power is determined by the heat balance across the power turbine.

$$P_{PT} = W_g (h_i - h_o) - Q_r - Q_m \qquad (13\text{-}3)$$

where

P_{PT} = power input to the load device, Btu/sec
W_g = weight flow of gas entering the power turbine, lb/sec
h_i = enthalpy of gas entering the power turbine, Btu/lb
h_o = enthalpy of gas exiting the power turbine, Btu/lb
Q_r = radiation and convection heat loss from the power turbine casing Btu/sec
Q_m = mechanical loss of bearing of power turbine, Btu/sec

In the above cases Q_r and Q_m can be reduced to functions of the air (or oil) flow and temperature differential.

TURBINE BLADE DISTRESS
(Erosion/Corrosion/Impact Damage)

Indications and Corrective Action

Turbine blade failures account for 25.5% of gas turbine failures[3]. Turbine blade oxidation, corrosion and erosion is normally a long-time process with material losses occurring slowly over a period of time. However, damage resulting from impact by a foreign object is usually sudden. Impact damage to the turbine blades and vanes will result in parameter changes similar to severe erosion or corrosion. Corrosion, erosion, oxidation or impact damage increases the area size of the turbine nozzle. The increased area size is seen as a decrease in turbine efficiency, which is a gas path measurement. It also causes an increase in N_1 stall margin; however, to detect this requires sophisticated tests and measurements. Parameter changes that can readily be seen by the operator are (at constant power): an increase in fuel flow, exhaust gas temperature, and compressor discharge pressure (Figure 13-1). While decreases in N_1 rotor speed and air flow also occur, these changes are not easily detectable, especially with small changes in nozzle area.

Considering the high pressure turbine on dual-spool machines, turbine blade and vane erosion results in a decrease in turbine ef-

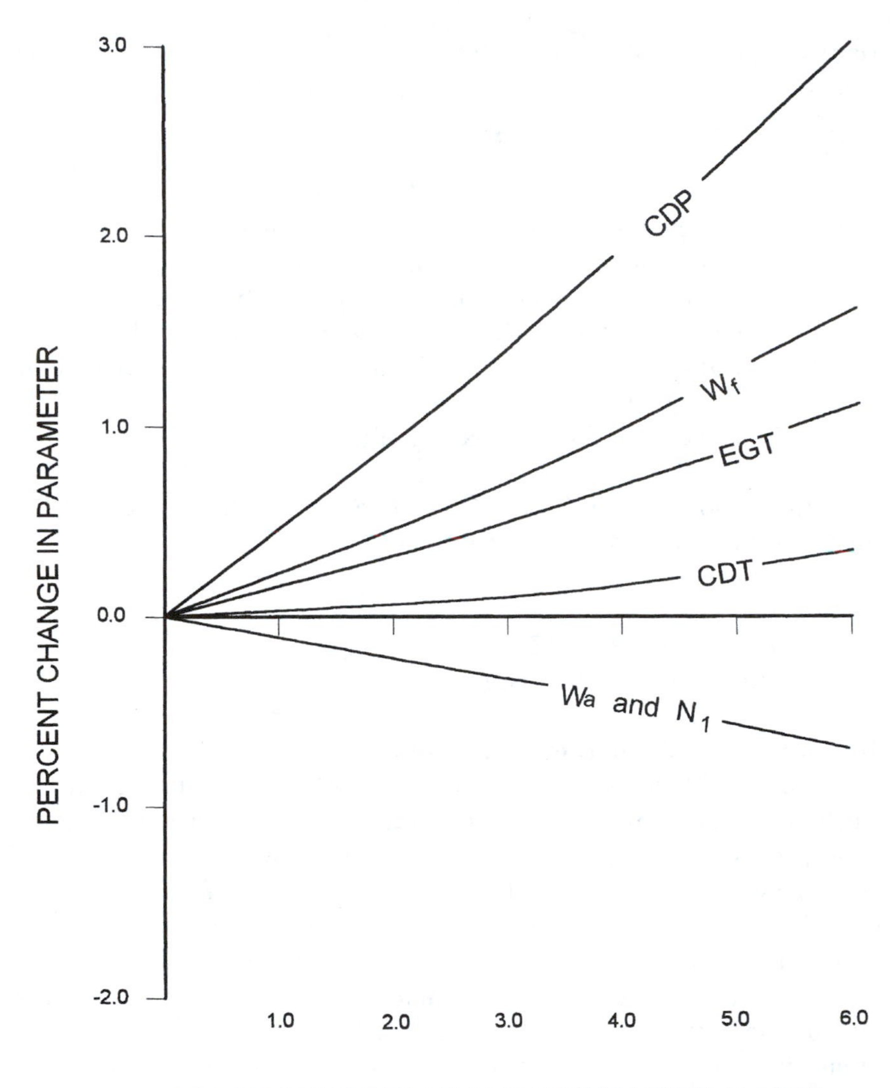

Figure 13-1. Single spool gas turbine estimated parameter changes with turbine nozzle area change.

> Note: Accurate airflow measurement requires a calibrated bellmouth at the inlet of the gas generator supported by temperature readings at this location. A bellmouth, when provided, is calibrated at the factory, usually with the gas generator. Since calibrated bellmouths are not normally provided, airflow measurements are rarely taken in the field. However, relative changes in airflow can be obtained by instrumenting the inlet bellmouth and establishing a baseline. Therefore, changes in airflow are indicated in this section in the hope that this parameter may eventually be utilized to evaluate engine health.

ficiency and an increase in N_2 stall margin. More readily seen parameter changes (at constant power) are increases in compressor pressure ratio of the low pressure compressor (P_3/P_2), exhaust gas temperature, and fuel flow, and decreases in compressor pressure ratio of the high pressure compressor (P_4/P_2), compressor discharge temperature, and N_2 rotor speed (Figure 13-2). Dccreases in N_1 rotor speed and air flow also occur, but these changes are not easily detectable, especially with small changes in nozzle area.

Considering the low pressure turbine on dual-spool machines, turbine blade and vane erosion result in a decrease in turbine efficiency and an increase in N_1 stall margin. More readily seen parameter changes (at constant power) are increases in N_2 rotor speed, fuel flow, exhaust gas temperature, and high pressure compressor discharge temperature (T_4), and decreases in N_1 rotor speed, low pressure compressor discharge pressure ratio (P_3/P_2), and low pressure compressor discharge temperature (T_3), as shown in Figure 13-3. Although a decrease in airflow also occurs, the change is not easily detectable.

To improve gas turbine package performance and increase overall power output, manufacturers have continually increased turbine inlet temperatures. While much of this increase has resulted from improvements in turbine materials, the implementation of turbine cooling has been a major contributing factor.

Turbine blade and nozzle cooling is accomplished by routing relatively cool air from the gas turbine-compressor back to the hot turbine components. This relatively cool (400°F-600°F) air is ported

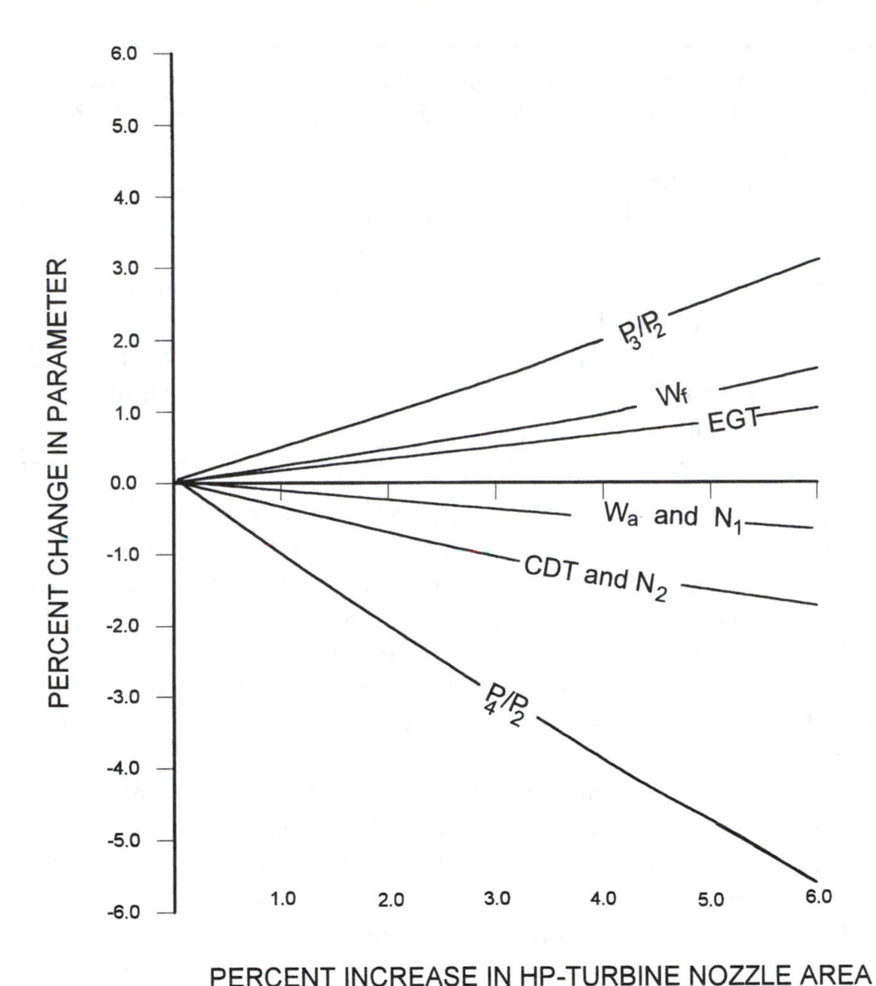

Figure 13-2. Dual spool gas turbine estimated parameter changes with high pressure-turbine nozzle area change.

through the turbine nozzles and turbine blades, and along the rims of the turbine discs to cool the metal temperatures exposed to the hot gases. These cooling paths are small "sieve-like" passages that are vulnerable to plugging by contamination in the gas turbine inlet air stream. Even a partially plugged cooling passage will result in 'local' temperature increases that can be significant. The resulting damage is often seen as localized burning or melting of the airfoil.

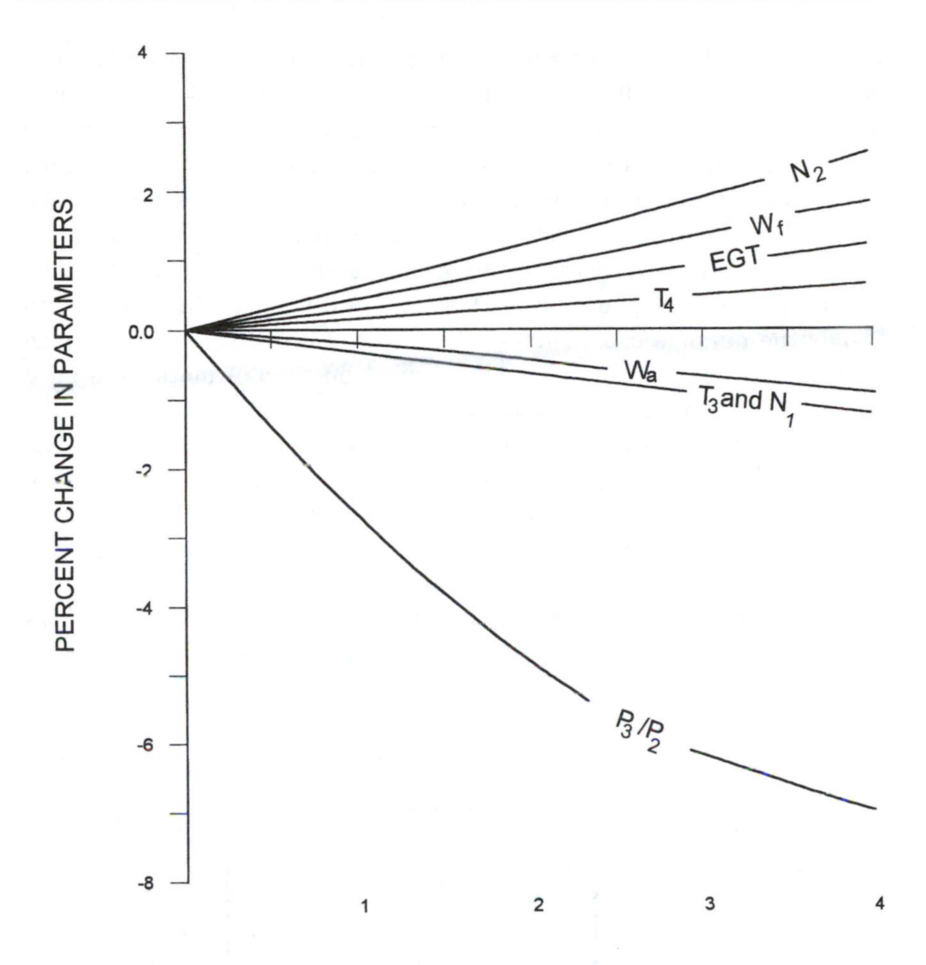

Figure 13-3. Dual spool gas turbine estimated parameter changes with low pressure-turbine nozzle area change.

Corrective Action

Geometry changes resulting from oxidation, corrosion, erosion, or impact damage can be corrected only during overhaul by removal and replacement of the damaged parts. Depending on the type of machine this could result in downtime of one day (aero-derivatives) to several weeks (heavy frame units)[4]. However, certain actions can be taken to maximize the time between overhaul and to minimize fired hour cost and overhaul cost. Many oxidation, corrosion or erosion problems initially discovered by gas path analysis can be verified by a boroscope inspection. Then an evaluation can be made as to how detrimental that fault is to continued operation of the unit. Once a fault has been isolated it can be tracked or trended using both gas path analysis and visual (boroscope) inspection. If the oxidation/corrosion/erosion is affecting the nozzle airfoils uniformly and the resulting inefficiency and power loss is tolerable, then the decision to repair the damage can be delayed while the progress of the damage is monitored. Figure 13-4 depicts varying degrees of metal loss on a turbine nozzle with estimated "urgency to repair."

As with nozzles, corrosion/erosion affects blades initially in the airfoil mid-span. Because blade loading is more severe than nozzle loading, less blade material loss can be tolerated. Furthermore, trailing edge losses are more tolerable than leading edge losses. Figure 13-5 depicts acceptable and unacceptable material loss.

Impact damage is usually first noticed as an increase in vibration. If the vibration is not severe and stabilizes after the initial

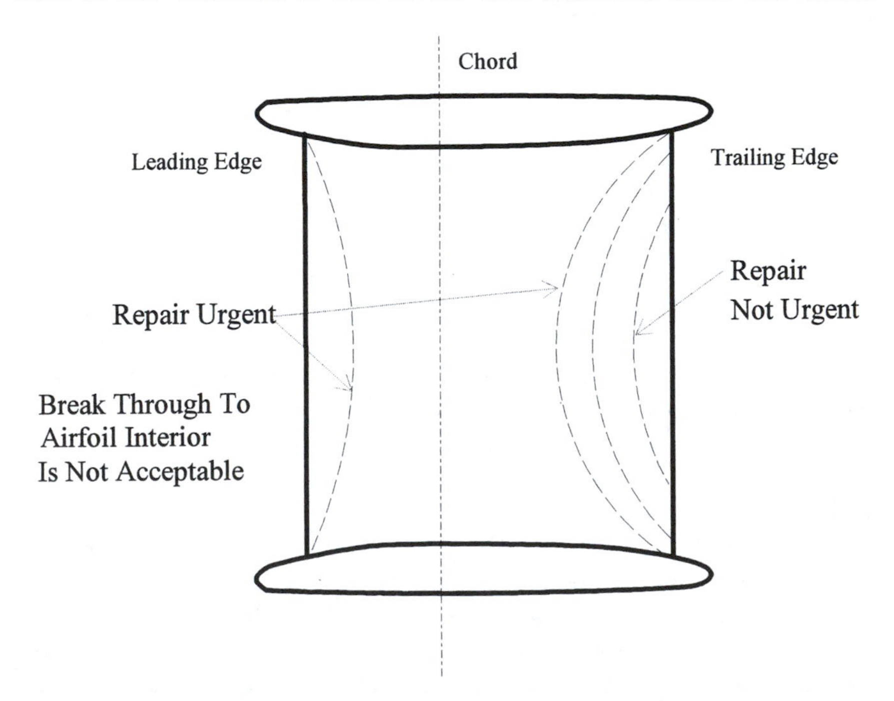

Figure 13-4. Turbine nozzle material loss and urgency of repair.

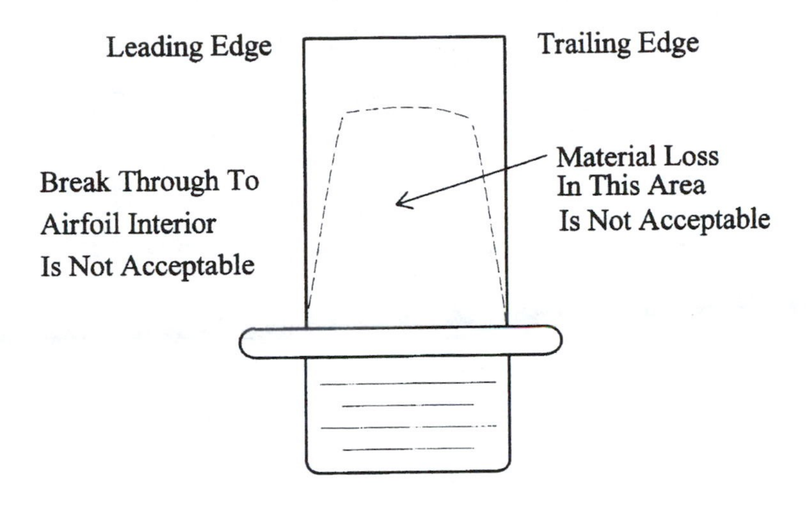

Figure 13-5. Turbine blade material loss.

increase, a gas path analysis may help define the extent of the damage. This should be followed by a visual or boroscope inspection. If a boroscope inspection verifies that:

1. the damage is the result of a blade or nozzle being impacted by an object originating upstream of the turbine section and the compressor or combustor is not the source of the object damage, or
2. the impact damage area is at or near the blade tip, and vibration is stable and not over limits,

then the unit can be run (until a replacement can be obtained or an overhaul scheduled) while being closely monitored. If blade damage is between the blade root and mid-span (Figure 13-6), regardless of vibration level reached, the unit should not be restarted and should be overhauled. Corrosion over the same parts of the turbine blade as shown in Figure 13-5 and 13-6 can be tolerated. However, if corrosion is evident under the tip of a shrouded blade, the unit should be shutdown and overhauled. If damage was the result of an object from within the turbine stage (such as a piece of turbine blade or nozzle airfoil), the unit should not be restarted.

Changes in blade airfoil dimensions will affect the blades natural frequency. Damage or material loss occurring at the blade tip will

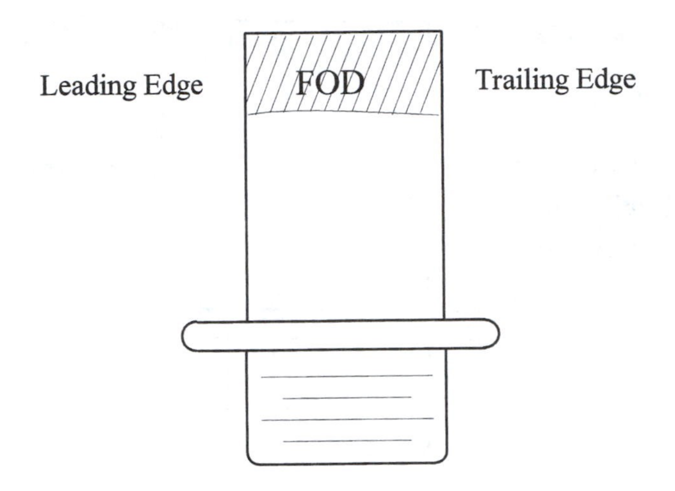

Figure 13-6. Turbine blade material loss.

increase the natural frequency. Conversely, damage occurring at or near the blade root weakens the blade structurally and decreases the blades natural frequency. Blade natural frequencies are above the normal running speed of a gas turbine. However, a reduction in the blades natural frequency could move it closer to the engines normal running speed.

Critical decisions, as discussed above, should be made only with the help of an engineer experienced with gas turbine failure signatures. Major manufacturers constantly test their designs and often run components (blade, nozzle, etc.) to failure. Throughout my years in this industry, I have witnessed numerous component failures—usually after the fact. However, on one occasion I inspected a first stage turbine blade failure in progress. After an increase in vibration was observed, the engine was boroscoped. A single turbine blade had failed at the blade mid-span. Normally this would be sufficient cause for immediate overhaul. However, as this was an experimental engine the decision was made to continue running until vibration reached 4 times acceptable running levels. The gas turbine was run for an additional 8 hours before overhauling the unit. Vibration had peaked and stabilized at 8 mils. The damage found in overhaul was no more severe than had been observed in my boroscope inspection.

COMPRESSOR FOULING

Indications And Corrective Action

Compressor fouling normally occurs due to foreign deposits on the airfoils. In petro-chemical operations, this is the result of the oil and other hydrocarbons in the atmosphere caused by neighboring process plant flares, etc. In coastal areas, this could be the deposit of salt in the compressor. In any event, compressor fouling is indicated by a drop in compressor efficiency, which is more readily seen as a drop in compressor discharge pressure at a constant speed and load. This will also manifest itself as a reduction in load capacity at a constant compressor inlet temperature. A compressor efficiency drop of 2 percent is indicative of compressor fouling. This can be calculated as shown.

$$\eta_c = \left(\frac{\left(\frac{CDP}{P_{in}}\right)^{\frac{k-1}{k}} - 1}{\left(\frac{CDT}{T_{in}}\right) - 1} \right) \tag{13-4}$$

For the field engineer or operator, compressor fouling is best indicated by a 2 percent drop in compressor discharge pressure at constant speed and load. Another indication of fouling is a 3 percent to 5 percent reduction in load capacity at constant compressor inlet temperature or ambient air temperature. The 1 psig decrease in compressor discharge pressure, which will accompany the load reduction is not easily detectable with field instrumentation.

As shown in Figure 13-7 for the single spool gas turbine, a decrease in compressor efficiency (at constant load) results in a decrease in rotor speed, compressor discharge pressure, and airflow, and a increase in fuel flow, exhaust gas temperature, and compressor discharge temperature.

Similarly, for the dual spool gas turbine, a decrease in low pressure compressor efficiency results in decreases in N_1 rotor speed, low pressure compressor discharge pressure (P_3), and airflow, and increases in fuel flow, exhaust gas temperature (T_7), N_2 rotor speed, and low and high pressure compressor discharge temperatures (T_3 and T_4) as shown in Figure 13-8. The decrease in high pressure compressor discharge pressure (P_4) is small and not easily detected.

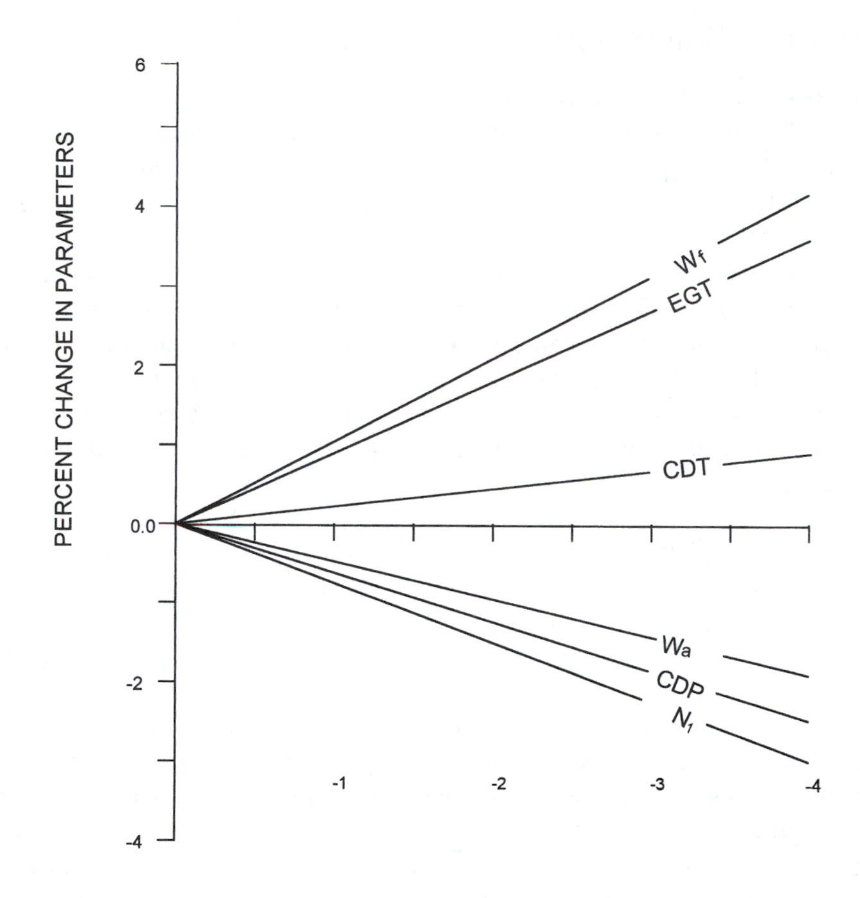

Figure 13-7. Single spool gas turbine estimated parameter changes with compressor efficiency changes.

Usually the low pressure compressor experiences most of the compressor fouling. However, ineffective compressor washing will move much of the debris from the low pressure compressor and re-deposit it in the high pressure compressor. This degrades the high pressure compressor efficiency. When this occurs the decrease in compressor efficiency (at constant power) results in a decrease in both rotor speeds and airflow and an increase in low pressure compressor

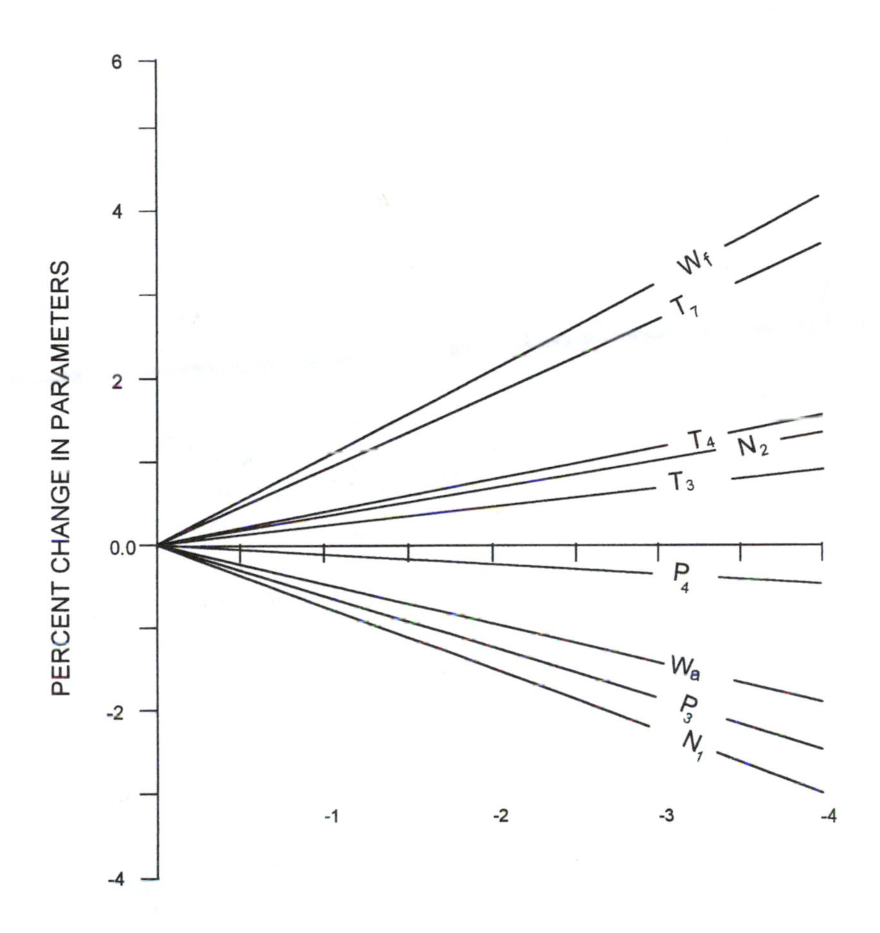

Figure 13-8. Dual spool gas turbine estimated parameter changes with low compressor efficiency changes.

discharge pressure (P_3), fuel flow, exhaust gas temperature (T_7), and both high (T_4) & low (T_3) compressor discharge temperatures (Figure 13-9).

The equipment manufacturer normally specifies cleaning agents. Typically, carbo-blast and water-wash are specified, with water-wash being the first method recommended. Care should be taken in cleaning compressors that are salt-laden, since the cleaning process will

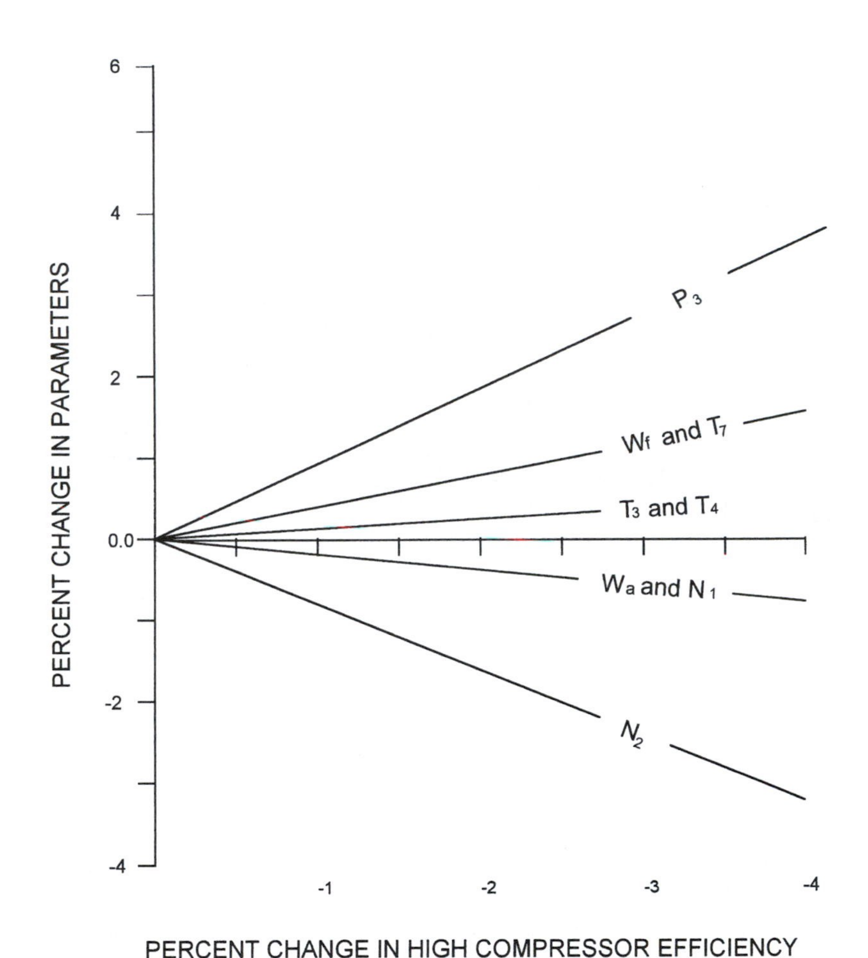

Figure 13-9. Dual spool gas turbine estimated parameter changes with high compressor efficiency changes.

dump huge quantities of salt into the turbine resulting in accelerated sulfidation corrosion. In some cases, it is recommended that the turbine also be washed after the compressor is washed.

Note: Impact damage to compressor blades can mimic compressor fouling.

COMBUSTOR DISTRESS AND PLUGGED FUEL NOZZLES

Plugged nozzles and/or combustor and transition piece failures will always result in distorted exhaust gas temperature patterns. Sometimes an increase in fuel flow can be seen, depending on the severity of the damage. Changes in burner efficiency produce no appreciable change in other parameters (Figure 13-10).

An obvious indicator of plugged nozzles or combustor damage is a distorted temperature pattern or profile accompanied by a change in this profile with power changes. This is a result of the swirl effect through the turbine from the combustor to the exhaust gas temperature-measuring plane (Figure 13-11).

Distortion in the temperature pattern or temperature profile not only affects combustor performance but can have a far reaching impact on the turbine. Turbine blades can be excited both thermally and structurally by radial variations in temperature or flow. For example, a dramatic failure of a gas turbine with only 800 operating hours was due to the failure of a portion of the combustor liner. The failed liner material lodged against the first stage turbine nozzle and blocked airflow to that quadrant of the turbine. The subsequent cycling, with each revolution of the turbine wheel, led to thermal fatigue failure of forth stage turbine blades at the blade root. The failed blades so unbalanced the rotor as to dynamically fail the bearings resulting in further damage and fire. The damage, after the liner had failed, could have been avoided if operations had been monitoring the exhaust gas temperature profile. Fortunately disaster was averted on a similar unit when the distortion in the temperature profile led to the shutdown and inspection of this unit. The combustion liner had failed in the same place and lodged against the first stage turbine nozzles.

FOREIGN/DOMESTIC OBJECT DAMAGE

Foreign Object Damage (FOD) is defined as material (nuts, bolts, ice, etc.) ingested into the engine from outside the engine envelope. Domestic Object Damage (DOD) is defined as objects from any other part of the engine itself. Statistically foreign or domestic object damage accounts for 10.5% of gas turbine failures.[3] Most of the impact incidents are the result of small loose parts within the engine. These parts are either left in the engine during the build or

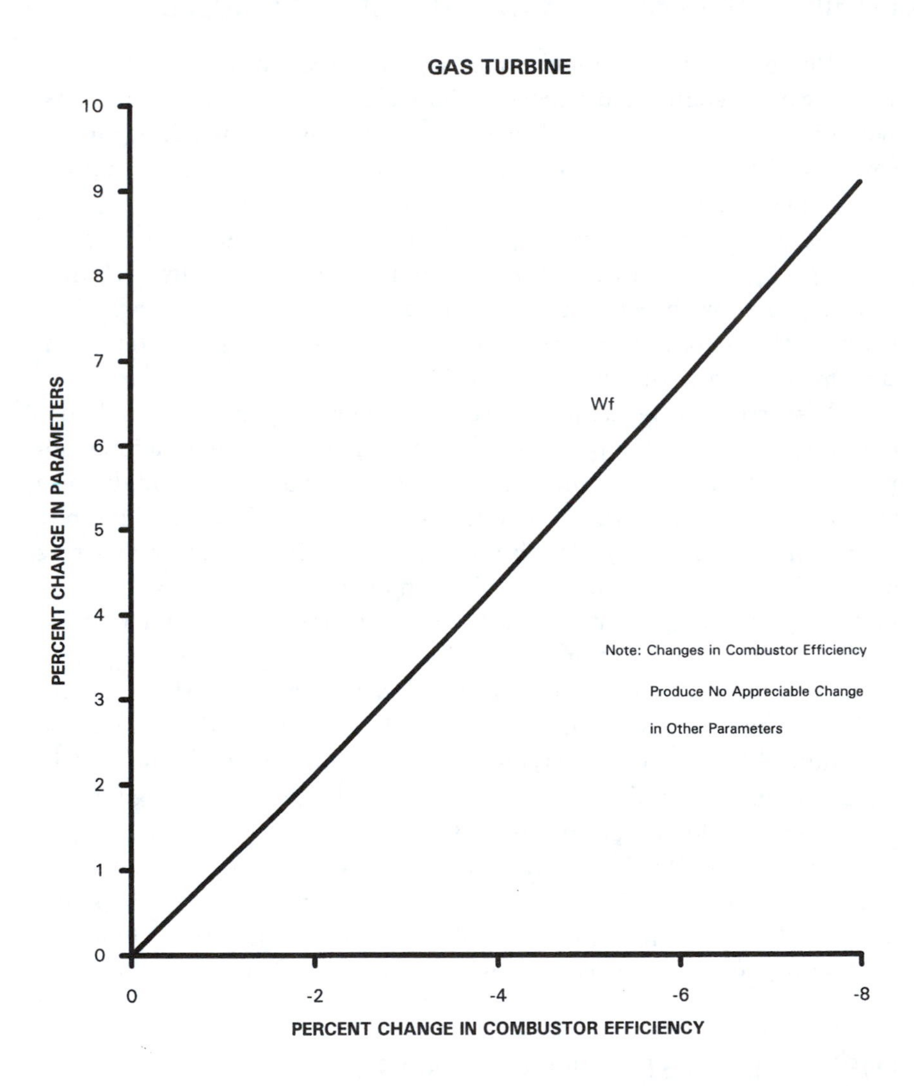

Figure 13-10. Estimated parameter changes with combustor efficiency change.

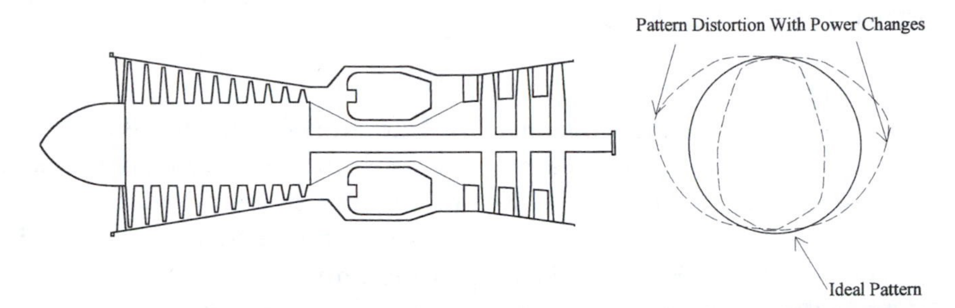

Figure 13-11. Plugged fuel nozzle.

overhaul, or break loose after operation is initiated. In a relatively small number of cases, damage is the result of tools left in the inlet or the formation of ice in the inlet downstream of the filters. There have even been several cases where the inlet silencers failed and bits of silencer debris entered the engine. In any event, FOD is almost always accompanied by an increase in vibration.

Compressor FOD symptoms are similar to compressor fouling. Severe Object Damage (OD) will also affect the combustors and turbine. Combustor damage or blockage is evident as temperature pattern distortion similar to plugged nozzles. Turbine FOD will cause a decrease in turbine efficiency similar to blade and vane erosion.

WORN AIR/OIL SEALS

Pressure balanced labyrinth and positive contact carbon-faced air/oil seals are the most common type used in gas turbines. Worn or damaged seals may result in oil leaking into the gas path. Sometimes this leakage will only occur during start-up and shut-down, when the air pressure is at its lowest and the oil pressure at its normal level. The first clue to oil leakage, depending on the size of the lubrication oil system, could be a decrease in lube oil level. For large, combined, lube oil systems, this may not be easily detected. Internal engine leaks can be detected by smoke in the exhaust during operation. Detection is also possible when the engine is shutdown by inspecting for oil streaks in the exhaust duct and along the compressor ID hubs. During operation, worn or damaged seals are more likely to leak hot air or gas into the oil system. This is normally detected by an increase in oil temperature and possibly oil frothing.

FUEL CONTROL PROBLEMS

Starting

Common starting problems are "hot starts" and "hung starts." Hot starts are so called because they produce excessive exhaust gas or turbine inlet temperatures. Sometimes hot starts are also associated with compressor surge. Hot starts are caused by too rich a fuel schedule. A "too rich" fuel schedule will rapidly increase the temperature, but drive the gas generator-compressor into surge. Other causes might be bleed valve malfunction or variable geometry malfunction or an "out of calibration" condition. Hot starts may also be the result of FOD in the compressor. The hung start is evidenced by a very slow acceleration and a slow increase in exhaust gas or turbine inlet temperature. Usually it results from insufficient starter motor torque or too lean a fuel schedule.

Failure to start due to a fuel mixture that is too lean or too rich is not an unusual problem. To determine that the fuel in the combustor has ignited, manufacturers utilize turbine temperature measuring sensors (thermocouples). Some gas turbine manufacturers also install ultraviolet (UV) flame detectors in the combustor. Depending on the method of control, the thermocouples or UV detectors may directly activate a fuel shutoff valve or the signal may be processed through the main control system. If complete shutoff is not achieved, it is probable that excess fuel will accumulate in the combustor and turbine. After repeated start attempts this excess fuel could produce a "hot start" situation.

A failed fuel shutoff valve spring was the cause of a turbine failure I investigated several years ago. This failure went undetected until the unit was shutdown. Without a tight shutoff, fuel gas leaked into the gas turbine over a period of several days. When operations attempted to bring the unit back on line, the first two attempts to start failed (the fuel gas mixture was too rich). On the third start attempt enough air had been mixed with the fuel to reduce the fuel mixture to the upper end of the explosive limit. The explosive mixture spread from the combustor through the turbine and into the exhaust duct. When the mixture ignited it resulted in an exhaust duct explosion with most of the damage occurring at the exhaust duct elbow section.

Running

Running problems associated with fuel controls consist of the inability to accelerate (or increase load), or accelerate too rapidly. There are also numerous running problems (such as miscellaneous trips, aborts, etc.) that are specific to the type of control method employed. Pneumatic and hydraulic controls are very susceptible to leaks and contamination. In addition pneumatic controls are sensitive to moisture in the lines, that can result in erratic operation. This often leads to erratic or otherwise unstable operation. Mechanical and electric controls are subject to wear and leaks in bellows and capillary sensors. In addition, accuracy deteriorates over a period of time. Computer or electronic controls are generally the most problem free, once the unit has been commissioned and all the programming logic fully and completely implemented.

Acceleration problems, either a lack or too much of, usually result from governor malfunctions (although the malfunctions may not be specifically with the governor mechanism). These malfunctions may result from faulty governor input signals such as speed, compressor discharge pressure, temperature, or fuel flow. Overspeed could be the result of a governor malfunction, electrical upsets (generator drives), or coupling failure. Overspeed is detrimental not only to the gas turbine but also to the driven load. Redundant systems (usually two electrical and one mechanical) are installed to protect against overspeed. The electrical systems should be checked periodically (simulation test) and the mechanical fly-bolt should be cleaned. It is not possible to check the operation of the mechanical fly-bolt mechanism in the field.

References

1. "Gas Turbine Parameter Interrelationships," Louis A. Urban, 1969.
2. American Society of Mechanical Engineers Power Test Code 22, 1974.
3. *Sawyer's Turbomachinery Maintenance Handbook*, First Edition, Vol 1, p 12-25.
4. "How Lightweight And Heavy Gas Turbines Compare," Anthony J. Giampaolo, *Oil & Gas Journal*, January 1980.

Chapter 14

Boroscope Inspection

OBJECTIVES AND EXPECTATIONS

When used in conjunction with gas path, vibration, and trending analysis techniques, boroscope inspections often provide the final step in the process of identifying an internal problem. However, it would be misleading to think that periodic boroscope inspections could be a substitute for the other analysis techniques. Boroscope inspections are useful in providing a general view of the condition of critical components, but they are limited by the gas turbine design, boroscope design, and capability of the inspector.

The problems discussed in this chapter, and the photographs used, are taken from actual gas turbines. All gas turbines have, at one time or another, experienced these or similar problems. Therefore, the reader is cautioned not to conclude that a particular problem is associated only with a specific gas turbine model or manufacturer. To insure that this treatise remains as generic as possible, the photographs will be identified as to engine component position and nominal gas turbine power output and not to manufacturer or model number (although readers well acquainted with particular models will readily recognize them from the photographs).

Boroscope inspections are visual examinations subject to the experience of the individual and the quality of the instrument being used. Before undertaking a boroscope inspection, the inspector needs to determine the suitable diameter and length of the boroscope for use in each boroscope port. Also the intensity of the boroscope light source must be determined for each location to be inspected. Depending on the skill and experience of the inspector, one boroscope, properly selected, will satisfy all of the necessary requirements. Where access to the viewed object is achieved through a straight path, the rigid boroscope (Figure 14-1) is a useful tool. This is especially true for inspecting the combustor. Their use is further enhanced with articulating, side viewing mirrors. However, the most versatile boroscopes are the flexible fiber optic scopes (Figure 14-2). These are available in diameters ranging from 0.3 millimeters to 13 millimeters and working lengths of approximately 250 millimeters to 6,000 millimeters. Boroscopes are also available with either two-way (up and down) and four-way (up, down, left, and right) articulation.

Special adapters are also available for the attachment of still or video cameras. Much like a camera, a boroscope is an optical instrument consisting of an object lens, a series of relay lenses, and an eyepiece lens[1]. In addition, an illumination system is also provided. Care should be taken in selecting the light source. For visual inspection only, the less expensive 150 watt light sources are adequate. For video use, a 300 watt white light source is recommended. The size of

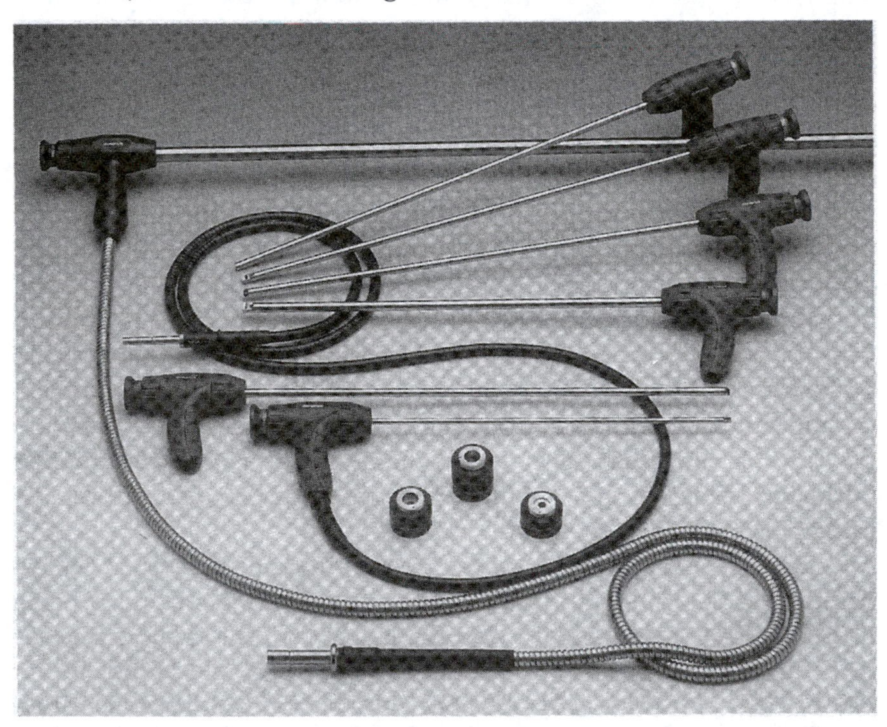

Figure 14-1. Courtesy of Olympus America Inc. The rigid boroscope is best suited for use in areas where the insertion distance and viewing direction is well defined. Rigid boroscopes are available in a range of diameters and working lengths.

the entry port determines the maximum diameter boroscope probe that can be used. It is always desirable to use the largest diameter probe that will fit easily through the opening to obtain the largest, clearest image and the brightest illumination. Also to be considered, primarily with rigid boroscopes, is the length of the boroscope probe and the object distance. The length of the boroscope probe is determined by the size of the gas turbine and the distance from the viewer to the object to be viewed. The object distance is the measurement from the window of the object lens to the surface being viewed. The location of the area to be inspected relative to the access port dictates whether the boroscope views directly ahead or at some angle. This is an important consideration when using a rigid-type boroscope. However, an articulating, flexible boroscope can accommodate almost any viewing angle.

A major consideration in the selection of a boroscope is the field of view (FOV). FOV falls into three categories: narrow, normal, and wide angle. Where

- Narrow FOV ≡ 10 to 40 degrees
- Normal FOV ≡ 45 degrees
- Wide Angle FOV ≡ 50 to 80 degrees.

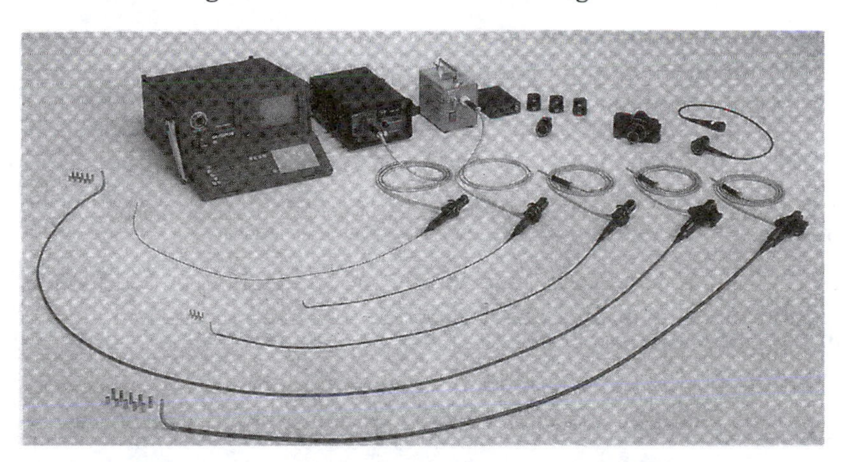

Figure 14-2. Courtesy of Olympus America Inc. The flexible boroscope is used for viewing those hard to get at areas. This instrument can also be used in most areas viewable with the rigid boroscope. Flexible boroscopes are available in a range of diameters and working lengths.

Depth of Field (DOF) is the minimum and maximum distance away from the objective lens at which the object is in sharp focus and is a function of the field of view. Fixed-focus type boroscopes do not have adjustable focus control, and objects will be in sharp focus only within a pre-determined range. As the FOV narrows the DOF also narrows.

With a boroscope it is possible to see the early signs of oxidation, erosion, corrosion, cracking, and the results of foreign object damage without disassembling the gas turbine. The inspector must be able to distinguish between component conditions that are potentially catastrophic and those that are just insignificant abnormalities. Later in this chapter visual comparisons will be drawn between gas turbine components as seen through a boroscope and the same or similar components as viewed prior to installation or after disassembly. The degree of damage that may be confirmed by way of a boroscope inspection will be discussed in detail. This information can then be used to confirm or refute problems initially identified through gas path or vibration analysis and attributed to a specific section of the gas turbine. One man can boroscope a gas turbine in one hour, whereas over 60 man-hours would be required to teardown a heavy frame unit to achieve the same level of inspection.

A BACKUP TO CONFIRMING SUSPECTED PROBLEMS

While a boroscope inspection can (and should) be done on a regular basis, it is most effective when it is used to confirm a problem highlighted by either vibration analysis or gas path performance analysis. For example, a shift in turbine temperature profile, with no apparent changes in either turbine or compressor efficiencies, could be indicative of a plugged fuel nozzle or a hole in the combustion liner or transition piece. Boroscoping the combustor will quickly confirm the problem.

An increase in turbine temperature could indicate an increase in turbine nozzle area (as a result of FOD or corrosion) or a decrease in compressor efficiency (due to compressor fouling). An inadequate compressor water wash procedure would further cloud the issue by not significantly improving compressor efficiency. This might lead the operator into prematurely removing the gas turbine from service for replacement (aero-derivative) or overhaul (heavy frame). Boroscoping the gas turbine would help identify the problem area.

Compressor vibration could be indicative of a failed compressor blade or imbalance created by contaminant flaking off an airfoil. The failed blade, blade portion, or loose contaminant could be so small that it has an insignificant influence on compressor efficiency. A boroscope inspection will provide the direction to either increase water washing efforts, overhaul the unit, or continue running the unit.

ASSESSING DAMAGE TO INTERNAL ENGINE COMPONENTS

For convenience in assessing engine health, some manufacturers have installed boroscope ports in strategic locations throughout their engines. Figure 14-3 is a schematic representation of a gas turbine with the typical location of boroscope ports indicated. However, manufacturers are not consistent as to the location of boroscope ports on different engine types and models. Therefore, an inspector must use every opening available or, as in some cases, openings that are easily created by removing an engine component (such as fuel nozzles, pressure or temperature probes, etc.). In general, boroscope ports are located in the combustor and in the turbine. The combustor boroscope port can be used to view the condition of the fuel nozzles, the combustion liner, the first stage turbine nozzles, and the first stage turbine

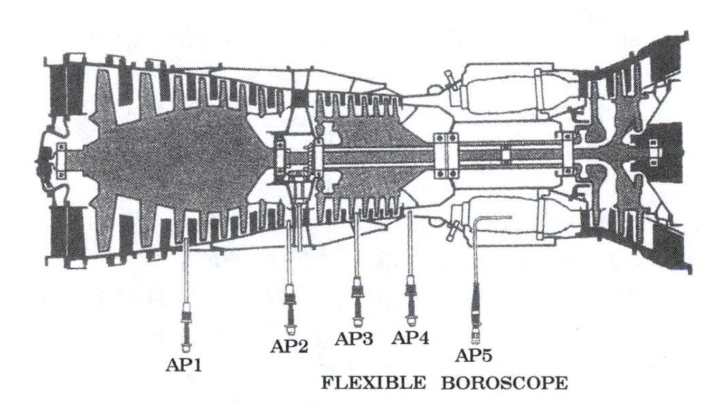

Figure 14-3. Courtesy of United Technologies Corporation, Pratt & Whitney Canada. Schematic cross-section of the FT8 aero-derivative gas turbine with the location of boroscope ports indicated.

blades. Mid-turbine boroscope ports can be used to view the turbine blades both upstream and downstream of the boroscope position.

Compressor Blades And Stators

Except for some aero-derivative gas turbines, boroscope ports are rarely provided within the compressor section of the gas turbine. However, the compressor inlet guide vanes and first stage compressor blades are accessible via the clean air compartment immediately upstream of the gas turbine. For a thorough examination, a boroscope can be inserted through the inlet guide vanes and all of the first stage blades inspected as the compressor rotor is turned. A boroscope view of the first stage compressor blades is shown in Figure 14-4. Note: the tar-like substance on the leading edge of the airfoil demonstrates the detail that can be achieved. This amount of contamination does affect compressor performance. Contamination such as this can be removed with a wash solution as recommended by the gas turbine manufacturer. However, if the wash/rinse length is inadequate, the residue will be distributed over a larger portion of the airfoil surface. This is shown in Figure 14-5. As the contamination accumulates the compressor efficiency would continue to deteriorate. The first stage compressor stators can also be partially viewed through the inlet guide vanes and the first stage blades. It is advisable to use extreme caution when inserting the boroscope probe into the blade path (be sure the rotor is locked so that it cannot be turned).

Figure 14-6 is a boroscope view of the last (eleventh) stage stators and blades of a 5,000 brake horsepower gas turbine. All of the boroscope photographs were taken using a 35mm single lens reflex camera attached to a flexible fiber optic boroscope. As there was not a boroscope port provided on this particular unit, a fuel nozzle was removed to inspect this area. After removing a fuel nozzle, the boroscope was snaked into the diffuser case toward the stators and blades. Note that with proper lighting the view is clear and sharp. While only a few stator vanes can be viewed from this position, by turning the rotor all the compressor blades can be viewed. To view all stator vanes, continue removing fuel nozzles around the circumference of the engine. Figure 14-7 is a photograph of the last (eleventh) stage compressor blades similar to that of Figure 14-6. The contamination visible in the compressor, specifically between the blades in Figure 14-7, would have been obvious when viewed through a boroscope. Due to the relative light weight of compressor blades, especially at the higher compression stages, blade failures may not create a noticeable amount of damage downstream of the compressor. Often only minimal damage is visible at the discharge of the compressor or the combustor inlet. Even damage resulting from severe compressor surge, which could wipe out several rows of compressor blades (and stator vanes if they are the cantilever design), seldom show a great amount of damage at the compressor discharge stators. However, a coating of a "plasma spray" consisting of compressor material will be

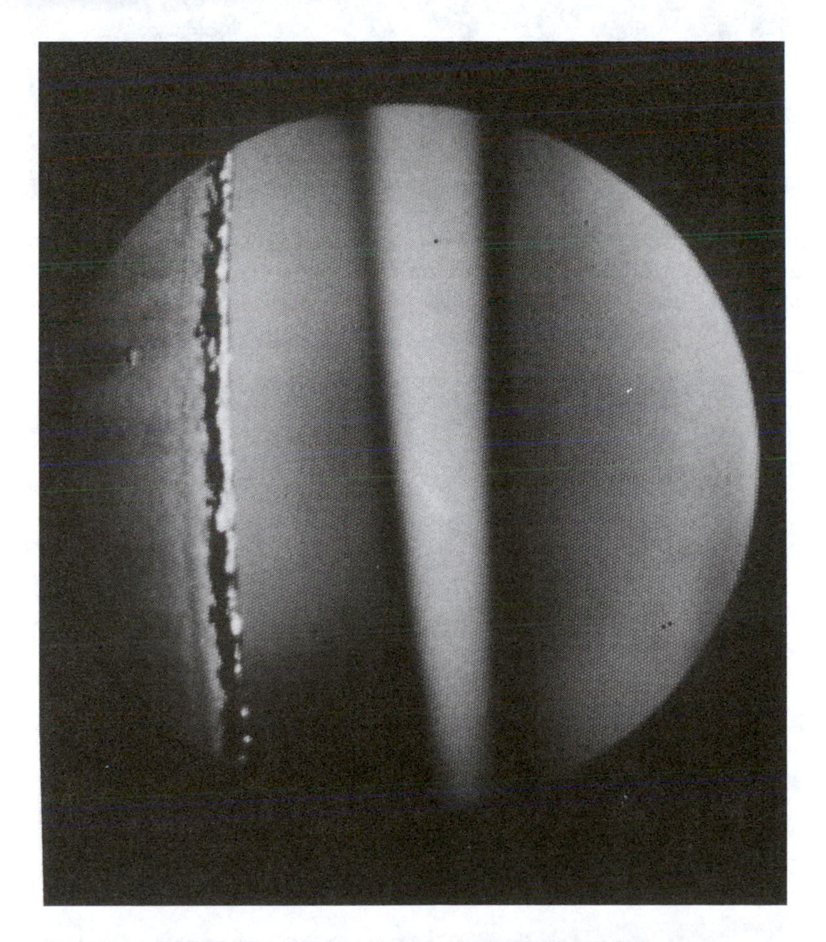

Figure 14-4. Tar-like deposit on a first stage compressor blade leading edge. 5,000 horsepower gas turbine-generator drive unit after 1,000 hours operation.

Figure 14-5. Carbon deposits distributed across the compressor blades. These deposits extend from the first to the sixth stage blades. 5,000 horsepower gas turbine-generator unit with over 16,500 hours operation.

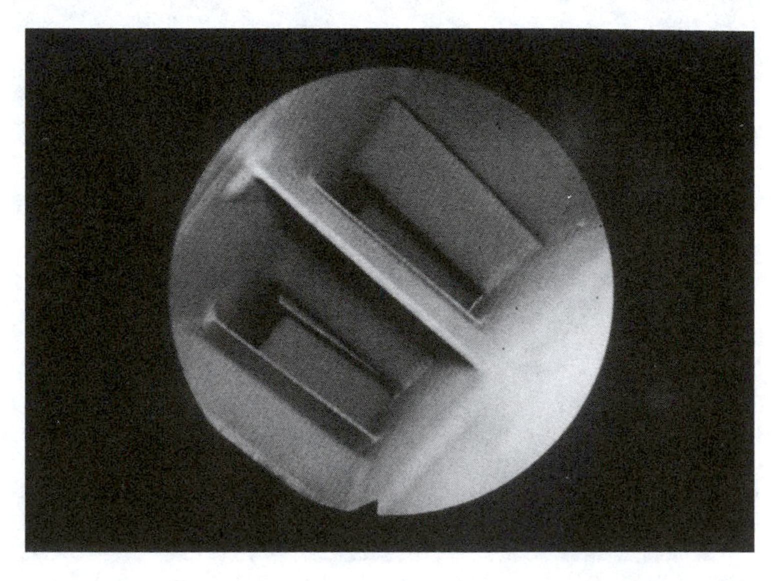

Figure 14-6. Boroscope view of last stage of compressor stators and blades. 5,000 horsepower gas turbine-generator unit after 1,000 hours operation.

Figure 14-7. View of last stage compressor blades with contamination evident on and between blades. 5,000 horsepower gas turbine-generator unit after 44,500 hours operation.

noticeable on the turbine blades and turbine nozzles. This is material from the compressor blades and stators, that melts as it travels through the combustor and is deposited as a "plasma spray" on the turbine airfoils.

Combustor And Fuel Nozzles

If the manufacturer has not made combustor boroscope ports available several other alternatives are possible.

- For heavy frame units the fuel nozzles can be removed and the combustor, first stage turbine nozzles and first stage turbine blades can be viewed. It may be necessary to fabricate a piece of rigid tubing to guide the boroscope through the combustor to the turbine nozzles and blades. The tubing must have an inside diameter slightly larger than the outside diameter of the boroscope and long enough to extend from the fuel nozzle plane to the first stage turbine nozzle plane.

- For gas turbines with a can-annular design combustor it will be necessary to remove at least one fuel nozzle from each combustor to view all the turbine nozzles.

- For gas turbines with an annular design it is sufficient to remove one fuel nozzle from each side of the gas turbine in order to view all of the first stage turbine nozzles and blades.

Figure 14-8 is a boroscope view of the internal annular combustor liner and fuel nozzles. This photograph was taken by inserting a flexible boroscope through one of the boroscope ports provided in the combustor section. This procedure would have confirmed the increase in turbine temperature spread and the decrease in combustor efficiency associated with the events leading up to the complete disintegration of the combustor shown in Figure 14-9, the hole in the combustion liner shown in Figure 14-10, and the hole in the transition piece shown in Figure 14-11. Liquid carryover in the fuel gas was the primary cause of the problems depicted in Figures 14-9 and 14-10. Identifying the problem early is preferable to finding it after extensive damage has been done. The combustion process is often accompanied by a high frequency vibration, which can excite the combustor and transition ducts. High frequency movement of the mating metal parts, can and often does, result in fretting and subsequent failure. Detecting the initial fretting is difficult. Once a crack starts it grows rapidly. Therefore, monitoring the turbine temperature or exhaust temperature profile is the best indicator of an internal duct failure.

Turbine temperature profile distortions are also caused by burnt, plugged or "coked-up" fuel nozzles as shown in Figures 14-12 and 14-13. Where boroscope ports are available, this condition (once highlighted by a temperature profile change) can be verified quickly and easily. Another alternative is to calculate the swirl effect and remove the suspected fuel nozzles. Frequent use of a boroscope will detect problems such as the small holes in a combustion liner (as shown in Figure 14-14) before the problem becomes serious enough to affect burner efficiency, temperature profile, or unit operation.

Turbine Blades And Nozzles

Boroscoping through the combustor boroscope ports or the fuel nozzle port, provides a broad view of the first stage turbine nozzles as shown in Figure 14-15. Moving the boroscope closer makes possible a good view of the turbine nozzle airfoil ID and OD platforms as shown in Figure 14-16 and 14-17. This view is sufficiently close to have detected the corrosion, erosion, and burning (oxidation) shown in

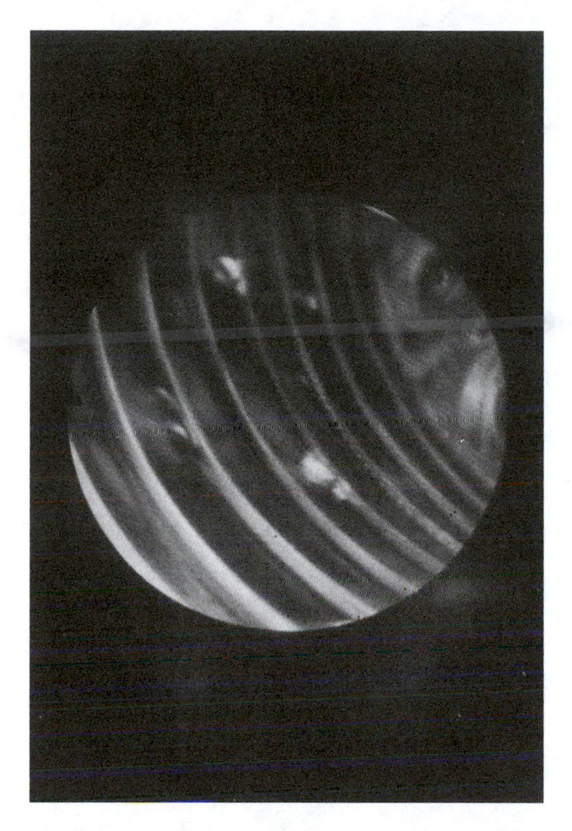

Figure 14-8. Internal boroscope view of annular combustor liner and fuel nozzle after 1,000 hours operation.

Figure 14-18 and 14-19, and foreign object damage and cracks shown in Figures 14-20 and 14-21, respectively. By performing periodic gas path analysis, the changes in exhaust gas temperature, fuel flow and rotor speed would have pointed to a diagnosis of an increase in turbine area. The material loss visible in Figures 14-18 and 14-19 is, in fact, an increase in turbine area. Even if the diagnosis was unclear, it would have prompted a visual inspection of the hardware.

A flexible fiber-optic boroscope can be inserted through a first stage turbine nozzle for a good view of the first stage turbine blades. By rotating the turbine shaft, all of the first stage turbine blades can be inspected, including the blade tips and rub strips as shown in Figures 14-22, 14-23, and 14-24.

With the above inspection techniques, blade surface condition as

Figure 14-9. Can-annular combustor failure due to liquid carry-over in the fuel gas. 30,000 horsepower gas turbine-pump drive unit with less than 24,000 operating hours.

shown in Figure 14-25 and 14-26 and tip condition as seen in Figure 14-27, can be evaluated. While a distorted temperature profile would have been an early indication of the burnt turbine blade tips, damage would have had to progress beyond the stage visible in the other photographs before a gas path analysis would have indicated a problem. Very early signs of damage due to erosion, corrosion, oxidation, or FOD can be detected to the degree shown in Figure 14-28.

Additional information regarding engine health can be obtained on gas turbines equipped with boroscope ports located between turbine stages. Utilizing these ports is best accomplished with a rigid-type, 90-degree boroscope. This port may be provided in either the

Figure 14-10. Can-annular combustor failure (burning and cracking) due to liquid mist in the fuel gas. 55,000 horsepower gas turbine-generator unit with less than 10,000 operating hours.

second- or third-stage turbine nozzle area depending on the total number of turbine stages. Through this port, the trailing edge of the upstream turbine blades and the leading edge of the downstream turbine blades can readily be viewed. Figure 14-29 shows a buildup of a foreign residue on the leading edges of the third stage turbine blades that would have been visible on a boroscope inspection.

Gas turbine performance and vibration trend analyses are important tools to be used in keeping gas turbine down-time and overhaul expenses to a minimum. Periodic boroscope inspections should also be used.

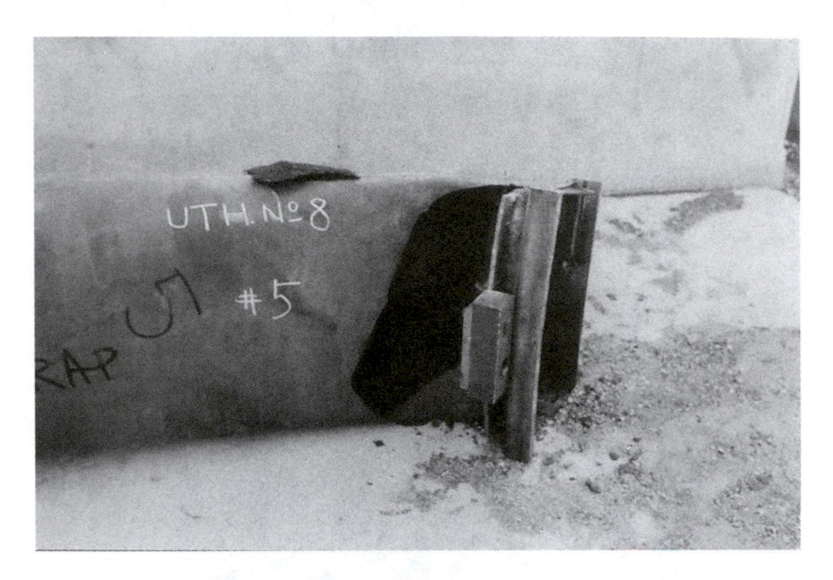

Figure 14-11. Can-annular combustor transition piece failure due to brittle fatigue due to fretting. 55,000 horsepower gas turbine-generator with less than 10,000 hours operation.

Figure 14-12. Can-annular combustor fuel nozzle coking. 30,000 horsepower gas turbine-pump drive unit. Operating hours unknown.

Figure 14-13. Burnt fuel nozzles from can-annular combustor. 55,000 horsepower gas turbine-generator with less than 10,000 operating hours.

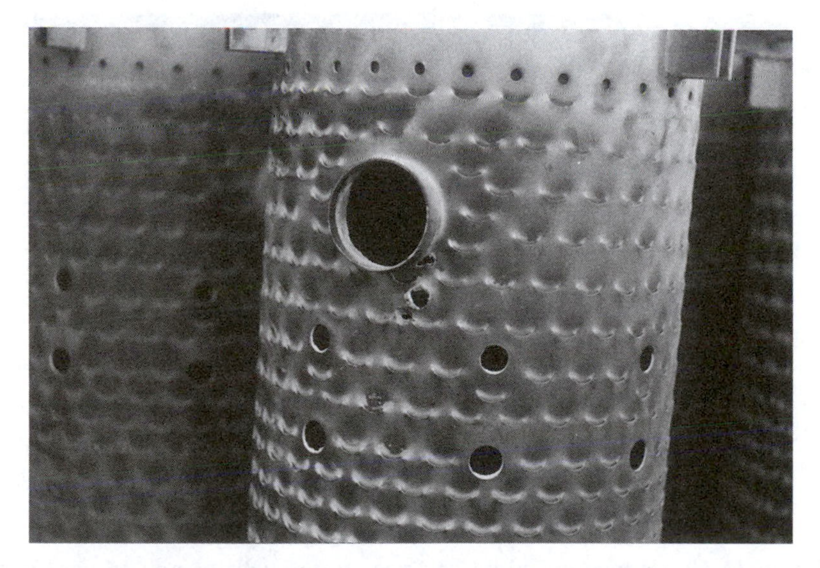

Figure 14-14. Can-annular combustor cracked and burnt at cross-over tube location. Distress was caused by cooling air distortion. 30,000 horsepower gas turbine-pump drive unit. Operating hours unknown.

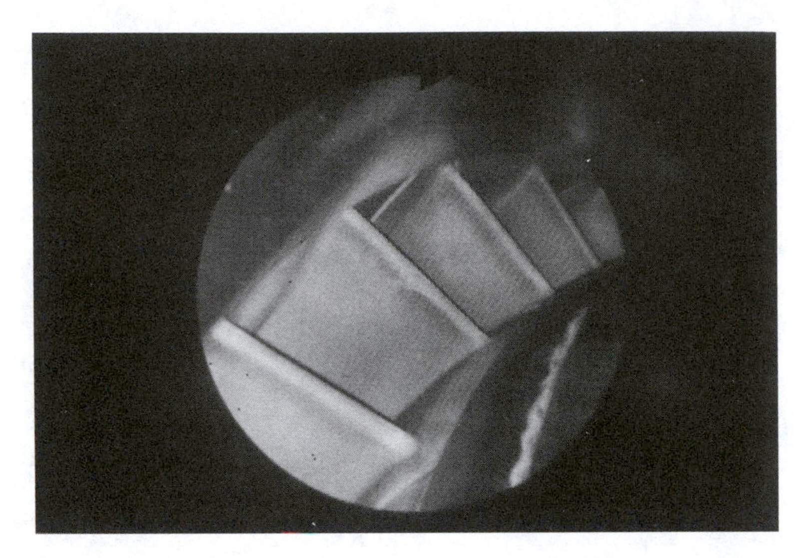

Figure 14-15. Boroscope view of first stage turbine nozzles. 5,000 horsepower gas turbine-generator unit with less than 1,000 hours operation.

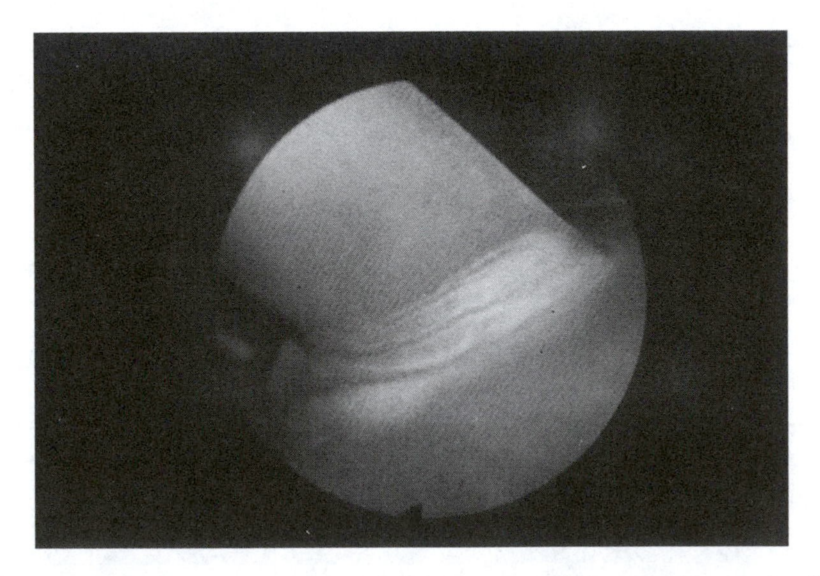

Figure 14-16. Boroscope view of first stage turbine nozzle root with water streak marks. 1,000 horsepower gas turbine-cogeneration unit shortly after installation.

Figure 14-17. Boroscope view of first stage turbine nozzle leading edge, platform, and concave surface. 5,000 horsepower gas turbine-generator unit shortly after installation.

CAUTION

Extreme care should be taken to insure that the boroscope tip does not extend into the path of the moving blades. Any contact between the boroscope, the blade, and the nozzle airfoil will result in severe damage to the boroscope and possibly the blade and nozzle. For this level of inspection the boroscope should be equipped with a locking feature that will prevent it from moving.

Figure 14-18. Cracked, burnt, eroded, and corroded first stage turbine nozzle concave trailing edge surface. This 4,500 horsepower gas turbine-generator unit operated for 17,000 hours with undersized water injection nozzle orifices in half the distribution ring.

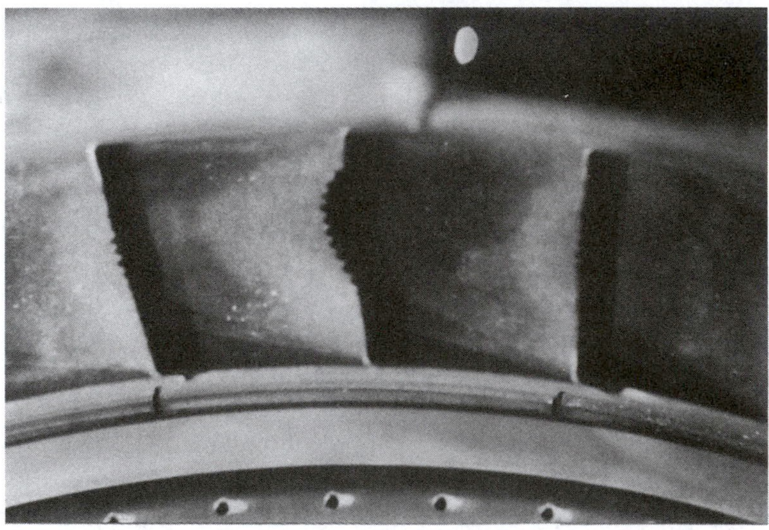

Figure 14-19. Burnt and eroded trailing edges of first stage turbine nozzles. 4,500 horsepower gas turbine-generator unit after 44,500 hours operation.

Figure 14-20. Foreign object damage to a second stage turbine nozzle trailing edge. 4,500 horsepower gas turbine-cogeneration unit after 16,820 hours operation.

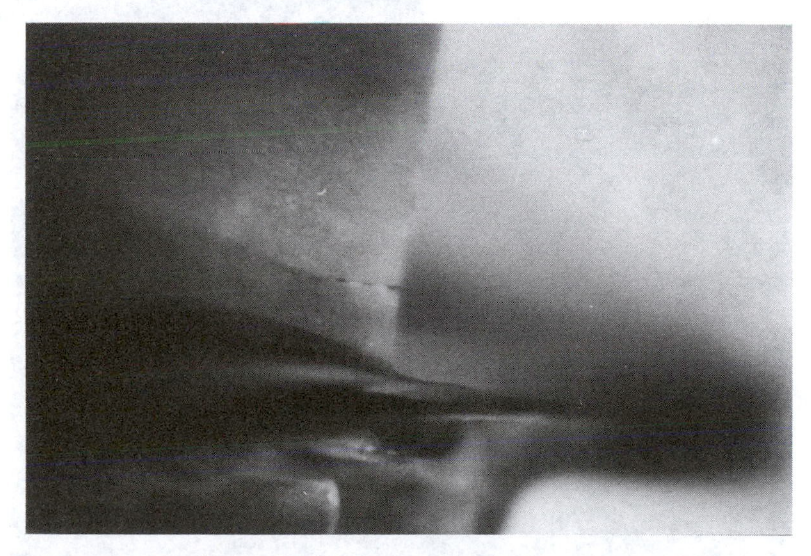

Figure 14-21. Trailing edge crack in the first stage turbine nozzle trailing edge. 5,000 horsepower gas turbine-compressor drive unit with less than 7,000 hours operation.

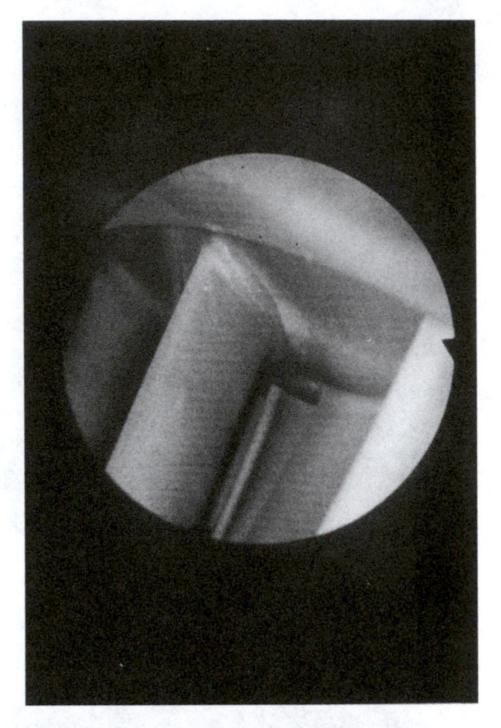

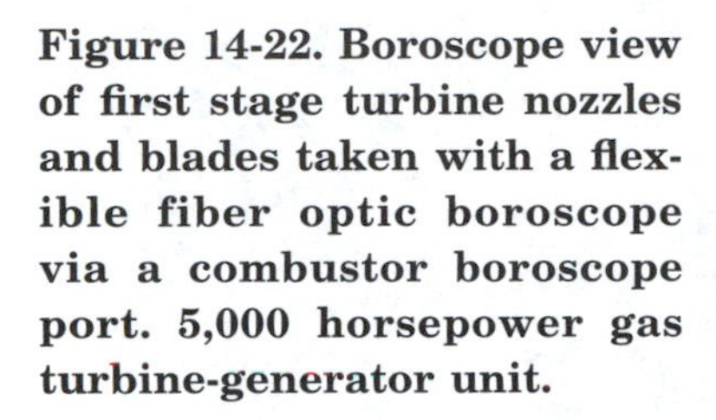

Figure 14-22. Boroscope view of first stage turbine nozzles and blades taken with a flexible fiber optic boroscope via a combustor boroscope port. 5,000 horsepower gas turbine-generator unit.

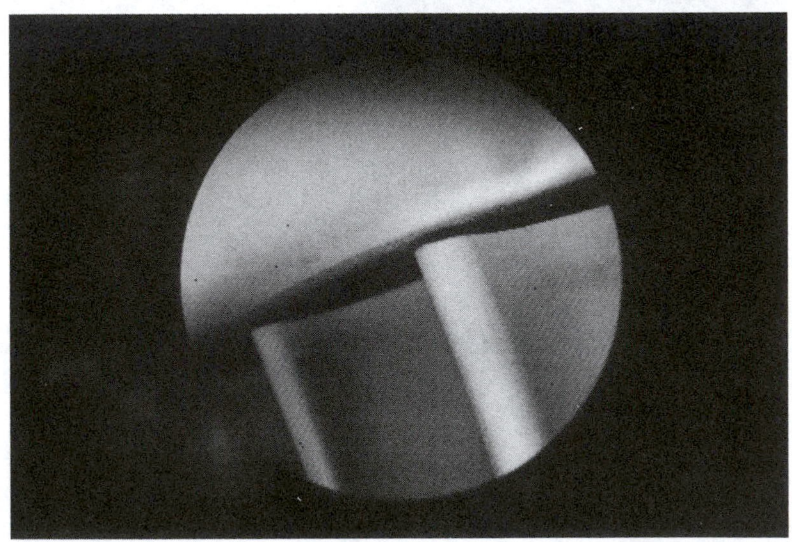

Figure 14-23. Boroscope view of first stage turbine blades taken with a flexible fiber optic boroscope via a combustor boroscope port. 5,000 horsepower gas turbine-generator unit with 1,000 hours operation.

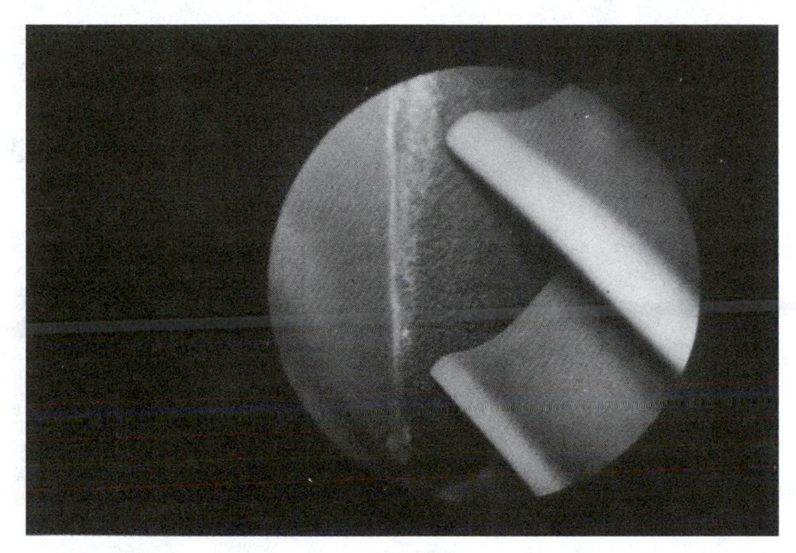

Figure 14-24. Boroscope view of first stage turbine blades and rub strip taken with a flexible fiber optic boroscope via a combustor boroscope port. 5,000 horsepower gas turbine-generator unit with 1,000 hours operation.

Figure 14-25. First stage turbine blade with burnt tip and "chalky-substance" built-up on concave surface. 5,000 horsepower gas turbine-generator drive unit with 9,000 hours operation.

Figure 14-26. Sulfidation corrosion attack on first stage turbine blade. 4,500 horsepower gas turbine-generator drive unit with under 45,000 hours operation.

Figure 14-27. First stage turbine blade trailing edge tip rub. 5,000 horsepower gas turbine-compressor drive unit with 8,000 hours operation.

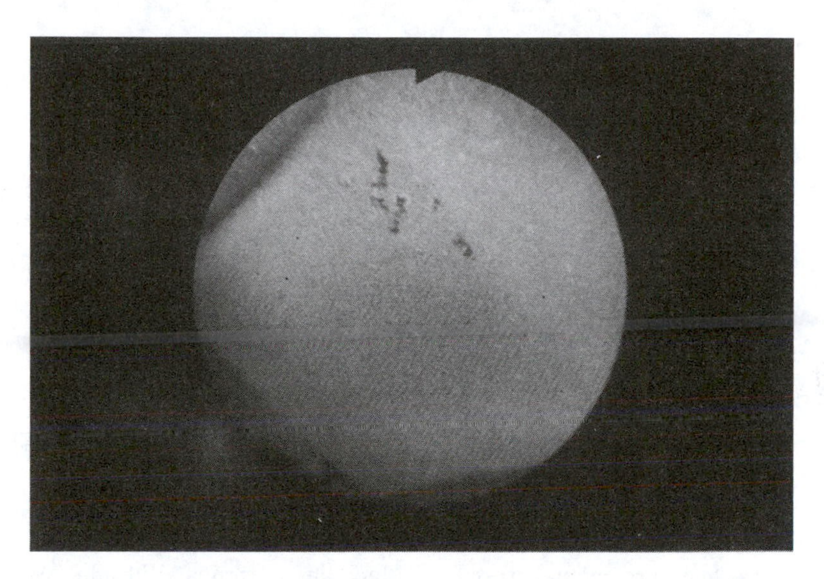

Figure 14-28. Early signs of sulfidation corrosion on a first stage turbine nozzle concave surface. 5,000 horsepower gas turbine-generator drive unit after 1,000 hours operation.

Figure 14-29. "Chalky-substance" build-up on third stage turbine blades. 4,500 horsepower gas turbine-generator drive unit with 17,000 operating hours since last overhaul and 80,000 total hours.

SOME SAFETY ADVICE

- Properly ground the light source.

- Do not use the light source in areas that are "wet" with either water or fuel.

- Do not use the light source where volatile gases could reach the projection lamp.

- Always turn off and disconnect the light source from the main power before opening the light source housing or changing lamps.

Reference

1. Boroscopes are manufactured by Machida, Olympus, Videodoc, Welch Allyn, and others.

Chapter 15

Case Study #1

Case study #1 addresses the conversion of coal to a gaseous fuel via a coal gasification process; and then application of combined cycle technology to the production of power utilizing the coal gas. This integrated gasification-combined-cycle (IGCC) system was installed at the Sierra Pacific Power Company's Piñon Pine Power Plant.

INTEGRATION OF AN ADVANCED TECHNOLOGY GAS TURBINE IN AN AIR BLOWN GASIFIER

By: Julian D. Burrow Engineer, MS6001FA GT/CC New Products
General Electric Industrial and Power Systems
1 River Road Schenectady, NY 12345
with acknowledgment to T.A. Ashley and R.A. Battista

and: John W. Motter
Director of Technology Management
Sierra Pacific Power Company
P.O. Box 10100
Reno, Nevada 89520

ABSTRACT

The Piñon Pine IGCC project is an 800-ton-per-day air-blown integrated gasification combined cycle (IGCC) plant to be located at Sierra Pacific Power Company's (SPPCo) Tracy Station near Reno, Nevada. The project is being partially funded under the DOE Clean Coal Technology IV program. SPPCo will own and operate the plant which will provide power to the electric grid to meet its customer's needs.

SPPCo has contracted with Foster Wheeler USA Corporation (FWUSA) for engineering procurement and construction management services for the project. FWUSA in turn has subcontracted with The M.W. Kellogg Company (MWK) for engineering and other services related to the gasifier island. General Electric has been selected to provide the MS6001FA gas turbine/air-cooled generator and steam turbine/TEWAC generator for the project. At Piñon Pine site conditions these units will have a combined cycle output of about 100MW.

The Piñon Pine project is the first air blown IGCC power plant to incorporate an "F"-technology gas turbine generator. This type of combustion turbine has a high combined cycle efficiency which greatly enhances the economic justification of an IGCC application. The Piñon Pine Power Project will be an early application of the MS6001FA which is nominally a 70 MW design, scaled from the MS7001FA.

This paper reviews the basic overall gasification and power plant configuration of The Piñon Pine Power Project and focuses on the integration of the MS6001FA gas turbine within this configuration. Specifically, aspects of the low Btu combustion system, compressor air extraction, fuel delivery system and control system are addressed.

INTRODUCTION

Role of IGCC Power Plants

Although the preponderance of recent domestic powerplant construction has been based on either simple or combined cycle natural gas fueled combustion turbines, coal-fired power plants are still a mainstay of power generation world-wide. The emissions produced by coal combustion have led to environmental concerns to the point where coal use for electricity generation has been threatened. Parallel to the installation of flue gas scrubbers in conventional coal-fired power plants, IGCC development is proceeding on innovative power plant concepts which are not only more acceptable from an environmental standpoint, but also feature higher efficiency. To apply combined cycle technology to power production from coal requires conversion of coal to a gaseous fuel via a coal gasification process. Coal gasification processes involves partial combustion of coal to provide energy for further conversion of the coal into a gaseous fuel primarily containing carbon monoxide, hydrogen and nitrogen.

In many regions of the United States, electric reserve margins have declined to low levels and increases in new power plant orders are predicted. The criteria used for selecting a power plant technology must include cost competitiveness, environmental superiority, module size, fuel flexibility, reliability and availability, and construction lead time. Based on these criteria, IGCC technology is a leading candidate for new capacity additions. Demonstration of this technology should provide a coal-based option with a cost of electricity that is competitive with more conventional technologies.

In order to meet the challenges of market-place and environment, a simplified IGCC system incorporating air-blown gasification with hot gas cleanup has been developed. Eliminating the oxygen plant and minimizing the need for gas cooling and wastewater processing equipment, reduces capital cost and improves plant efficiency.

Project Overview

Sierra Pacific Power Company (SPPCo) submitted a proposal requesting co-funding for the Piñon Pine Power Project in response to the United States Department of Energy (DOE) issuing its Program Opportunity Notice for Round IV of the Clean Coal Technology program. This proposal was selected for co-funding by the DOE in Fall 1991, and a Cooperative Agreement between the DOE and SPPCo was executed in August 1992. SPPCo's proposal was for design, engineering, construction, and operation of a nominal 800 ton-per-day, air-blown, IGCC project. The project was to be constructed at SPPCo's existing Tracy Station, a 400 MW, gas/oil-fired power generation facility. The Tracy Station is located on a rural 724-acre [293 hectare] plot about 20 miles [32Km] east of Reno at an elevation of 4280 feet [1305m] above sea level. The DOE will provide funding of $135 million in matching funds for construction and operation of the facility. (Figure 15-1, Tracy Power Station's Location within SPPCo's Service Territory).

The Kellogg-Rust-Westinghouse (KRW) ash-agglomerating flu-

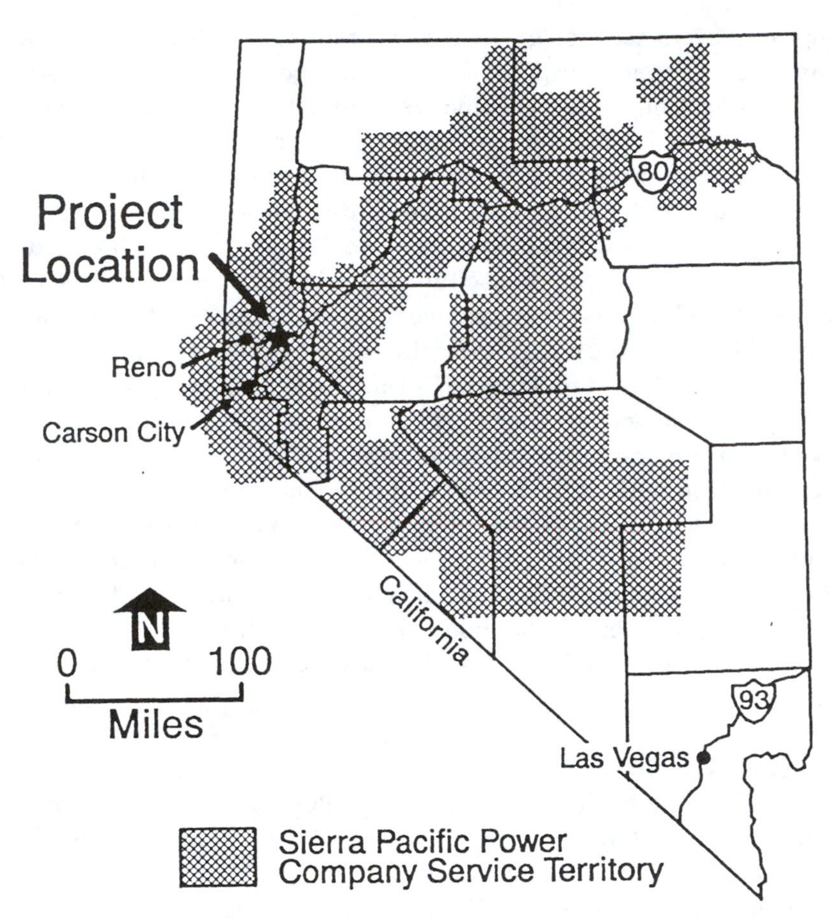

Figure 15-1. Tracy Power Station's Location within SPPCo's Service Territory

idized-bed gasifier will be the basis of the Piñon Pine project. This gasifier, operating in the air-blown mode, and coupled with hot gas cleanup (HGCU) will provide a low heating value fuel gas to power the combustion turbine. High temperature exhaust from the combustion turbine will then supply the energy required to generate steam in a heat recovery steam generator (HRSG) for use in a steam turbine. Both the combustion turbine and the steam turbine will drive generators to supply approximately 100 MW (net) of power to the electric grid.

The objectives of the Piñon Pine Power Project are to meet the power needs of the SPPCo's customers and to demonstrate the technical, economic and environmental viability of a commercial scale IGCC power plant. The project will demonstrate that power plants based on this technology can be built at capital costs and with thermal efficiencies that significantly reduce electric power costs over more conventional technologies. The project will also demonstrate the effectiveness of hot gas clean-up in achieving low environmental impacts (reduced SO_2 and NO_x emissions) and the operation of a low-Btu fuel gas turbine. The performance to be demonstrated will include all major sub-systems of the IGCC including coal and limestone feed systems, pressurized air-blown fluidized-bed gasifier, hot product gas filtering and desulfurization with a regenerative sulfur sorbent, sulfator system, gas turbine and steam cycle and integrated control systems. Another objective is to assess the long term reliability, maintainability, and environmental performance of the IGCC technology in a utility setting at a commercial scale.

Project Schedule and Status

The project is scheduled to be mechanically complete in late 1996 and in commercial operation by early 1997. Since execution of the Cooperative Agreement in 1992, project activities have been concentrated in the areas of regulatory approval, permitting, design and engineering, and recently procurement.

Construction of new utility powerplants in Nevada must follow approval of the unit as part of an approved "least cost" electric resource plan by the Public Service Commission of Nevada. That approval was obtained in November, 1993.

Environmental and permitting work has been proceeding on the project since inception. As part of the National Environmental Policy Act (NEPA) activities an Environmental Impact Statement (EIS) was required for the plant. Following comprehensive site-specific studies and analysis, a draft EIS was prepared, public and agency comments were received and the final EIS was issued in September, 1994. A Record of Decision (ROD) is expected by October, 1994. SPPCo expects to submit its "Continuation Application" to the Department of Energy in November 1994, following the ROD and approval by the SPPCo Board of Directors.

All other permits, including an air quality PSD permit to construct, have been received or are expected to be granted in the near future. Prior to commencement of construction in February, 1995

SPPCo expects to be granted a Nevada Utility Environmental Protection Act (UEPA) permit in January, 1995.

Design and preliminary engineering of the plant is substantially complete, and detailed engineering is approximately 40% complete. Procurement and vendor engineering is well underway (Figure 15-2. Piñon Pine Project Schedule).

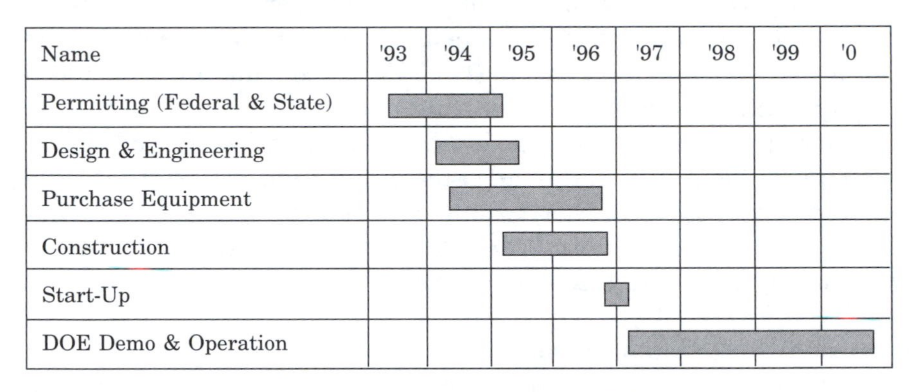

Figure 15-2. Piñon Pine Project Schedule.

Major Participants

SPPCo will own and operate the demonstration plant which will provide power to the electric grid to meet its customer needs. Foster Wheeler USA Corporation (FWUSA) will provide engineering and construction management services. The M.W. Kellogg Company (MWK) will provide engineering of the gasifier and hot gas cleanup systems. The General Electric Company was awarded contracts for both the gas turbine/generator and the steam turbine/generator for the project following competitive solicitations.

PLANT CONFIGURATION AND PERFORMANCE

Process Description

The Piñon Pine Power Project will demonstrate the performance of a coal-based IGCC power plant. The power plant will include a gasifier island, coupled to a combustion turbine and steam turbine based power island. The gasifier island has an air-blown ash-agglomerating KRW gasifier to produce low-Btu syngas. Coal is supplied to the gasifier by a pneumatic coal feed system, fed by lockhoppers. The syngas is transferred to a hot gas conditioning system for removal of sulfur compounds, particulates and other contaminants to meet environmental and combustion turbine fuel requirements. The power island includes a GE model MS6001FA combustion turbine capable of operating on natural gas, syngas or propane and a heat recovery steam generator (HRSG). Superheated, high pressure steam generated in the HRSG and the gasification and desulfurization sections supplies the steam turbine.

Description of Major Components

Key equipment items and systems which will be part of the unique technology of the Piñon Pine Power Project are the pressurized (20 bars) KRW gasifier with in-bed desulfurization, primary cyclone, external regenerable sulfur removal, fine particulate filters, and some aspects of the gas turbine. Advanced KRW gasification technology produces a low-Btu gas (heating value approximately 130 Btu/SCF [4838KJ/m^3]) which is used as fuel in the combined-cycle power plant, and includes hot gas removal of particulates and sulfur compounds from the fuel gas, resulting in low atmospheric emissions. Desulfurization will be accomplished by a combination of limestone fed to the gasifier and treatment of the gas in desulfurization vessels using a zinc-based sulfur sorbent such as Phillips proprietary Z-Sorb III®. Particulates will be removed by a pair of high-efficiency cyclones and a barrier filter. These operations will be carried out at the elevated temperature of approximately 1000°F [538°C] to eliminate the thermodynamic inefficiency and capital cost of cooling and cleaning the gas at low temperature, which is associated with other IGCC systems. Since water vapor is not condensed in the hot gas clean-up process, water effluents will be reduced and will consist of only feed water treatment effluent and boiler and cooling tower blow down.

Sub-bituminous coal will be received at the plant from a unit train consisting of approximately 84 railcars of between 100- and 110-ton capacity, arriving approximately once a week. In addition, the project will also demonstrate the ability to operate on higher sulfur, eastern coals. Currently, Southern Pacific Railroad facilities are on site; the railroad line is a main east-west supply route. Upgrading and extending the spur on SPPCo. land will be required for the proposed project. Coal will be received at an enclosed unloading station and transferred to a coal storage dome. The unloading station will consist of two receiving hoppers, each equipped with a vibrating-type unloading feeder that will feed the raw coal conveyor systems. All material handling systems will be enclosed and supplied with dust collection systems for environmental control.

The steam turbine will be a straight condensing unit with an uncontrolled extraction and an uncontrolled admission. The guaranteed rating is 46226 kW, with the VWO expected output of 50785 kW. The steam conditions are 950 psia [65.5 Bars], 950°F [510°C] inlet, with exhaust at 0.98 psia [67mBars]. At the rating, the uncontrolled extraction pressure is 485 psia [33.4Bars], and the uncontrolled admission is 54 psia [3.7Bars]. The turbine will have 26 inch [0.66m] last stage buckets in an axial exhaust configuration. The turbine will be fully packaged, baseplate mounted, with a combined lubrication and control system mounted on a separate console. The control will be a GE Mark V Simplex system. The generator is a fully packaged and baseplate mounted GE TEWAC unit rated at 59000 KVA at 0.85 power factor.

The gas turbine is a General Electric MS6001FA. It is an evolutionary design, being aerodynamically scaled from the larger MS7001FA machine. It shares the same thermodynamic cycle with previous GE "F"-class machines. Key characteristics of the "FA" cycle pertinent to the Sierra Pacific application are the 2350°F [1288°C] firing temperature and high output for this frame size, elevated exhaust temperature making possible high combined cycle efficiencies and the ability of the combustion system to burn a wide spectrum of fuels.

The MS6001FA has a compressor or "cold-end" drive flange, turning an open-ventilated GE 7A6 generator (82000 KVA @ 0.85PF) through a reduction load gear. The exhaust is an axial design directing the gas flow into the HRSG. These mechanical features have facilitated an optimum plant arrangement for Sierra Pacific's combined cycle heat recovery application.

Plant Performance

Key aspects of the predicted plant performance are summarized in the tables opposite.

Table 15-1. Expected Plant Performance on SYNGAS*

Heat Input (10^6 Btu/Hr) LHV/HHV[KJ/Hr]	807.3[850.9]/836.6[881.9]
Combustion Turbine Power (MW)	61
Steam Turbine Power (MW)	46
Steam Turbine Conditions (psia[Bars]/°F[°C])	950[65.50]/950[510]
Station Load (MW)	7
Net Power Output (MW)	100
Heat Rate(Btu/kWhr) LHV/HHV[KJ/kWhr]	8096[8534]/8390[8844]

*At 50°F[10°C] and 4280'[1305m] elevation, evaporative cooler off.

Table 15-2. Expected Plant Performance vs. Temperature.

Ambient Temperature	5°F*[–15°C]	50°F[10°C]	95°F**[35°C]
Expected Performance - Coal			
Net Power Output (MW)	102	100	96
Heat Rate (Btu/kWhr) LHV/HHV	7956/8245	8096/8390	8154/8450
(KJ/kWhr) LHV/HHV	8387/8691	8534/8844	8596/8908
Expected Performance - Natural Gas			
Net Power Output (MW)	91	88	84
Heat Rate (Btu/kWhr) HHV	8103	8144	8207
(KJ/kWhr) HHV	8541	8584	8717

*5°F[–15°C] Data do not include spiking.
**95°F[35°C] Data include evaporative cooling.)

GAS TURBINE DESIGN DETAILS

Flange-to-Flange

In a coal gasification system, partial combustion takes place in the gasifier and is completed in the gas turbine combustors. Particulates and sulfur as are removed by cyclones and a HGCU unit between these stages. The final fuel arriving at the machine, although of low heating value, requires relatively few changes to the gas turbine flange-to-flange design. Full benefit can be taken from technology advances in traditionally fueled gas turbines. As in GE's other "F"-class gas turbines the MS6001FA uses an axial flow compressor with eighteen stages, the first two stages being designed to operate in transonic flow. The first stator stage is variable. Cooling, sealing and starting bleed requirements are handled by ninth and thirteenth stage compressor extraction ports. The rotor is supported on two tilt-pad bearings. It is made up from two subassemblies, the compressor rotor and turbine rotor, which are bolted together. The sixteen bladed disks in a through-bolted design give good stiffness and torque carrying capability. The forward bearing is carried on a stub shaft at the front of the compressor. The three stage turbine rotor is a rigid structure comprising wheels separated by spacers with an aft bearing stub shaft. Cooling is provided to wheel spaces, all nozzle stages and bucket stages one and two.

Main auxiliaries are motor driven and arranged in two modules. An accessory module contains lubrication oil, hydraulic oil, atomizing air, natural gas/doped propane skids and bleed control valves. The second module would normally house liquid fuel delivery equipment but in this IGCC application, holds the syngas fuel controls. A fire resistant hydraulic system is used for the large, high temperature syngas valves.

IGCC Integration/Design Customization

Fuel Delivery

The fuel system is designed for operation on syngas, natural gas or doped propane. Only natural gas or doped propane can be used for start up with either fuel being delivered through the same fuel system. Once the start is initiated it is not possible to transfer between natural gas and doped propane until the unit is on-line.

A full range of operation from full speed no load to full speed full load will be possible on all fuels and transfers, to or from syngas, can take place down to thirty percent of base load. This maintains a minimum pressure ratio across the fuel nozzle preventing transmission of combustor dynamics or cross-talk between combustors. Co-firing is also an option with the same thirty percent minimum limit apply to each fuel. A low transfer load also minimizes the amount of syngas flaring required.

The gasifier has a fixed maximum rate of syngas production. At low ambient temperatures, fuel spiking is required to follow gas turbine output. The MS6001FA gas turbine can accept a limited amount of natural gas mixed with the syngas, to boost the Btu content. In this application the maximum expected is about 8% and an orifice in the natural gas mixing line will limit supply to 15% in the event of a valve malfunction. This protects the gas turbine from excessive surges of energy release in the combustor. On hot days power augmentation can be achieved with steam injection or evaporative coolers.

The natural gas/doped propane fuel system, uses a modulating stop ratio valve to provide a constant pressure to a "critical flow" control valve. The position of the valve is then modulated, by the Mk. V Speedtronic® control to supply the desired fuel flow to the combustors. When not in use, the natural gas/doped propane system is purged with compressor discharge air.

Syngas is supplied from the gasifier at a temperature and pressure of approximately 1000°F [538°C] and 240 psia [16.5Bars] respectively (Table 15-3, Typical Syngas Composition).

Table 15-3. Typical Syngas Composition

CONSTITUENT	PERCENTAGE BY VOLUME
Hydrogen	14.58
Nitrogen	48.68
Carbon Monoxide	23.91
Carbon Dioxide	5.45
Methane	1.35
Water	5 45
Ammonia	0.02
Argon	0.56

The exact figures depend on specific operating conditions. Specific heating value of the syngas can vary, therefore the entire delivery system was sized to accommodate the resulting volumetric flow changes. Piping and valve materials were upgraded to nickel based super alloys and higher grade stainless steels for their improved high temperature properties and corrosion resistance. The low Btu content and elevated temperature of the syngas imply that large volumes of gas are required. Flow from the gasifier is accommodated in a 16-inch [41cm] diameter pipe with a 12-inch [30cm] diameter manifold supplying syngas to the six combustors via 6-inch [15cm] diameter flexible connections. By contrast, the corresponding figures for natural gas are 4 [10cm], 3 [7.6cm] and 1.25 inches [3.1cm] respectively. With combustor case bleed, syngas, natural gas, steam injection, compressor discharge purge bleed and inlet bleed heat manifolds all positioned closely on the unit, careful attention must be paid to manifold arrangement. (Figure 15-3, View of MS6001FA Manifold Arrangement)

Syngas at 1000°F [538°C] burns spontaneously when in contact with air. The fuel system is designed to eliminate syngas-to-air interfaces. Both steam and nitrogen purging and blocking systems are being evaluated. General Electric has used nitrogen as a purge medium in previous IGCC applications. While operating on syngas, nitrogen would be supplied at a pressure higher than the syngas and would use a block and bleed strategy to prevent contact with the purge air supply. Operating on natural gas, the syngas would be blocked by an additional stop valve and a stop ratio valve. The nitrogen purge valves open to flush syngas from the entire system downstream of the stop valve, and then close. Compressor discharge air is then used to purge the line from the control valve to the combustor while nitrogen provides a block and bleed function between the control valves. Correct sequencing of the valves maintains separation between syngas and air. A steam based system would be functionally similar. (Figure 15-4.)

Combustion System

The MS6001FA uses a reverse flow can-annular combustion arrangement common to all GE heavy duty gas turbines. There are six combustors, equally spaced and angled forward from the compressor discharge case. Compressor discharge air flows over and through

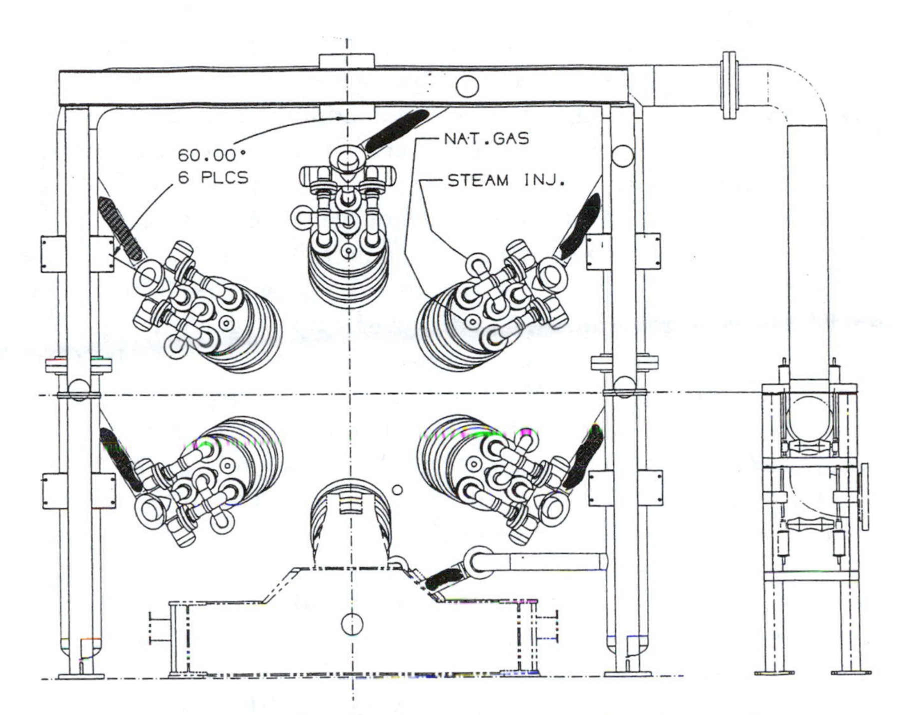

Figure 15-3. Combustor Case Extraction Manifold.

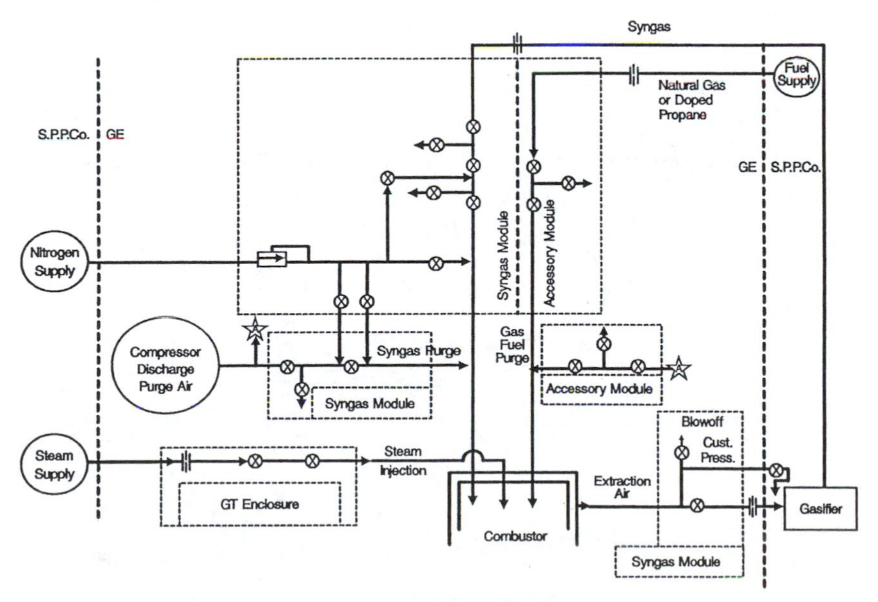

Figure 15-4. Fuel Delivery Schematic

holes in the impingement sleeve providing cooling to the transition piece while the remainder of the flow passes through holes in the flow sleeve. The air then flows along the annulus formed by the case and the liner. The combustor uses a conventional film cooling design based on the experience gained from previous MS7001FA IGCC applications. Fuel and diluents are introduced through nozzles in the end cover. The combustor for the IGCC application is a diffusion type and differs markedly from the standard Dry Low NO_x offering. This application features a dual gas end cover with large syngas nozzles and natural gas/doped propane nozzles for the alternate fuel. Steam injection nozzles are used for power augmentation at high ambient temperatures and NO_x control. The combustion cases, apart from the usual provision for spark plugs, flame detectors and cross-fire tubes, also have extraction ports to supply compressor discharge air to the gasifier. (Figure 15-5, X-section of Combustor.)

Design verification testing is taking place at the General Electric Development Laboratory in Schenectady, NY. Gas delivered by trailer will be blended to simulate the syngas composition predicted for the MWK gasifier. Test hardware will be fully instrumented to monitor temperatures, flow fields and dynamic effects within the combustor.

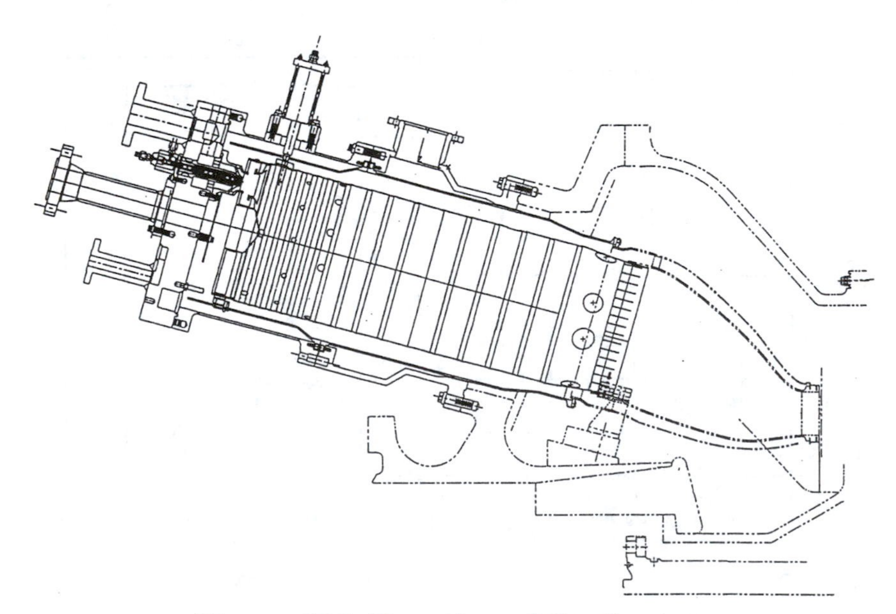

Figure 15-5. X-section of Combustor.

Two phases of testing are planned. The first will establish an initial database for the IGCC combustor operating on nominal composition syngas and natural gas in a generic test stand. Studies also cover the effect of extracting up to twenty percent of compressor discharge flow from the combustion case, cofiring, minimum heating value trials, ammonia conversion rates and demonstration of NO_x control capabilities. This first phase of testing will allow opportunity for hardware modifications and will result in a validated production design. The second phase will concentrate on production design verification, operation on different fuels, fuel transfers and operating procedures. This second test sequence will utilize a new test stand and facility, custom built to simulate the MS6001FA combustion operating environment. Initial field runs will take place at the Sierra Pacific Piñon Pine facility in the third quarter of 1996, with December 1996 as the date for commercial operation and demonstrations to begin. (Figure 15-6, IGCC Combustion Test Schedule)

Thermal oxidation of fuel nitrogen and oxidation of nitrogen species, such as ammonia, are two contributors to overall NO_x output. Thermal NO_x emissions on syngas are guaranteed at 32 ppm @15% O_2 unabated. With mixed flow (spiking) or co-firing and steam dilution the guarantee is 42 ppm @ 15% O_2. Laboratory testing will determine what lower levels are achievable with steam injection for 100% syngas and natural gas/syngas mixtures. The aim is to demonstrate

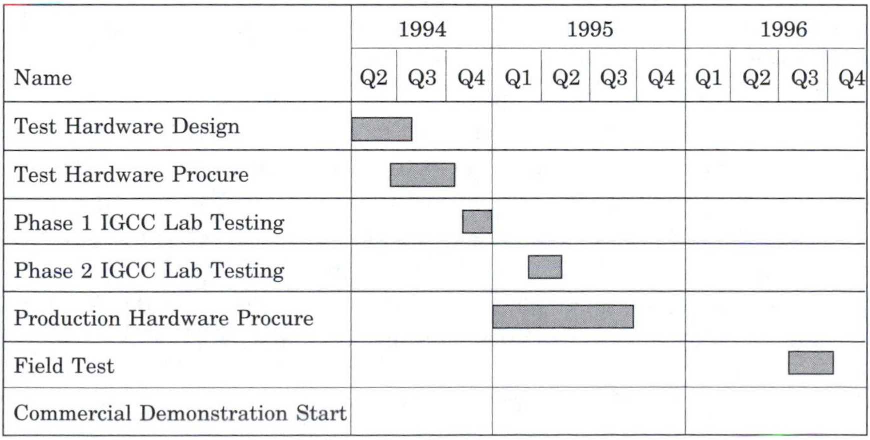

	1994			1995				1996			
Name	Q2	Q3	Q4	Q1	Q2	Q3	Q4	Q1	Q2	Q3	Q4
Test Hardware Design											
Test Hardware Procure											
Phase 1 IGCC Lab Testing											
Phase 2 IGCC Lab Testing											
Production Hardware Procure											
Field Test											
Commercial Demonstration Start											

Figure 15-6. IGCC Combustion Test Schedule.

9 ppm thermal NO_x emissions in the laboratory using 100% syngas with head end steam injection. An additional benefit of the compressor air-blown gasifier is the retention in the cycle of N_2 which helps with NO_x control. Syngas has an ammonia content which contributes to the overall NO_x output. Testing will also determine the ammonia to NO_x conversion rates.

Compressor Air Extraction

The most efficient way to provide air to the gasifier is to bleed compressor discharge air from the gas turbine. The MS6001FA has the capacity to supply high bleeds, in excess of 20% of compressor flow. This allows plant simplification in that an air blown gasifier can be used, supplied with compressor discharge air. Additionally, no turbo-machinery changes are required. Extraction flow can be supplied to the gasifier at all conditions from full speed no load to full load. This is particularly important in the event of a load trip because the gas turbine can still supply the gasifier even though fuel to the gas turbine has been cut back. This allows an orderly turn-down of the gasifier to match gas turbine fuel demand.

Compressor discharge air is extracted through casing ports. The ports incorporate a metering orifice designed to limit the flow to a level that does not cause operability problems for the gas turbine compressor. The extraction pressure is increased by a boost compressor which delivers the flow to the gasifier. As a supplement to the inlet bleed heat surge protection function already existing in the control system, the gas turbine is protected by a blow-off valve from excessive back-pressure caused by gasifier or boost compressor problems.

It is important in the design of the extraction and syngas delivery system to minimize pressure drops. Extraction and delivery pressures losses are a very powerful driver in the sizing of the boost compressor. High losses can double the size of the boost compressor, raising initial cost and reducing net power output over the life of the plant. To minimize the losses GE is developing a unique low pressure drop fuel control system for this application. (Figure 15-7, View of MS6001FA Combustor Case Extraction Manifold.)

Control System

The overall plant is controlled by a Distributed Control System (DCS). The Mk. V Speedtronic® turbine control systems interfaces with the DCS.

The natural gas/doped propane fuel system is standard apart from a customer signal which indicates which fuel is in use. Logic within the control system makes the necessary specific gravity adjustment for fuel flow measurement.

A new, patented syngas control system is being provided for the Sierra Pacific IGCC application. It provides an integrated control system for the syngas boost compressor and the gas turbine that will minimize fuel supply delivery pressure losses. The boost compressor raises the compressor extraction pressure prior to the gasifier. After the gasifier, fuel is metered to the gas turbine through the stop ratio and gas control valves. Pressure drops through these valves are minimized by controlling them to the full open position during normal control operation. By this method the system operates at the minimum pressure drop and controls the gasifier flow via the boost compressor, thereby controlling the gas turbine output. When the fuel flow is not meeting demand, a signal is sent to the gasification island control panel, which translates this input to the necessary response from the

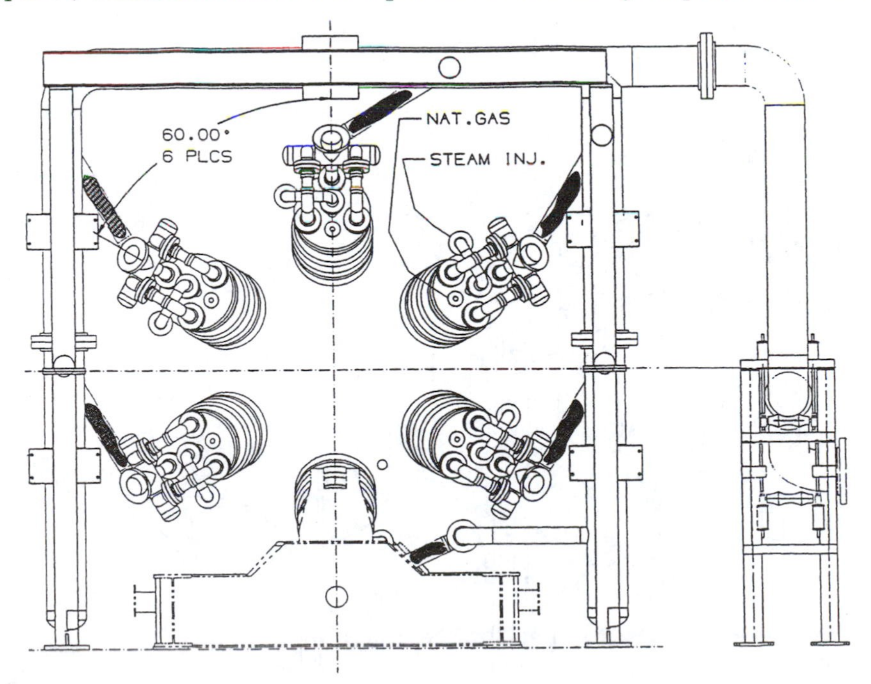

Figure 15-7. View of MS6001FA Combustor Case Extraction Manifold.

boost compressor/gasifier. Syngas output is increased or decreased as demanded while the stop ratio and gas control valves remain wide open.

Features in the Mk. V Speedtronic® control will handle fuel transfers, co-firing, purge/extraction valve sequencing, steam injection, surge protection plus all standard gas turbine control functions.

Verification testing of the modified control system will take place at General Electric's Schenectady, NY, controls laboratory. A revised Mk. V Speedtronic® control will be connected to a simulation of an "FA" gas turbine control system. The results will demonstrate and validate the new control software. (Figure 15-8, Control System Development Schedule)

SUMMARY

The integrated gasification combined cycle (IGCC) is gaining acceptance as a clean coal technology with potential for continued

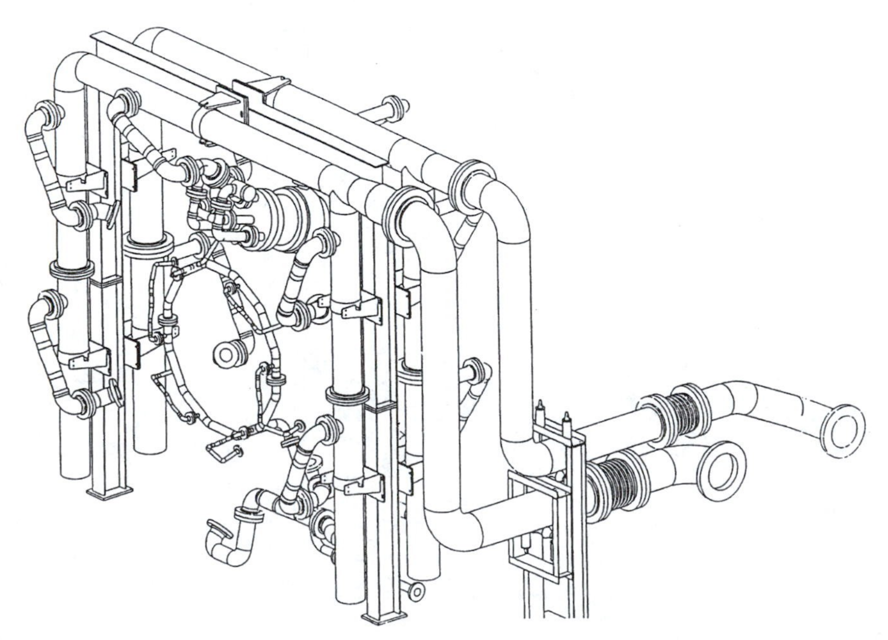

Figure 15-8. Control System Development Schedule.

development and performance improvement. The Piñon Pine project represents the next step in demonstrating this technology on a commercial scale. A unique feature of this IGCC application is the ability of the GE MS6001FA compressor to deliver the high extraction flows required to supply an air-blown gasifier. While the turbo-machinery is unchanged, the combustion design is modified, based on experience gained from previous "F" class machines that have been integrated into IGCC applications. The combustion system is flexible and can operate efficiently on a variety of fuels including the low Btu syngas and natural gas to be used at Piñon Pine. The highly efficient "F" technology MS6001FA confers excellent combined cycle performance. Because of its physical design, the MS6001FA gives a logical, economical combined cycle powerplant arrangement. Additionally, the air-blown gasifier, with hot gas clean-up, allows significant plant simplification and improvement in overall plant efficiency. This project will demonstrate reduced NO_x and SO_x emissions resulting from the hot gas clean-up technology.

DOE DISCLAIMER

This report was prepared by Sierra Pacific Power Company and General Electric pursuant to a Cooperative Agreement partially funded by the U.S. Department of Energy, and neither the Sierra Pacific Power Company nor any of its subcontractors nor the U.S. Department of Energy, nor any person acting on behalf of either:

(A) Makes any warranty or representation, express or implied with respect to the accuracy completeness, or usefulness of the information contained in this report, or that the use of any information, apparatus, method, or process disclosed in this report may not infringe privately owned rights.

(B) Assumes any liabilities with respect to the use of, or for damages resulting from the use or, any information apparatus, method or process disclosed in this report.

Reference herein to any specific commercial product, process, or service by trade name, trademark, manufacturer, or otherwise, does not necessarily constitute or imply its endorsement, recommendation, or favoring by the U.S. Department of Energy. The views and opinions of authors expressed herein do not necessarily state or reflect those of either the U.S. Department of Energy or the Sierra Pacific Power Company.

References

1. William M. Campbell, Martin O. Fankhanel, and Gunnar B. Henningsen, "Sierra Pacific Power Company's Piñon Pine Power Project," presented at the Eleventh EPRI Conference on Coal Gasification Power Plants, San Francisco, California (October 1992).
2. William M. Campbell, James J. O'Donnell, Satyan Katta, Thomas Grindley, Gary Delzer, and Gyanesh Khare, "Desulfurization of Hot Fuel Gas with Z-Sorb III Sorbent," presented at the Coal-Fired Power Systems '93 — Advances in IGCC and PFBC Review Meeting, Morgantown, West Virginia (June 1993).
3. E.B. Higginbotham, L.J. Lamarre, and M. Glazer, "Piñon Pine IGCC Project Status," presented at the Second Annual Clean Coal Technology Conference, Atlanta, Georgia (September, 1993),
4. John W. Motter, John D. Pitcher, and William M. Campbell, "The Piñon Pine IGCC Project Design and Permitting Issues," presented at the 55th Annual Meeting of the American Power Conference, Chicago, Illinois (April 1993).
5. Douglas M. Todd, "Clean Coal Technologies for Gas Turbines," presented at CEPSI Conference, New Zealand (September 1994).
6. James C. Corman and Douglas M. Todd, "Technology Considerations for Optimizing IGCC Plant Performance," presented at the International Gas Turbine and Aeroengine Congress and Exposition, Cincinnati, Ohio (May 1993).
7. W. Anthony Ruegger, "The MS6001FA Gas Turbine," presented at CEPSI Conference, New Zealand (September 1994).
8. Rodger O. Anderson, "MS6001FA, The 70MW-50Hz Solution," presented at the PowerGen Europe Conference (1994).

Chapter 16

Case Study #2

This case study examines how two refineries discovered savings within the processing plants by utilizing the waste heat energy from their gas turbines to increase refinery profitability.

Benefits of Industrial Gas Turbines for Refinery Services

Bo Svensson, ABB STAL, Finspong, Sweden
Henry Boerstling, ABB Power Generation Inc., North Brunswick, NJ

ABSTRACT

Asea Brown Boveri (ABB) supplied the industrial gas turbines for cogeneration application in two refineries outside of Athens, Greece. The Motor Oil Refinery in Corinth and the Hellenic Refinery in Aspropyrgos are the basis for this paper. These two plants have been in operation since 1985 and 1990, with a total of 140,000 and 55,000 operating hours, respectively. Topics discussed will include the integration of cogeneration systems into the refinery operations, technical details of the cogeneration systems, their applicability to the users' specific needs, and the savings realized by the owners.

INTRODUCTION

For years, the oil and gas industry has targeted the well-heads as the source for maximizing profit. Increasing petroleum exploration and oil recovery projects around the world exemplify such pursuit.

Today, with the maturing of the petro-chemical industry, the quest for profit has been moving away from the oil and gas fields and into every comer of the refinery processes.

Driven by a complex environment of regulation and profit motivation, more and more refineries are discovering savings in forgotten or overlooked areas within their processing plants. In a sense, the quest is centering on how to use the by-product waste to increase refinery profitability. The preferred solution is clearly in-plant cogeneration systems. The refinery cogen plants are providing savings to the industry while satisfying growing environmental concerns.

Refinery requirements for electric power and heat are being met with cogeneration systems consisting of one or more gas turbine-generators, Heat Recovery Steam Generators (HRSGs) which utilize the gas turbine waste heat. The generated steam is being used either for additional power generation and/or process heat for refinery application. However, in the volatile climate in which the refineries operate, the industry must tailor its products to market demand; thus the gas turbine units must contend with varying types of fuel supply (from liquid to gaseous, or both simultaneously) and quality (calorific value, viscosity), as well as the changing electrical and process heat demands throughout its process cycle. Often, this frequent and large fuel fluctuation presents a major obstacle for better efficiency and lower operating cost of the cogen system. However, these changing parameters do not represent a problem for the GT35 and GT10 gas turbines. The GT35 and GT10 units can utilize multiple fuels of changing quality and calorific value, including the waste gases derived from the refining process, for the gas turbine combustion process. ABB has a sizable fleet of industrial gas turbines installed throughout the world for refinery application. List of ABB's petrochem applications are exhibited in Figure 16-1.

This paper discusses the benefits of two gas turbine installations in Greek refineries utilizing the GT10 and GT35 units. These two medium size gas turbines are ideally suited for industrial cogeneration, because they are compact, heavy-duty industrial machines with proven reliability and efficiency track-records for operations in harsh environments.

These two industrial units have since been up-graded in both performance and emission control. The GT10B is presently ISO rated at 24.6 MW with a simple cycle Heat Rate of 9,970 Btu/kWh (LHV).

Figure 16-1. ABB Gas Turbines for Petro-Chem Services.

Order Year	Customer	Location	Application
GT35			
1968	TPAQ Refinery	Turkey	Cogen
1977	ELF	Norway	Power Gen
1979	BP Raffinaderi AB	Goeteborg, Sweden	Cogen
1981	NPO Refinery	Iraq	PowerGen
1982	Motor Oil Hellas	Corinth, Greece	Cogen
1984	BP Development Ltd	Ula, Norway	Power Gen
1986	Finos Petroleos De Angola	Luanda, Angola	PowerGen
1989	Shell Norske AS	Draugen Field, Sweden	Cogen
1990	Thai Taffeta Co	Rayong, Thailand	Cogen
1992/3	ESSO Refinery	Thailand	Cogen
GT10			
1984	Imperial Chem Industries	Runcorn, GB	Cogen
1986	Arcadian Corp	Geismar, LA/USA	Cogen
1988	Hellic Aspropyrgos Refinery	Aspropyrgos, Greece	Cogen
GT8			
1984	Vulcan Chemicals	Geismar, LA/USA	Combi/Cogen
1984	Shell Nederland Raffinaderij	Rotterdam/NL	Cogen
1986	SUN Refinery	Marcus Hook, PA/USA	Cogen
1987	CHEVRON Refinery	Richmond, WA/USA	Cogen
GT13			
1970	Chemische Werke Huels	Huels, Germany	Commbi/Cogen

While the GT35 is ISO rated at 16.9 MW with a simple cycle Heat Rate of 10,665 Btu/kWh (LHV). Both gas turbines are equipped with ABB's proprietary EV (Dry Low NO_x) burners providing NO_x emissions of 25 ppmvd with natural gas fuel. The combustor with EV burners is depicted in Figure 16-2.

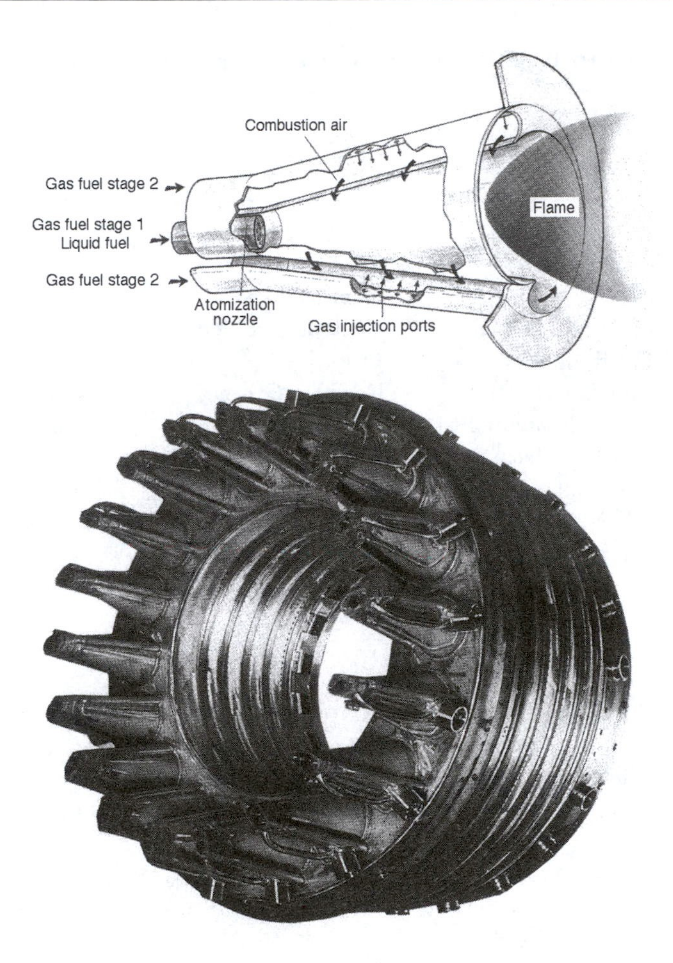

Figure 16-2.

REFINERY PROCESSES

The Corinth Motor Oil Refinery

The Motor Oil Refinery is located near Corinth, about one-hour drive north of Athens along the coast. The refinery was installed in 1972. It is a medium size refinery complex with sophisticated conversion units (i.e., fluid catalytic cracking, alkylation, isomerization, lube oil processing units), and the traditional distillation, reforming and hydrotreating units. Presently, the refinery processes 7,500,000 tons of crude oil per year, and produces 150,000 tons/year of lubrication oil.

The Motor Oil Refinery decided to install its own in-plant cogeneration plant in early 1980's with the following objectives:

1. Energy conservation by utilizing the refinery flare gas (11,000 ton/year);
2. Increase refinery productivity by avoiding the costly process shut-downs due to interruption of electrical power supply;
3. Increase refinery process steam availability and reliability by utilizing the gas turbine exhaust energy;
4. Reduce refinery operating costs;
5. Reduce refinery pollution.

In 1985, the refinery cogeneration plant, consisting of two ABB STAL GT35 Gas Turbine-Generator units, was placed in commercial operation. The two gas turbines exhaust into a single two pressure Heat Recovery Steam Generator (HRSG) with supplementary firing capability. The combined electrical output of the two units is 27 MW and a total 52 tons/hr of high pressure steam for power generation and 16 tons/hr of low pressure process steam for refinery purposes.

The primary combustion fuel is refinery flare gas originating from different refinery process streams as waste by-products. It consists essentially of propane and butane in varying proportions ranging from 60% to 100% propane, and 40% to 0% butane by volume. This by-product flare gas also contains a varying concentration of H_2S, up to a maximum of 10,000 ppm. 20-25 ppm of H_2S concentration is considered a highly corrosive environment for gas turbine application. Because the GT35 operates at a low hot blade path temperature profile (exhaust temperature of 710°F) which is below the melting point of sulfur, it is not affected by the high temperature corrosion phenomenon; thus making it an ideal machine for refinery flare gas application.

The flare gas varies in qualities, pressures and temperatures. Its heating value approximates that of natural gas, i.e. 1,145 Btu/scf. The cogen plant back-up fuel is gasified LPG (liquefied petroleum gas) with a heating value of 2,500 Btu/scf.

To use this flare gas for gas turbine combustion, it is initially de-slugged by a liquid trap which separate the liquid phase from the gas stream. It is then processed through a low pressure compressor. The condensates from the compressed fuel gas is further removed by a gas-liquid separator. Finally, the combustion fuel gas is brought to a pressure of 330 psia by a high pressure compressor. The fuel gas is fed to the two gas turbines at a temperature of 203°F and pressure of 300 psia. Operating parameters of the cogeneration plant are shown in Figure 16-3.

Figure 16-3. Corinth Motor Oil Company.
Corinth, Greece
GT35 Operating Parameters

Gas Turbine Units	2 × GT35, ABB STAL
Initial Unit Output	11.5 MW @ 95°F
Up-Graded Unit Output	13.5 MW @ 95°F
Fuels	Refinery Flare Gas/LPG
HRSG	1 × Supplementary
High Pres. Steam	52 tons/hr @ 683 psig/788°F
Low Pres. Steam	16 tons/hr @ 36 psig/280°F

The Corinth Cogeneration Plant contributes to the refinery profitability. First, it utilizes its by-product (cost-free) flare gases as the primary combustion fuel for the gas turbines. By having its own cogeneration system, it dramatically improves the refinery profit profile by eliminating the costly refinery downtime due to power outages from the local power grid. The cogen plant also eliminates the ever-present threat of electrical rate hikes. Lastly, the cogeneration plant has reduced NO_x emissions to an air quality well below the acceptable limits imposed by the Authority. The cogen plant's noise emission is 52 dBA at 400 feet. The cogen plant is viewed positively by the Greek Authority in the battle against pollution.

The Corinth cogen plant has met all the objectives set by the Refinery. The utilization of the refinery flare gas has resulted in an accelerated payback period for this cogen plant of 2.6 years. According to the Motor Oil Company, the net savings from the cogen plant for 1992 were US $8,000,000. Profitability is shown in Figure 16-4.

Hellenic Aspropyrgos Refinery

The Hellenic Refinery is located in Aspropyrgos, Greece. It was installed in 1958, with an initial through-put of 8,500,000 tons per year. This state-owned refinery has been modernized over the years both in capacity output and plant efficiency.

The Refinery combined cycle/cogeneration plant was placed in commercial operation on January 19, 1990. The cogeneration system consists of two GT10 gas turbine-generator units, two dual pressure Heat Recovery Steam Generators (HRSG) of forced circulation type with by-pass stack, and one ABB condensing steam turbine-generator unit. Saturated steam is also produced for general refinery purposes. Figure 16-5 shows the general arrangement of the Aspropyrgos Cogeneration Plant.

The GT10 unit electrical output is limited to 17 MW per Customer's specification. The steam turbine-generator is rated at 15 MW. The combined electrical output of the cogeneration system is 49 MW. The generated high pressure steam of 612 psia/760°F from the HRSGs operates the condensing steam turbine-generator set. Each HRSG also produces 18,520 lbs/hr of low pressure steam at 68 psia/342°F for refinery consumption.

Like the Corinth Motor Oil Refinery, the Aspropyrgos Cogeneration Plant operates normally on refinery flare gas. The primary fuel for the two GT10s has a heating value range of 18,360 Btu/lb to 23,580 Btu/lb; however, heating values of as high as 29,520 Btu/lb has been recorded. The gas turbines are also capable of firing propane, diesel oil, and a mixture of the various refinery by-product gas streams. The two GT10s start up on diesel oil.

Similar to the Motor Oil Refinery, the factors that motivates the Hellenic Refinery to install its own in-plant electrical power and steam plant are the high purchase price of the electrical power and the unreliability of the local electrical power supply, coupled with the need to improve refinery efficiency and plant profitability. Since its commercial operation, the Aspropyrgos Refinery has retired several gas fired heaters and boilers. The Refinery average electrical consumption is 33,300 kW, and, it exports its excess electrical generation; thus further improves the Refinery profitability. Cogen plant reliability has been close to 100% up to its first Major Inspection.

The first Major Inspection of the two GT10s was performed after 25,000 hours of operation (scheduled Major Inspection is 20,000 operating hours). Unit #2 was inspected and immediately returned to service. More corrosion was found on Unit #1 then anticipated. It was overhauled, the first two stages of turbine vanes and blades were

Figure 16-4. Corinth Motor Oil Refinery, Corinth, Greece

	1985	1986	1987	1988	1989	1990	1991	1992
Refinery Power Demand, MWh	156,507	210,028	213,754	211,493	212,424	216,327	223,703	222,074
Net In-Plant Power Generation, MWh	149,616	183,128	194,733	172,285	203,526	203,937	215,520	210,530
Purchase Power from Grid, MWh	6,891	26,900	19,021	39,208	8,898	12,390	8,183	11,544
Power Coverage by Cogen Plant, %	0.96	0.87	0.91	0.81	0.96	0.94	0.96	0.95
Availability of GTs, %								
GT#1	90.8	97.8	98.6	78.0	95.4	97.9	99.4	98.2
GT#2	87.4	96.7	98.6	91.7	94.8	91.3	99.0	99.0
INCOMES (1000×$)								
Saving In Energy Cost	$8,019	$12,299	$14,738					
Credit for Generated Steam	$4,155	$2,152	$3,150					

TOTAL SAVINGS	**$12,173**	**$14,450**	**$17,888**
EXPENSES (1000x$)			
Additional Fuel Cost	$8,182	$4,650	$6,744
Purchase Power Cost	353.2	1095.1	1311.4
Maintenance Costs			
Materials	46.3	87	125
Labor	39	42	45
Personnel	251	276	320
TOTAL EXPENSES	**8871.4**	**6149.9**	**8545.7**
GROSS REVENUE (1000×$)	**3301.9**	**8300.2**	**9342.4**

TOTAL INVESTMENT IN 1984	US 18,000,000
TOTAL REVENUE, '85-'87	US $20,944,500
PAYBACK PERIOD	2.6 Years

Source: Corinth Motor Oil Refinery Co.

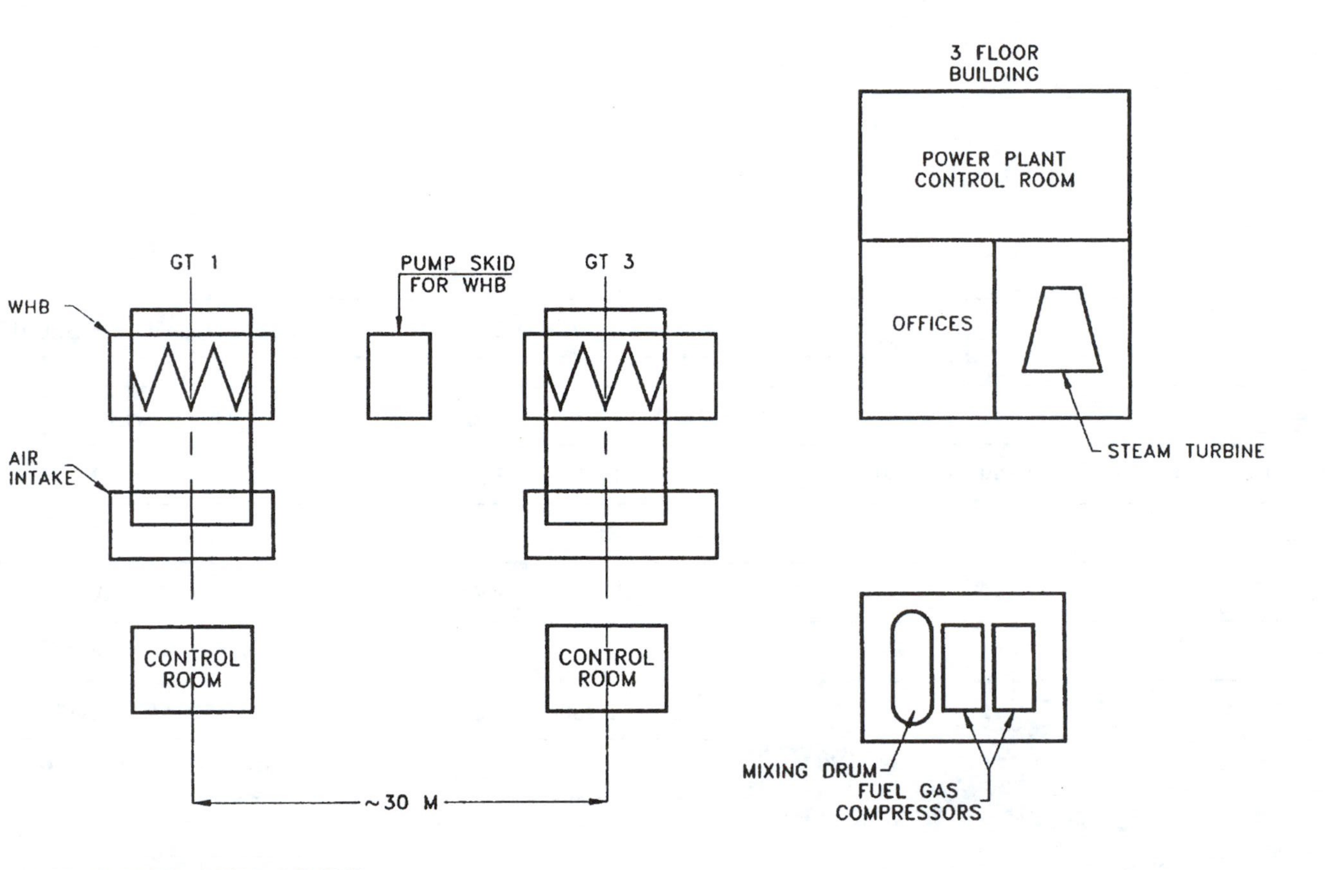

Figure 16-5. Aspropyrgos GT10 Project.

replaced and cracks in the combustor were field repaired.

It is necessary to point out that the two machines operate in a very polluted environment. Traces of the lead, zinc and natrium were found in the gas turbine compressors during the Major Inspection. Furthermore, ABB Field Service personnel have observed, on occasions, that implosion doors of the air inlet plenums were found to be opened by operators during the plant operation. The subsequent metallurgical analysis of the damaged parts indicated that the rapid and frequent change of fuel quality is also contributed to the damages and cracks in the combustor.

The spare parts and maintenance cost of the two GT10 turbines up to the first Major Inspection was US $140,000 for the two units (for the first three years). For a total of 50,000 operating hours at a combined output of 34,000 kW, this service cost comes to less than 0.2 mils/kWhr. The estimated total cost of parts and service of the Major Inspection of Unit #2 was US $1,120,000; which is equivalent to 2.6 mils/kWhr.

Using the same payback calculation technique that was employed in the Corinth Motor Oil Project, the payback of the refinery cogen plant is estimated at 3.5 years. Figure 16-6 describes the plant operating characteristics.

Figure 16-6. Hellenic Aspropyrgos Refinery Aspropyrgos, Greece.

GT10 Combi/cogen Operating Parameters

Gas Turbine Units	2 × GT10, ABB STAL
Unit Output	17 MW @ 95°F
Fuels	Refinery Flare Gas/Propane/ Diesel Oil
Steam Turbine-Generator	15 MW, ABB Type WG1000
HRSG	
High Pres. Steam	30 tons/hr @ 609 psia/770°F
Low Pres. Steam	65 tons/hr @ 55 psia/338°F

CONCLUSIONS

The maturing oil and gas industry is finding its refinery processes and in-plant utilities to be rich sources of plant improvement and profitability. Instead of the flaring the by-product waste gases, it is now harnessing them to power the refinery electrical generation and process steam production. In so doing, the refineries are becoming more self-sufficient; thus achieving better plant reliability, lowering refinery utility costs and increasing operating profit and improving the environment.

Chapter 17

Case Study #3

Case study #3 discusses the addition of a thermal energy storage plant to enhance output of this combined-cycle-cogeneration plant by cooling the gas turbine inlet air to 43 degrees F.

Gilroy Energy: Designed, Operated and Maintained for Customer Satisfaction

David B. Pearson, Plant Manager/Cogeneration Operations Manager, Bechtel Power Corp., Gilroy Energy Cogeneration Facility, Gilroy, Calif.

Gilroy Energy Co. (GEC), a 120-megawatt, combined cycle cogeneration plant in Gilroy, Calif., is designed, operated and maintained to provide maximum reliability and availability for its customers and owner, while operating within the prevailing regulatory controls and environmental considerations of the area.

The plant is a qualified facility (QF) in accordance with the Public Utilities Regulatory Policies Act (PURPA) of 1978. It provides electric power to Pacific Gas & Electric Co. under a modified Standard Offer 4 Power Purchase Agreement. The agreement allows PG&E curtailment privileges of 4,590 hours annually. It cycles off-line on a daily basis.

Thermal energy from the plant is supplied to Gilroy Foods, Inc., a large food processing facility that services the "salad bowl" region of Central California.

The plant began commercial operation in March, 1988. It was designed and constructed by Bechtel Power Corp., which also operates and maintains the facility under a long-term O&M agreement. The O&M agreement is managed by U.S. Operating Services Co. (USOSC), a Bechtel affiliate.

PLANT CONFIGURATION

The major components of the facility are the gas turbine generator, the heat recovery steam generator and the steam turbine generator. (See Figure 17-1).

The gas turbine generator is a General Electric Frame 7EA model rated at 89 MW. It is a "Quiet Combustor" unit with 10 combustion cans and 6 fuel nozzles per can. It utilizes steam injection for NO_x control. The turbine is controlled with a G.E. MK IV microprocessor based control system.

The waste heat from the gas turbine is absorbed by an unfired heat recovery steam generator (HRSG) provided by Henry Vogt Machine Co. It has a generating capacity of 292,860 lbs per hour at a design pressure of 1570 psi and 101,000 lb/hr of 82 psig steam. It has a boiler surface of 937,000 sf.

The steam turbine is a dual extraction condensing unit manufactured by AEG Kanis of Nurnberg, Germany. It is rated at 36 MW. Process steam to the food facility is supplied by controlled extraction from the turbine during normal operations. Steam for gas turbine NO_x control is supplied from a 300 psig uncontrolled bleed line.

During periods of plant shutdown for scheduled non-generating hours and maintenance, steam is supplied by dual auxiliary boilers. These are Nebraska Boiler Co. products rated at 86,890 lbs. per hour. They operate at 170 psi and 378 degrees.

The condenser, supplied by Graham Manufacturing, was designed to maintain feedwater O_2 less than 5 ppb without the use of a feed water deaerator. Up to 200 gpm of makeup water is scrubbed of O_2 by the associated vacuum makeup deaerator. The condenser handles up to 307,000 lb/hr of exhaust steam using 22,000 gpm cooling water flow.

A Thermal Energy Storage Inlet Air Cooling System was recently installed to enhance output of the facility during high electrical demand periods. It is designed to control gas turbine inlet air temperature at 43 degrees ±1 degree. The system is comprised of a 750,000-gallon ice storage tank, refrigeration system utilizing freon-22 and inlet air cooling coils. The system was designed and constructed under a turnkey contract by Henry Vogt Machine Co.

During off-peak periods, the turbine inlet air is cooled by an evaporative cooling fogging system supplied by Mee Industries.

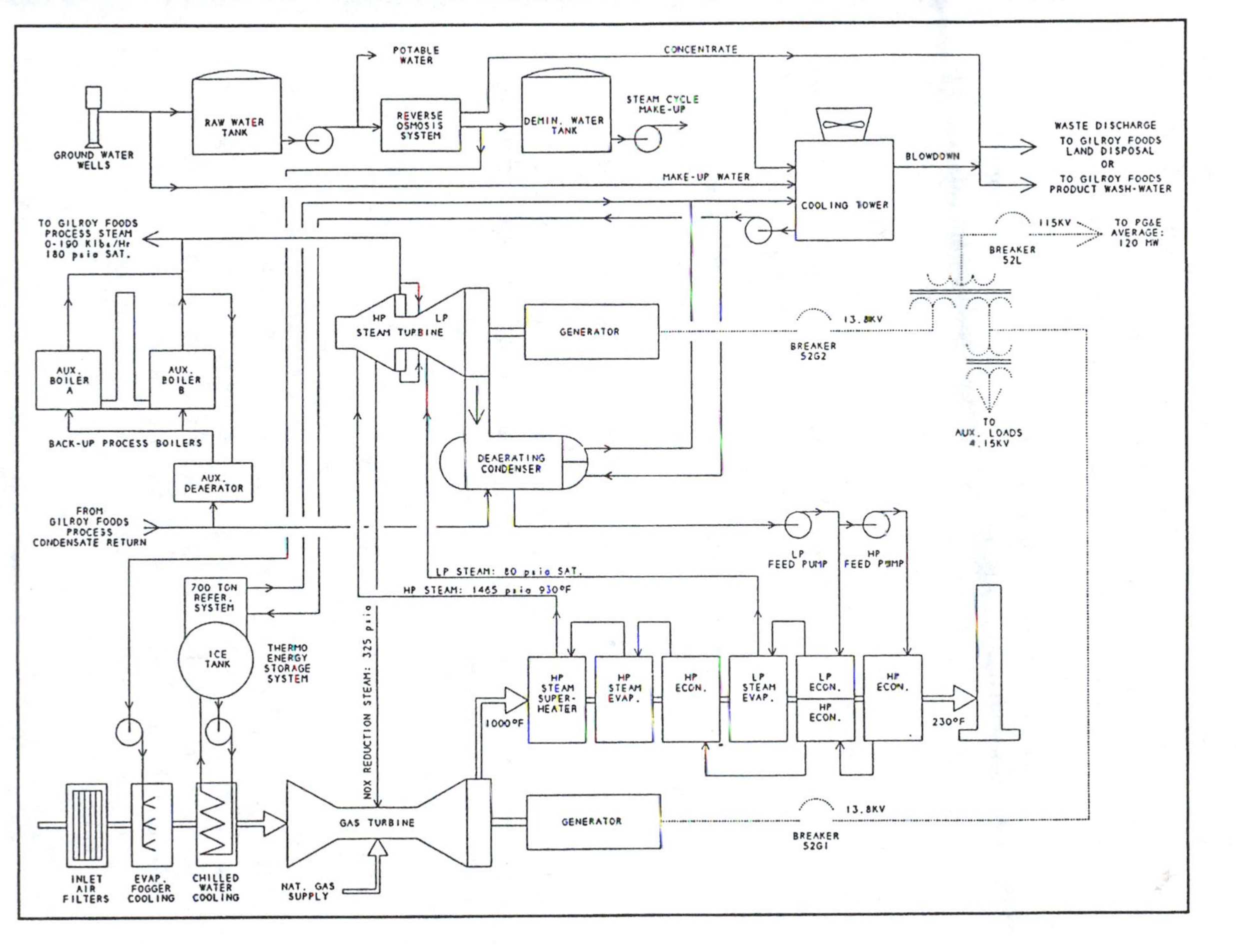

Figure 17-1. Plant Process Flow Diagram.

The facility, originally supplied with two mixed bed demineralizer trains, has contracted with Ionics Ultrapure Water to furnish a 200 gpm reverse osmosis unit with mixed bed resin bottles for polishing to supply the demineralized water needs. This system eliminates the requirement for on-site sodium hydroxide and greatly reduces sulfuric acid usage. It also eliminates demineralizer maintenance, resins and operations.

The plant control system is a Bailey Network 90 Distributed Control System (DCS). This system was provided to monitor and control the operation of the HRSG, steam turbine generator, auxiliary boilers, and balance of the plant equipment. It also monitors the operation of the gas turbine generator. The DCS provides graphic displays, trend indications, and alarm information.

The plant was outfitted with the best available control technology (BACT) for emissions reduction as recognized by the State of California. NO_x emissions are controlled by the Quiet Combustion System with steam injection, and the HRSG is equipped with a carbon monoxide catalytic reduction system for the control of CO emissions.

The plant maintains a continuous emissions monitoring system (CEM) to monitor the output criteria pollutants. This unit continuously samples stack gases and relays the results to a dedicated control room computer for calculating emissions in the format requested by the local air district.

OVERVIEW OF OPERATING PLAN

According to the terms of the Power Purchase Agreement, the utility provides an annual nonbinding estimate of hours to be operated during curtailable periods. This estimate is used to budget and acquire the fuel supplies for the coming year, coordinate annual maintenance activities, and budget operating costs.

The plant operates within the following two separate and distinct scheduling periods:

Scheduling Period 1—January 1 through April 30—The plant is dispatchable and operates per "Specific Operating Orders" from the utility. In 1994, for example, the plant was scheduled to operate base load in January and February, and was scheduled off-line in March and April.

Scheduling Period 2—May 1 through December 31—The plant is curtailable each night from midnight to 6:00 a.m., seven days a week.

The ability to adapt to changing utility power requirements and steam host process demands throughout the year is the key component of service to our customers.

Priorities of the GEC project are: to furnish process steam to the thermal host, provide "firm" net power generation for sale to the utility, maximize plant reliability, minimize fuel cost, plant heat rate, and operating overhead, and be environmentally responsible in all actions.

This plant has been designed, constructed and is operated in a manner to accomplish these goals while complying with strict standards, which are intended to protect personnel, property, and the environment.

An essential component in the success of a cycling operation is plant conformance to numerous and sometimes conflicting regulations, contracts and changing business environments. Adherence to Federal Energy Regulatory Commission (FERC) regulations as a qualifying facility, state and regional air permits, steam sales contract, power purchase agreements, equipment maintenance requirements and California Public Utilities Commission (CPUC) rules while maintaining high availability and low heat rate demand exceptional commitment, dedication and teamwork by all parties.

The result is that the staff must serve more than one customer. The owner, the steam host, the various regulating agencies, and the utility must all be satisfied at the same time for the project to be considered successful.

OPERATING RELIABILITY

Consistent with maximizing plant reliability and availability, the GEC facility design incorporated standby equipment such as spare pumps, redundant control loops and controls, and manual backup field equipment.

Special design consideration was required to minimize the stresses of thermal cycling in major equipment items such as the gas and steam turbines, heat recovery steam generator and vacuum deaeration condensing system consistent with a life expectancy of 30 years.

The plant's operation reflects these considerations. During the last six years the plant has generated 4,647,000 megawatt hours of power for the utility. GEC has accumulated 1,150 successful starts on 41,750 hours of operation.

These performance indicators show that the design has met the challenge. Since inception they are:

Operating Availability	89.7%
Peak Periods Availability	98.0%
Forced Outage Factor	1.25%

The availability factor of GEC is impacted by the way maintenance is scheduled and executed. The annual inspections are scheduled to coincide with the utility's low usage period, typically March and April. Because the plant is scheduled off anyway, we will usually extend our maintenance schedule to minimize overtime expenses. For example, a combustion inspection at a base load plant can be performed in 36-40 hours working around the clock. We will perform the same inspection in a week using a single shift.

MAINTENANCE POLICY

The maintenance associated with the plant is accomplished through use of the operator/technicians and contract maintenance support. All maintenance is tracked and equipment histories recorded through use of a computerized maintenance management system that also tracks inventory of spare parts. This same system is used for purchase of all materials.

The technicians are highly trained in the detailed operation of the plant. They are also highly skilled mechanics capable of disassembly, inspection and repair of mechanical and electrical equipment. The technicians are assigned specific systems for which they are responsible and accountable. Each is required to be the system's expert;

develop operating and maintenance instructions; perform routine, preventive and corrective maintenance; develop and present training to others; and manage any contractors that may be required to perform work on that system. They perform outage and parts planning required for work and coordinate any tests or vendor activities. They are truly "system owners" and that pride of ownership is reflected in the way the plant is maintained.

The plant's management staff is acutely aware of the importance of a solid spare parts program. In the event of an unplanned failure, the plant must be able to respond swiftly to return to service. This is accomplished most effectively by replacing any parts that may be suspect and troubleshooting those parts after the plant is returned to operation. The plant carries an inventory of spare parts in excess of one and one-half million dollars.

The facility is staffed with 20 full-time employees. With the exception of an instrument technician and a utility man, there is no dedicated maintenance crew at the plant. Routine maintenance, including welding, equipment repair and grounds maintenance is performed by the operating group. Major equipment overhauls and specialty services are contracted when expertise or manpower requirements preclude the use of site personnel. This arrangement eliminates many of the boundaries typical of a more conventional organization with separate maintenance and operations departments. As a result, the operator/technicians become more knowledgeable about the plant and develop an attitude of plant and equipment ownership.

COMMITMENT TO PARTNERSHIP

A host of technical disciplines were brought together to insure the delivery of a reliable energy source that would meet the requirements of PG&E and Gilroy Foods, Inc., the plant's customers.

In addition, an underlying philosophy of customer satisfaction and team involvement has been integrated into the daily operation of the plant. The plant staff assumes the customers of PG&E and Gilroy Foods, Inc. as their own and they work to insure these customers are well served.

The dedication to meeting PG&E's needs, as well as PG&E's dependence upon the plant is apparent in several respects. For one thing, although GEC is an independent power producer, the plant is included by PG&E as part of its master control center. In addition, continuing contacts between the operating employees of PG&E and GEC coordinate dispatchable operations and maintenance activities. Also, PG&E and GEC help one another with specific situations of either an emergency or routine nature.

Our commitment to customer satisfaction also includes our suppliers. It is our intent to develop win-win relationships with suppliers whenever possible. For example, we have an ongoing, partnering agreement with General Electric that provides significant resources and services to us and gives them a commitment for planned work and materials. We currently have similar agreements in place with several of our vendors.

While it is true there are contracts between companies that create obligations and responsibilities, the contracts do not preclude flexibility in meeting the needs of all parties. "Partners" are always looking for ways to create mutual benefit. There is a commitment to know, understand and work with your "partner." This attitude is evident in our relationship with both our customers and our suppliers.

Chapter 18

Case Study #4

Case study #4 demonstrates the use of microturbines, fueled with landfill gas, in providing electric power and heat for a 262,000 square foot high school. The microturbines, operating at 96,000 rpm, produce 360 kW of electricity and 3.48 million Btu/hr, at 550^{0}F, of exhaust energy.

A Case Study in the Development of a Landfill Gas-to-energy Project for the Antioch, Illinois, Community School District

Mark Torresani
Solid Waste Engineering
RMT, Inc.
Madison, Wisconsin
Ben Peotter
Solid Waste Engineering
RMT, Inc.
Madison, Wisconsin

ABSTRACT

The HOD Landfill, located within the Village of Antioch in Lake County, northeastern Illinois, is a Superfund site consisting of approximately 51 acres of landfilled area. On September 28, 1998, the United States Environmental Protection Agency (USEPA) issued a Record of Decision (ROD) for the site requiring that specific landfill closure activities take place. The Remedial Action (in response to the ROD), which was completed in January 2001, included the installation of a landfill gas collection system with 35 dual gas and leachate extraction wells. This system collects approximately 300 cubic feet per minute of landfill gas. In 2001, RMT and the Antioch Community School District began exploring the option of using this landfill gas to generate electricity and heat for the nearby high school. In April 2002, the Antioch Community School District applied for and received a $550,000 grant from the Illinois Department of Commerce and Community Affairs' (DECCA's) Renewable Energy Resources Program (RERP) to construct a cogeneration system to use the landfill gas to produce electricity and heat at the high school. On December 24, 2002, construction of the system began.

The design and construction of the energy system posed a number of challenges, including resolving local easement issues, meeting local utility requirements, connecting to the existing school heating system, crossing under a railroad, and meeting the USEPA's operational requirements. One-half mile of piping was installed to transfer approximately 200 cubic feet per minute of cleaned and compressed landfill gas to the school grounds, where 12 Capstone MicroTurbines™ are located in a separate building. The 12 Capstone MicroTurbines™ produce 360 kilowatts of electricity and, together with the recovered heat, meet the majority of the energy requirements for the 262,000–square foot school. The system began operating in September 2003.

This use of landfill gas proved beneficial to all parties involved. It provides energy at a low cost for the high school; clean, complete combustion of waste gas; decreased emissions to the environment by reducing the need for traditional electrical generation sources; public relations opportunities for the school and community as being the first school district in the U.S. to get electricity and heat from landfill gas; and educational opportunities in physics, chemistry, economics, and environmental management for Antioch High School students, as a result of this state-of-the-art gas-to-energy system being located at the school.

INTRODUCTION

In 2001, RMT contacted Antioch Community High School (ACHS) to inquire as to their interest in using landfill gas as an energy source. In 2002, RMT and ACHS entered into an agreement to develop the landfill gas into the primary energy source for the high school. The

$1.9 M project became operational in October 2003, designating the high school as the first in the country to use landfill gas for both heat and electrical production. The landfill gas is now used to power 12 microturbines, providing 360 kilowatts of electricity and enough heat to meet the majority of the energy requirements for the 262,000–square foot school.

The design of the energy system included tying into the existing gas collection system at the landfill, installing a gas conditioning and compression system, and transferring the gas 1/2 mile to the school grounds for combustion in the microturbines to generate electricity and heat for the school. A schematic layout of the landfill gas-to-energy system is shown on Figure 18-1. This work presented many challenges, including resolving easement issues, meeting local utility requirements, connecting to the existing heating system, crossing railroads, cleaning the landfill gas, and meeting the USEPA's operational requirements to control landfill gas migration. RMT staff worked with the local government, school officials, and the USEPA, in addition to leading the design efforts and managing the construction activities throughout the project. RMT also provided public relations assistance to ACHS by attending Antioch Village Board meetings to describe the project and to answer any questions from concerned citizens and Village Board members. The HOD Landfill is located

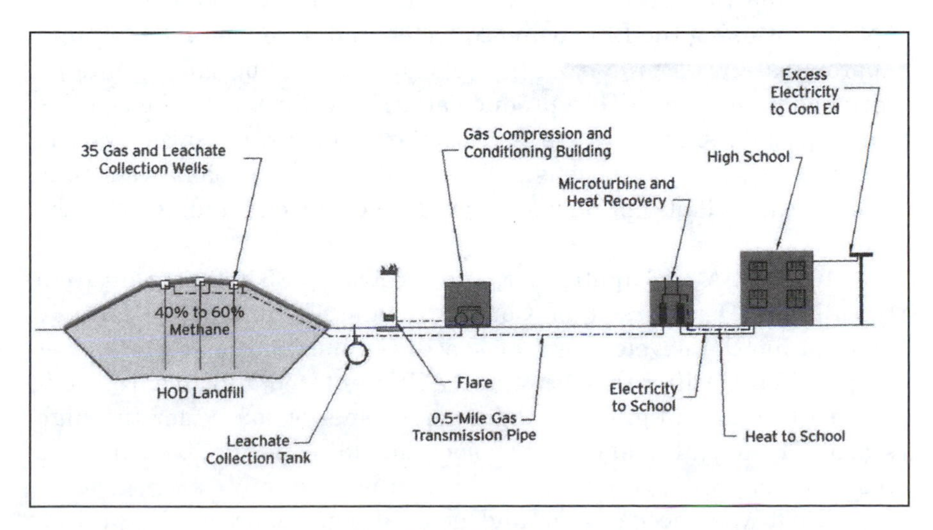

Figure 18-1. Layout of Landfill Gas-to-energy System.

within the eastern boundary of the Village of Antioch in northeastern Illinois. The closed 51-acre municipal and industrial solid waste disposal facility was active from 1963 to 1984. During that time, the landfill accepted approximately two million tons of municipal waste.

On September 28, 1998, the United States Environmental Protection Agency (USEPA) issued a Record of Decision (ROD) for the site, with concurrence from the Illinois EPA, which required that specific landfill closure activities be performed. The final Remedial Design (RD), including a landfill gas and leachate management system and final cover, was approved by the USEPA on August 9, 2000.

The final RD included 35 dual gas extraction wells that were located to allow for athletic fields as an end use option. Construction activities for the RD were essentially completed in April of 2001. At that time, the landfill gas and leachate collection systems began operating.

Initial operation of the gas management system indicated that approximately 300 cubic feet per minute (cfm) of landfill gas (LFG) was available for potential use as an energy source. Local businesses and industries were identified as potential users of this LFG. Ultimately, the ACHS was identified as the only user that was able to use the energy from 300 cfm of LFG.

During the fall of 2002, the ACHS and RMT began to explore options for using the LFG being collected and flared at HOD Landfill approximately 1/2 mile from the school. Potential options evaluated included using the LFG to produce electricity, for use in the school's existing boilers, and for use in a combined heat and power system. Through these evaluations, it was determined that the only economically viable option was to produce electricity and heat for the school.

In 2002, ACHS applied for, and received, a $550,000 grant from the Illinois Department of Commerce and Community Affairs to be used for the development of the LFG combined heat and power project. Shortly after this, RMT and ACHS entered into an agreement to turn the landfill gas into the primary energy source for the high school. The overall cost of this project, including design, permits, and construction, was approximately $1.9 million.

RMT was the designer and general contractor on the project. Specifically, RMT's team was responsible for:

- Designing the system
- Administering contracts, including coordinating access rights, railroad access, and obtaining all appropriate permits
- Creating a health and safety plan
- Managing construction
- Coordinating utility connections

In addition, RMT attended public meetings to address concerns from the Village Board and residents, and worked with the Village of Antioch and ACHS to resolve any conflicts that arose during the project.

PROJECT DESIGN

This project included 12 Capstone MicroTurbines™, to turn landfill gas into the primary energy source for the 262,000–square-foot ACHS. This is the first landfill gas project in the U.S. to be owned by and to directly provide heat and power to a school.

The challenges surrounding the design of this system, are discussed below.

HOD Landfill Gas Collection System Tie-in

The collection system at the HOD Landfill, which includes 35 landfill gas extraction wells, a blower, and a flare, must remain operational to control landfill gas migration. Therefore, the construction of the new cogeneration system required connection to the existing system to allow for excess LFG to be combusted in the flare. Additional pipes and control valves were included in the system to route the gas to the conditioning and compression building and to allow the existing blower and flare to remain operational, while providing the correct volume of landfill gas to the microturbines.

Gas Cleaning and Compression

The gas that is collected from the landfill is conditioned through a series of chillers that drop the gas temperature to -10°F to remove moisture and siloxane compounds. A schematic diagram indicating the LFG compression and conditioning system is shown on Figure 18-2. An activated carbon unit is also included to remove additional impurities.

The landfill gas is compressed to 95 pounds per square inch (psi) to meet the input fuel requirements of the Capstone MicroTurbinesTM. The gas cleaning and conditioning system is located at HOD Landfill in a building adjacent to the blower and flare. The gas conditioning and gas compression systems are shown on Figures 18-3 and 18-4.

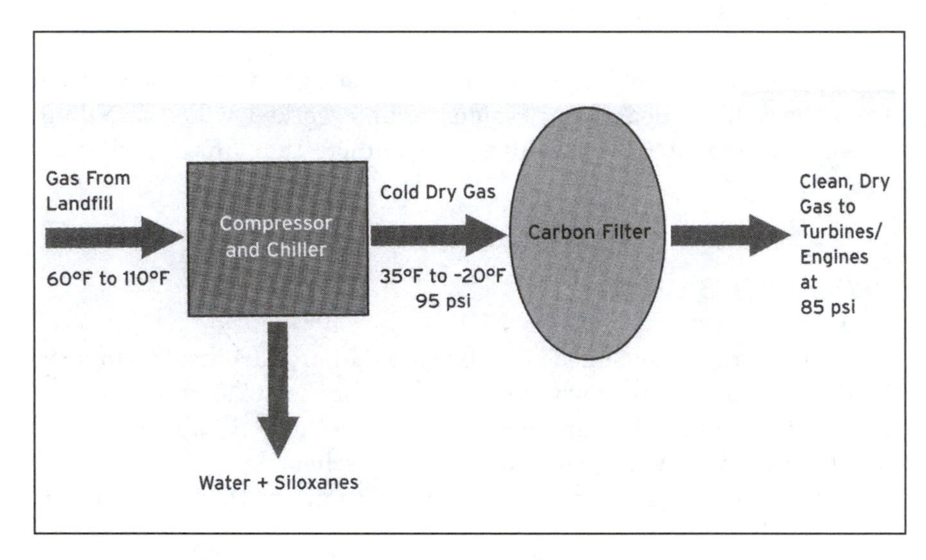

Figure 18-2. Landfill Gas Conditioning System

Figure 18-3. Gas Conditioning System.

Figure 18-4. Gas Compression System

Gas Piping to the Microturbines Located at the School

High-density polyethylene (HDPE SDR 9) pipe 4 inches in diameter and 1/2 mile long was installed 4 to 12 feet below ground, running from the HOD Landfill to the microturbines at the school. The use of horizontal drilling techniques allowed the pipe to cross beneath a stream, a road, public utilities, athletic fields, and a railroad, with minimal disturbance of the ground surface. This was extremely important for the community and the school's athletic programs.

Electric Generation

Twelve Capstone MicroTurbines™ are located at the school to provide the electricity and heat from the LFG. Each Capstone MicroTurbineTM fueled by the landfill gas produces up to 30 kW of three-phase electricity at 480 volts, using 12 to 16 cubic feet per minute (cfm) of landfill gas for a total of 360 kW of electricity—enough to power the equivalent of approximately 120 homes. The microturbine system incorporates a combustor, a turbine, and a generator. The rotating components are mounted on a single shaft supported by air bearings that rotate at up to 96,000 RPM. The generator is cooled by airflow into the gas turbine. Built-in relay protection (over/under voltage and over/under frequency) automatically trips off the microturbines in the event of a utility system outage or a power quality disturbance. Excess electricity not used by ACHS is sold to Commonwealth Edison. A 12-turbine system was selected to provide a system that will remain functional as LFG production from HOD Landfill decreases. Based on the initial LFG collection rates, up to 18 turbines could have been installed. The final payback for this project, based on conservative assumptions for future energy costs, is approximately 8 years. The installed microturbines are shown on Figure 18-5.

Heat Generation

Each Capstone MicroTurbineTM produces exhaust energy of around 290,000 Btu/hr at 550°F. The exhaust from the microturbines is routed through heat exchangers that heat the liquid, which then circulates through underground insulated steel pipes running beneath a parking lot to the school's boiler system. Because heat is being transferred to the school through insulated 4-inch–diameter pipes, locating the turbines next to the school was critical in preventing excess heat loss. When waste heat recovery is not required by ACHS, the microturbine exhaust is automatically diverted around the exchanger, allowing continued electrical output. During extremely cold weather, the school boiler system automatically uses natural gas to supplement the heat output of the microturbines.

SUMMARY

This project serves as a model of how a landfill with relatively small quantities of LFG can be used to produce clean efficient energy. By using the electricity and heat created during power production, microturbines become more practical for landfill gas utilization. The main advantages of microturbine technology over other more traditional internal combustion engines are the clean, quiet operation and the ability to add and remove microturbines as gas flow increases or decreases.

The project's design and construction can be a model for other communities that are interested in the beneficial reuse of nearby

Figure 18-5. Installed Microturbines.

landfill gas resources. It is an example of how to deal with the numerous community concerns related to developing an alternative energy system based on landfill gas. Determining suitable equipment for system design, construction, and operation, while considering local community needs and requirements, is critical to a successful project.

This project is a prime example of how innovative partnerships and programs can take a liability and turn it into a benefit. The solution has created a win-win situation for all involved, including HOD Landfill, ACHS, the Village of Antioch, the State of Illinois, Commonwealth Edison, and the USEPA. Each key player is seeing significant benefits of the energy system:

- Low energy costs for the high school
- Use of waste heat for internal use in the high school
- Clean, complete combustion of waste gas
- Decreased emissions to the environment through reduced need for traditional electrical generation sources
- Reduction in greenhouse gas emissions
- Public relations opportunities for ACHS and the community as the first school district in the U.S. to get electricity and heat from landfill gas
- Educational opportunities in physics, chemistry, economics, and environmental management as a result of this on-campus, state-of-the-art gas-to-energy system

Chapter 19

The Gas Turbine's Future

Gas turbine production worldwide has varied over the past fifteen years from a low of $20 billion per year to a high of just over $40 billion per year. The total value of production in 2004 was $21.9 billion. Of this amount $14.9 billion was for aviation gas turbines and $7.1 billion was for non-aviation gas turbines.

In the aviation category $3.7 billion was for military applications while $11.2 billion was for commercial aviation. The US Military F35 Joint Strike Fighter powered by the Pratt & Whitney Aircraft (Division of United Technologies Corp.) F135 is a 40,000 pound thrust jet engine (Figure 19-1). This jet engine will power the F35 for all three types of service—conventional takeoff/landing (CTOL), carrier variant (CV), and short takeoff/vertical landing (STOVL).

In commercial aviation the Airbus A380, a 555- to 800-passenger aircraft, will be powered by the General Electric—Pratt & Whitney Alliance GP7000 or the Rolls Royce Trent 900 jet engines. These engines will each produce between 70,000-80,000 pounds of thrust.

In the non-aviation gas turbine category $6.04 billion was for electric generation, $0.78 billion for mechanical drive applications, and $0.28 billion for marine applications. The Queen Mary 2, the largest cruise ship at 1,132 feet long (about three times the size of the Titanic), is powered by two General Electric LM2500+ gas turbines and four diesel engines. There are presently 16 gas turbine powered cruise ships in operation.[1]

Large power plants utilizing techniques such as the atmospheric fluidized bed combustion systems, advanced pressurized fluidized bed combustion systems, integrated gasification-combined-cycle systems (reference Case Study #1), and integrated gasification-fuel cell combined-cycle systems and small microturbine electric generator sets will make up the new generation of gas turbine applications.

Figure 19-1. Courtesy of United Technology Corporation, Pratt & Whitney Aircraft. The F 135 propulsion system team consists of Pratt & Whitney, the prime contractor with responsibility for the main engine and system integration; Rolls Royce, providing lift components for the STOVL F 35B aircraft, and Hamilton Standard, provider of the F 135 control system, external accessories and gearbox.

While many of these techniques—implemented to facilitate the use of solid fuels (coal, wood waste, sugar cane bagasse) and improve power and efficiency—are new, others are not. The integration of technologies [such as, combined cycle (reference Case Study #2), inlet air cooling (reference Case Study #3), compressor inter-cooling, water or steam injection, recuperators and the use of water or steam to internally cool turbine airfoils] to enhance gas turbine performance and output is more than a century old. With the rapid improvements in gas turbine technology brought about by the war effort sixty years ago, these power and efficiency improvement techniques were set aside. Now their time has come—again!

Advances achieved in aircraft engine technology (such as; airfoil loading, single crystal airfoils, and thermal barrier coatings) have been transferred to the industrial gas turbine. Improvement in power output and efficiency rests on increases in airfoil loading and turbine inlet temperature. The union of ceramics and superalloys will provide the material strength and temperature resistance necessary to facilitate increased turbine firing temperature. However, increasing firing temperature will also increase emissions. To ensure continued acceptance of the gas turbine, emissions must be eliminated or significantly reduced. To reduce emissions, catalytic combustors are being developed (see Figure 19-2). The catalytic combustors will reduce NO_x formation within the combustion chamber. They will also reduce combustion temperatures and extend combustor and turbine part life.

While increased power (in excess of 200+ megawatts) will be provided without increasing the size of the gas turbine unit, the balance of plant equipment (such as; pre- and inter-coolers, regenerators, recuperators, combined cycles, gasifiers, etc.) will increase the overhaul size of the facility. Designers and engineers must address the total plant not just the gas turbine. The size of

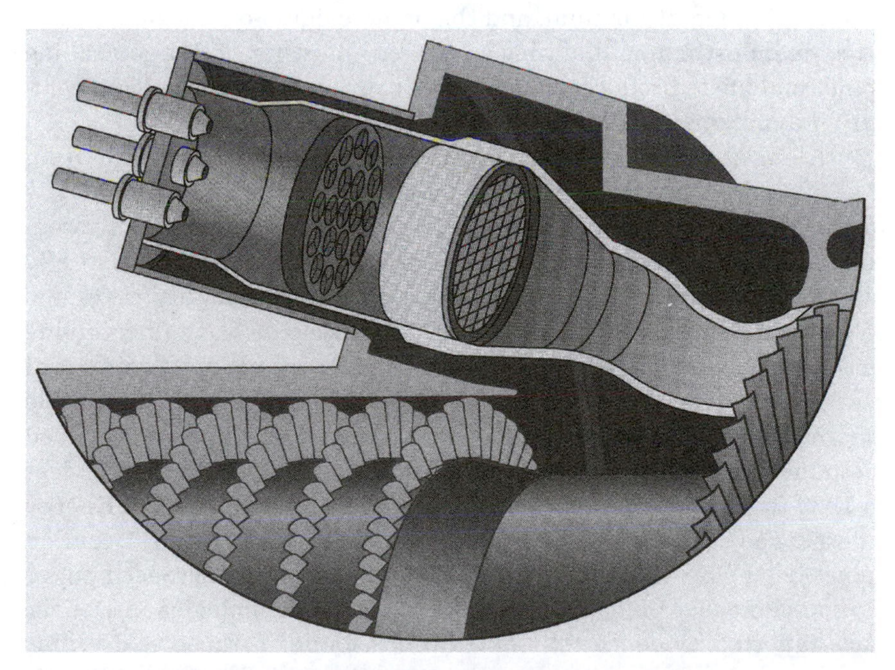

Figure 19-2. Courtesy of Catalytica Combustion Systems. Artist rendition of the catalytic combustor. The catalytic element enables combustion well below the temperatures where NO_x can form.

the various process components must be optimized to match each component's cycle with the gas turbine cycle, as a function of ambient conditions and load requirements. Computer (microcomputer or programmable logic controller) control systems must be designed to interface and control these various processes during steady-state and transient operation.

Though the supply of natural gas and oil, globally [proven gas reserves of 6.34×10^{15} cubic feet (179.5×10^{12} cubic meters) and proven oil reserves of 1.19×10^{12} barrels][2], will satisfy consumption needs for many years to come, a major effort is being made to produce gas turbines capable of burning all types of fossil fuel, biomass and waste products[3,4].

Researchers are intensifying their efforts to supply hydrogen, processed from fossil and non-fossil resources, for use in petroleum-based equipment. The objective is to produce recoverable, cost effective, and environmentally benign energy. One technique being studied at Purdue University's College of Science requires only water, a catalyst based on the metal rhenium, and the organic liquid organosilane. (The study team estimates that about 7 gallons of water and organosilane could combine to produce 6.5 pounds of hydrogen, which could power a car for approximately 240 miles.)[5]

Along these lines efforts are underway to obtain methane from coal. Coal-bed methane accounts for one-twelfth the natural gas production in the US. For example, one estimate puts the total coal resource in the Powder River Basin (in Wyoming & Montana) at 800 billion tons. More than 300 million tons are mined annually. A volume of coal can contain 6-7 times as much methane as the same volume in a conventional sandstone reservoir. The Powder River Basin's coal seams hold nearly 40 trillion cubic feet of technically recoverable gas, or about 20% of the total gas reserves in the US. Coalbed methane gas production has soared from just 100 billion cubic feet in 1989 to more than 1,600 billion cubic feet in 2003. The US Energy Information Agency reported that there was fewer than 4 trillion cubic feet of proven reserves in 1989, that figure increased to 19 trillion cubic feet in 2003. Scientists at Luca Technologies in Denver have reported evidence that natural gas found in these coal seams may be generated by microbes digesting the coal. These microscopic creatures, called Archaea, have been shown to produce 900 cubic feet of natural gas per ton of coal in only 5 months.[6] Note the current global coal reserve is approximately $90.9x10^{10}$ tons[2] (over 8.18×10^{14} cubic feet of methane).

Transportation is one of the main sources of energy demand—20% to 30%—and all of it is petroleum based. Ultimately the gas turbine will be required to burn hydrogen, which is the leading candidate to replace gasoline. Presently hydrogen is made most often be reforming methane. The use of hydrogen eliminates the fuel bound nitrogen that is found in fossil fuels. With combustion systems currently capable of reducing the dry NO_x level to 25 ppmv; and catalytic combustion systems demonstrating their ability to reduce emissions to "single digits"; the use of hydrogen fuel promises to reduce the emission levels to less than 1 ppmv. The gas turbine will need to achieve this low level in order to compete with the fuel cell. Today the largest fuel cell manufactured is rated at 200 kilowatts. This technology is available and it will be developed.

Hospitals, office complexes, and shopping malls will become the prime sales targets for 1 to 5 megawatt power plants. Residential homes, small businesses, and schools will turn to the smaller gas turbines—the microturbines. Operating these small plants, on site, may prove to be more economical than purchasing power through the electric grid from a remote power plant. This is especially true with the current restructuring of the electric utility industry. In the very small gas turbines, manufacturers are returning to the centrifugal design in both the compressor and the turbine. Centrifugal compressors and radial inflow turbines are used in most microturbines. The advantage is that this design can be produced as three distinct parts (compressor, turbine, and shaft). The compressor and turbine components are then pressed onto the shaft forming one part. This small gas turbine generator design entered the low power market in the mid-to-late 1990s with units from Capstone (Figure 19-3), Allied Signal, Ingersoll Rand, and Elliott Energy Systems, Inc. The capital cost of the microturbine is already at $1.00 per watt and it is targeted to reach $0.50 per watt. This is possible primarily due to the design of the compressor and turbine components. These components are relatively easy and inexpensive to cast. With precision casting methods, finish machining is limited and in some cases unnecessary, which further reduces cost. These units operate on gaseous or liquid fuel and generate from 20 to 500 kilowatts.

To reduce the expense of units under 5 megawatts, the next

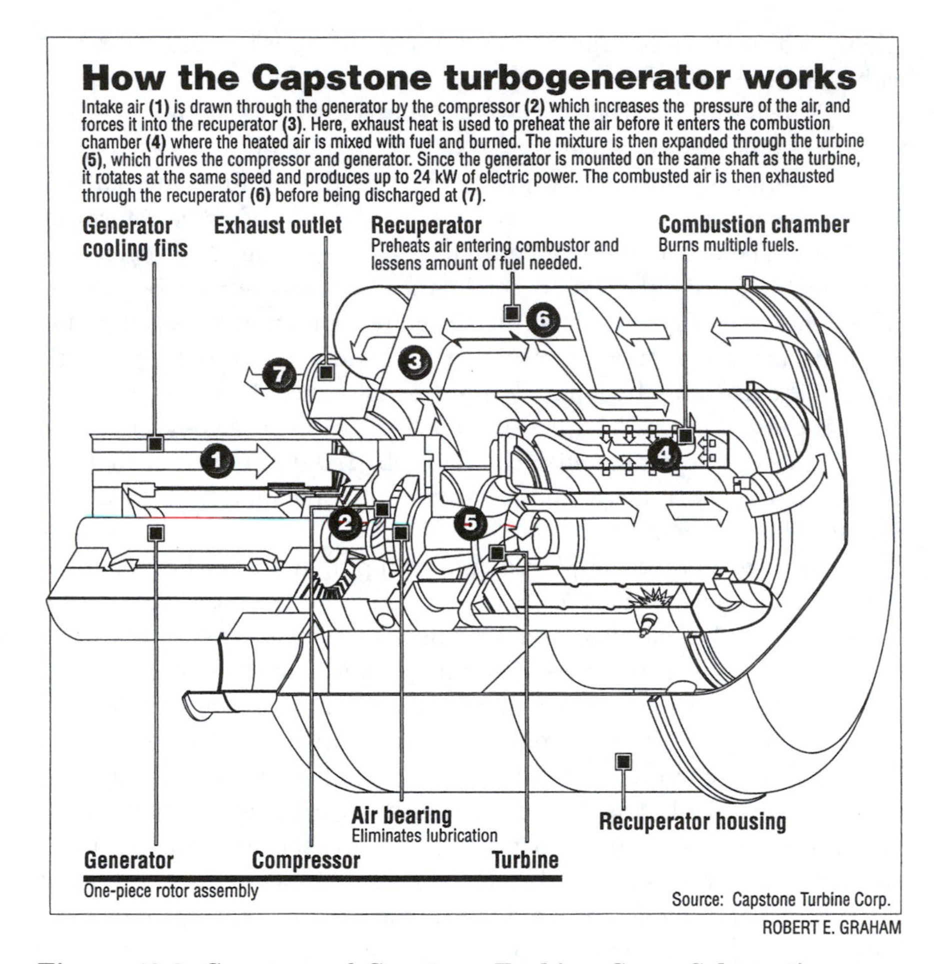

Figure 19-3. Courtesy of Capstone Turbine Corp. Schematic cross-section of the ISO base load 24 kilowatt turbogenerator. This unit, configured with the centrifugal compressor and turbine wheels mounted on the same shaft as the electric generator, weights 165 pounds.

breakthrough must be in the production of axial blade and disc assemblies as a single component. The greatest single obstacle to reducing gas turbine costs is the manufacturing process—machining and assembly. A typical gas turbine, over 1 megawatt, has over 4000 parts. About one third of these parts are made from exotic materials that incur high development cost. Each of the 4000 parts must be handled several times from initial machining to final installation. Maintaining this machine is every bit as complicated, requiring the same technical skills, as building a new unit.

Producing the blade and disc assembly, as a single part will reduce the quantity of parts handled and relax the requirement for tight dimensional tolerances between parts. Today manufacturers have access to advanced design tools such as computational fluid dynamics (CFD) to optimize compressor and turbine aerodynamic designs and computer aided drafting and computer aided machining (CADCAM) software.

The gas turbine still has not found its place in land transportation use (automobiles and trucks). This application has been stifled in the past due to high production and maintenance cost. However, a hybrid electric car application may utilize the microturbine generator as a range extender. In this capacity it will provide a constant power source to continuously charge the on-board battery pack.

The gas turbine (turbofan, turbojet, and turboprop) will continue to be the major player in aircraft applications for the next 100 years. In the next 10 years, alone, over 76,000 gas turbines—turbojet and turbofans—will be built[7]. Also, the gas turbine will be used to a greater extent in marine applications, primarily in military applications but also in cruise ships and fast-ferries. However, the greatest advances will be in land based, stationary, power plant applications. It is here the gas turbine will demonstrate its growth potential, flexibility, and adaptability.

Future energy growth can be projected from assumptions of future economic, population, energy supply, and energy price estimated growth—approximately 2% per annum growth average. The trend clearly shows energy needs increasing two-fold to seven-fold by the middle to the end of the 21st Century.

Considering that America uses 25 % of the world's energy consumption today. To supply just the world's energy demand by 2050, we'll need to build at least 4 times the current capacity (10,224.4 million tons oil equivalent in 2004).[2,8]

References

1. New Horizons," Lee S. Langston, *Power & Energy*, June 2005
2. GEOHIVE - Global Statistics, web site www.GEOHIVE.com
3. "Electrification Will Enable Sustained Prosperity," Henry R. Linden, Illinois Institute of Technology, *Power Engineering*, October 1996.

4. Making Biomass Energy A Contender," George Sterzinger, *Technology Review*, October 1995
5. "Hydrogen Production Method Could Aid Fuel Cell Push," *InTech Magazine*, August 31, 2005
6. "Deep Seam Farming," Jeffrey Winters, Power & Energy, June 2005
7. "Turbine Market Forecast, 1995-2004" *Turbomachinery International Handbook*, 1995.
8. Green Atoms" Romney B. Duffey, *Power & Energy*, June 2005

Appendix A-1

Gas Turbine Manufacturers

The following list the original equipment manufacturers (OEM) of gas generators, gas turbines and gas turbine packages. The information provided in these pages, used in conjunction with Chapters 2, 3, 4, 7-9 and 12 is instrumental in identifying comparable gas turbine units. It should be noted that gas generators (especially aero-derivatives) are packaged by several companies. Oftentimes the packager provides a different power turbine. This explains the different output and heat rate of the same gas generator provided by different packagers.

This information, provided courtesy of **Diesel and Gas Turbine Publications**, is reproduced from the 2013 Diesel & Gas Turbine Publications Global Sourcing Guide.

2013 Basic Specifications — Gas Turbine Engines

Manufacturer	Page Reference	Model Number	Type	Fuel	Gross Output MW	Continuous Output at ISO Conditions		Heat Rate		Pressure Ratio	Mass Flow		Turbine Inlet Temp (°C)	Exhaust Temp (°C)	Output Shaft Speed (r/min)	
			EG=Electric Generator GG=Gas Generator MD=Mechanical Drive MN=Marine Propulsion	L=Liquid G=Gaseous		(bhp)	(kW)	(btu/hph)	(kJ/kWh)		(lb/s)	(kg/s)			min	max
ALSTOM	290	GT24-60 Hz	EG	L/G	230.7			8531	9000.00	35.4:1	1113	505		597		3600
		GT26-50 Hz	EG	L/G	326			8467	8933.00	35.0:1	1526	692		603		3000
		GT13E2-50 Hz	EG	L/G	202.7			8980	9474.00	18.2:1	1376	624		501		3000
		GT11N2-50 Hz	EG	L/G	113.6			10,247	10,811.00	15.9:1	882	400		526		3610
		GT11N2-60 Hz	EG	L/G	115.4			10,066	10,619.00	15.9:1	882	400		526		3600
ANSALDO ENERGIA	300	AE64.3A	EG	L/G			75,000		10,028.0	16.7	470.00	213		574	3000	3600.0
		AE94.2	EG	L/G			170,000		10,366.0	11.5	1179.00	535		552	3000	3000.0
		AE94.2K	EG	L/G			170,000		9863.0	12	1190.00	540		545	3000	3000.0
		AE94.3A	EG	L/G			310,000		9045.0	19.5	1653.00	750		576	3000	3000.0
CAPSTONE	*	C30 Low Pressure	GG	G		38	28	10,200	14,400	Variable	0.69	0.3	-20 to +50	275	45,000	96,000
		C30 High Pressure	GG	G		40	30	9800	13,800	Variable	0.69	0.3	-20 to +50	275	45,000	96,000
		C65 High Pressure	GG	G		87	65	8800	12,400	Variable	1.08	0.5	-20 to +50	309	45,000	96,000
		C65 ICHP High Pressure	GG	G		87	65	8800	12,400	Variable	1.08	0.5	-20 to +50	309	45,000	96,000

		C65 CARB High Pressure	GG	G		87	65	9100	12,900	Variable	1.13	0.5	-20 to +50	311	45,000	96,000
		C200 Low Pressure	GG	G		255	190	8200	11,600	Variable	2.93	1.3	-20 to +50	280	30,000	61,000
		C200 High Pressure	GG	G		268	200	7700	10,900	Variable	2.93	1.3	-20 to +50	280	30,000	61,000
		C200 CARB Low Pressure	GG	G		255	190	8200	11,600	Variable	2.93	1.3	-20 to +50	280	30,000	61,000
		C200 CARB High Pressure	GG	G		268	200	7700	10,900	Variable	2.93	1.3	-20 to +50	280	30,000	61,000
		C600 Low Pressure	GG	G		764	570	8200	11,600	Variable	8.79	4.0	-20 to +50	280	30,000	61,000
		C600 High Pressure	GG	G		805	600	7700	10,900	Variable	8.79	4.0	-20 to +50	280	30,000	61,000
		C800 Low Pressure	GG	G		1019	760	8200	11,600	Variable	11.72	5.3	-20 to +50	280	30,000	61,000
		C800 High Pressure	GG	G		1073	800	7700	10,900	Variable	11.72	5.3	-20 to +50	280	30,000	61,000
		C1000 Low Pressure	GG	G		1274	950	8200	11,600	Variable	14.65	6.7	-20 to +50	280	30,000	61,000
		C1000 High Pressure	GG	G		1341	1000	7700	10,900	Variable	14.65	6.7	-20 to +50	280	30,000	61,000
		C30	GG	L		39	29	10,200	14,400	Variable	0.69	0.3	-20 to +50	275	45,000	96,000
		C65	GG	L		87	65	8800	12,400	Variable	1.08	0.5	-20 to +50	309	45,000	96,000
		C65 ICHP	GG	L		87	65	8800	12,400	Variable	1.08	0.5	-20 to +50	309	45,000	96,000
		C200	GG	L		255	190	7700	10,900	Variable	2.93	1.3	-20 to +50	280	30,000	61,000
		CR30	GG	G		40	30	9800	13,800	Variable	0.69	0.3	-20 to +50	275	45,000	96,000
		CR65	GG	G		87	65	8800	12,400	Variable	1.08	0.5	-20 to +50	309	45,000	96,000
Continues		CR65 ICHP	GG	G		87	65	8800	12,400	Variable	1.08	0.5	-20 to +50	309	45,000	96,000

2013 Basic Specifications — Gas Turbine Engines

Manufacturer	Page Reference	Model Number	Type (EG=Electric Generator, GG=Gas Generator, MD=Mechanical Drive, MN=Marine Propulsion)	Fuel (L=Liquid, G=Gaseous)	Gross Output MW	Continuous Output at ISO Conditions		Heat Rate		Pressure Ratio	Mass Flow		Turbine Inlet Temp (˚C)	Exhaust Temp (˚C)	Output Shaft Speed (r/min)	
						(bhp)	(kW)	(btu/hph)	(kJ/kWh)		(lb/s)	(kg/s)			min	max
CAPSTONE Continued	*	CR200	GG	G		268	200	7700	10,900	Variable	2.93	1.3	-20 to +50	280	30,000	61,000
		CR600	GG	G		805	600	7700	10,900	Variable	8.79	4.0	-20 to +50	280	30,000	61,000
		CR800	GG	G		1073	800	7700	10,900	Variable	11.72	5.3	-20 to +50	280	30,000	61,000
		CR1000	GG	G		1341	1000	7700	10,900	Variable	14.65	6.7	-20 to +50	280	30,000	61,000
		C30	GG	G		40	30	9800	13,800	Variable	0.69	0.3	-20 to +50	275	45,000	96,000
		C30 Hazloc	GG	G		40	30	9800	13,800	Variable	0.69	0.3	-20 to +50	275	45,000	96,000
		C65	GG	G		87	65	8800	12,400	Variable	1.08	0.5	-20 to +50	309	45,000	96,000
		C65 ICHP	GG	G		87	65	8800	12,400	Variable	1.08	0.5	-20 to +50	309	45,000	96,000
		C65 Hazloc	GG	G		87	65	8800	12,400	Variable	1.08	0.5	-20 to +50	309	45,000	96,000
		C200	GG	G		268	200	7700	10,900	Variable	2.93	1.3	-20 to +50	280	30,000	61,000
		C200 Hazloc	GG	G		268	200	7700	10,900	Variable	2.93	1.3	-20 to +50	280	30,000	61,000
		C600	GG	G		805	600	7700	10,900	Variable	8.79	4.0	-20 to +50	280	30,000	61,000
		C800	GG	G		1073	800	7700	10,900	Variable	11.72	5.3	-20 to +50	280	30,000	61,000
		C1000	GG	G		1341	1000	7700	10,900	Variable	14.65	6.7	-20 to +50	280	30,000	61,000
DAIHATSU DIESEL MFG. CO., LTD.	304	DT-4	EG	L		441		18,110	8.00	6.60			570		1800	

		DT-4W	EG	L		883		18,110	8.00	13.20			570		1800	
		DT-6	EG	L		662		17,890	8.00	10.50			610		1800	
		DT-10	EG	L		1103		19,190	8.00	18.30			520		1800	
		DT-10A	EG	L		1324		17,890	8.00	18.30			560		1800	
		DT-14	EG	L		1546		17,890	8.00	22.70			580		1800	
		DT-20	EG	L		2206		17,890	8.00	33.10			570		1800	
		DT-10AW	EG	L		2648		17,890	8.00	36.60			560		1800	
		DT-14W	EG	L		3089		17,890	8.00	45.40			580		1800	
		DT-20W	EG	L		4412		17,890	8.00	66			570		1800	
DRESSER-RAND	299	DR-61G4	EG	G			33 175		9290	23	202	92	855	526		3600
		DR-63G PC	MD	G		59,436	44,322	6042	8549	28	280	127	859	457	1600	3780
		VECTRA 30G	EG	G			22,767		9941	18	150	68	832	547		6200
		VECTRA 40G	EG	G			30,460		9257	22	190	86	827	526		6200
		VECTRA 40G4	MD	G		45,902	34,230	6316	8937	24	198	90	855	541	3100	6510
		DR-63G PG	MD	G		66,822	49,830	6054	8566	30	259	134	908	486	1782	3930
		KG2-3E	EG	L/G		2588	1930	21,005	22,160	5	33	15		549		1800
		VECTRA 30G	MD	G		31,469	23,467	6816	9645	18	150	68	832	547	3100	6510
		DR-61G	EG	G			23,394		9785	18	153	70	832	533		3600
		VECTRA 40G	MD	G		42,102	31,359	6347	8981	22	190	86	827	526	3100	6510
		DR-61GP	EG	G			30,742		9300	23	192	87	836	515		3600
		VECTRA 40G4	EG	G			33,209		9212	24	198	90	855	541		6200
		DR-63G PC	EG	G			43,738		8610	28	279	127	859	453		3600
		DR-63G PG	EG	G			50,447		8659	31	308	140	908	471		3930

* This engine builder is not represented in this 2013 Edition of the Global Sourcing Guide with a section description of its products.

2013 Basic Specifications — GAS TURBINE ENGINES

Manufacturer	Page Reference	Model Number	Type EG=Electric Generator GG=Gas Generator MD=Mechanical Drive MN=Marine Propulsion	Fuel L=Liquid G=Gaseous	Gross Output MW	Continuous Output at ISO Conditions		Heat Rate		Pressure Ratio	Mass Flow		Turbine Inlet Temp (°C)	Exhaust Temp (°C)	Output Shaft Speed (r/min)	
						(bhp)	(kW)	(btu/hph)	(kJ/kWh)		(lb/s)	(kg/s)			min	max
EBARA CORPORATION	*		EG	L		69,063	51,500			20	371	168		600	750	3600
			MD	L		69,063	51,500			20	371	168		600	750	3600
GE OIL & GAS	*	LMS100	EG	L/G		131,682.6	98,196	5885	8327	40	456	207		417		3600
		PGT16	EG	L/G		18,473.1	13,775	7298.1	10325	20.10	103.80	47		498		7900
		PGT20 SAC	EG	L/G		23,449.9	17,487	7227.0	10225	19.70	138.00	63		479		6500
		PGT20 DLE	EG	L/G		24,054.5	17,937	7238.0	10240	19.80	137.30	62		491		6500
		PGT25 SAC	EG	L/G		30,113.4	22,456	7001.5	9906	17.90	151.90	69		522		6500
		PGT25 DLE	EG	L/G		30,103.1	22,448	7039.6	9960	17.90	151.00	69		529		6500
		PGT25+SAC	EG	L/G		40,598.2	30,274	6411.8	9072	21.50	185.80	84		500		6100
		PGT25+DLE	EG	L/G		40215	29,988	6432.3	9101	21.50	184.70	84		507		6100
		PGT25+G4 SAC	EG	L/G		43,900.6	32,737	6433.9	9103	23.00	198.40	90		512		6100
		PGT25+G4 DLE	EG	L/G		43,583.5	32,500	6432.3	9101	23.00	197.30	90		513		6100
		LM6000 PC SAC FIXED IGV	EG	L/G		58,196.8	43,397	6093.8	8621	27.90	276.90	126		454		3600
		LM6000 PC SAC OPEN IGV	EG	L/G		58,367.6	43,525	6098.1	8628	28.20	278.90	127		452		3600
		LM6000 PC SAC VARIABLE IGV	EG	L/G		58,470.1	43,601	6089.4	8615	28.10	278.20	126		454		3600

		LM6000 PD	EG	L/G		57,633	42,977	6107.4	8641	28.30	274.90	125		455		3600
		LM6000 PF	EG	L/G		57,633	42,977	6107.4	8641	28.30	274.90	125		455		3600
		LM6000 PC SAC FIXED IGV	MD	L/G		59,384.5	44,283	5971.9	8449	27.90	276.90	126		454		3600
		LM6000 PC SAC OPEN IGV	MD	L/G		59,558.8	44,413	5976.1	8455	28.20	278.90	127		452		3600
		LM6000 PC SAC VARIABLE IGV	MD	L/G		59,663.4	44,491	5967.6	8443	28.10	278.20	126		454		3600
		LM6000 PD	MD	L/G		58,809.2	43,854	5985.3	8468	28.30	274.30	125		455		3600
		LM6000 PF	MD	L/G		58,809.2	43,854	5985.3	8468	28.30	274.30	125		455		3600
		LMS100	MD	L/G		13,4370	100,200	5767.6	8160	40.00	456.10	207		417		3600
		PGT16	MD	L/G		19,143.1	14,275	7042.7	9964	20.10	103.80	47		498		7900
		PGT20 SAC	MD	L/G		24,300.4	18,120.8	6974.1	9867	19.70	138.00	63		479		6500
		PGT20 DLE	MD	L/G		24,926.9	18,588	6984.7	9882	19.80	137.30	62		491		6500
		PGT25 SAC	MD	L/G		31,205.6	23,270	6756.4	9559	17.90	151.90	69		522		6500
		PGT25 DLE	MD	L/G		31,194.9	23,262	6793.2	9611	17.90	151.00	69		529		6500
		PGT25+SAC	MD	L/G		42,070.7	31,372.1	6187.4	8754	21.50	185.80	84		500		6100
		PGT25+DLE	MD	L/G		41,673.6	31,076	6207.2	8782	21.50	184.70	84		501		6100
		PGT25+G4 SAC	MD	L/G		45,492.8	33,924	6208.7	8784	23.00	198.40	90		512		6100
		PGT25+G4 DLE	MD	L/G		45,164.3	33,679	6207.2	8782	23.00	197.30	90		513		6100
		MS5001	EG	L/G		36,046.7	26,880	8842.2	12,510	10.50	276.00	125		483		5100
		MS5002E	EG	G		42,577.5	31,750	7110.5	10,060	17.00	225.50	102		508		5714
Continues		MS6001B	EG	L/G		57,127.5	42,600	7888.0	11,160	12.30	322.30	146		546		5160

* This engine builder is not represented in this 2013 Edition of the Global Sourcing Guide with a section description of its products.

2013 Basic Specifications — GAS TURBINE ENGINES

Manufacturer	Page Reference	Model Number	Type (EG=Electric Generator, GG=Gas Generator, MD=Mechanical Drive, MN=Marine Propulsion)	Fuel (L=Liquid, G=Gaseous)	Gross Output MW	Continuous Output at ISO Conditions		Heat Rate		Pressure Ratio	Mass Flow		Turbine Inlet Temp (°C)	Exhaust Temp (°C)	Output Shaft Speed (r/min)	
						(bhp)	(kW)	(btu/hph)	(kJ/kWh)		(lb/s)	(kg/s)			min	max
GE OIL & GAS Continued	*	MS7001EA	EG	L/G		12,0378	89,766	7523.3	10,644	12.90	662.00	300		544		3600
		MS9001E	EG	L/G		17,4467	130,100	7357.9	10,410	12.80	926.60	420		538		3000
		MS7121(EA)	EG			120,378	89,766	7523.3	10,644	12.90	662.00	300		544		3600
		MS9171(E)	EG			174,467	130,100	7357.9	10,410	12.80	926.60	420		538		3000
		MS5002C	MD	L/G		37,950.9	28,300	8700.9	12,310	8.80	274.00	124		517		4670
		MS5002C POWER CRYSTAL	MD			39,520	29,470	8714.0	12,330	9.10	270.00	122		540		4670
		MS5002D	MD	L/G		43,717.3	32,600	8411.1	11,900	10.80	311.70	141		509		4670
		MS5002D POWER CRYSTAL	MD			45,553	33,980	8413.0	11,904	10.40	308.00	140		534		4670
		MS5002E	MD	G		42,912.7	32,000	7052.6	9978	17.00	225.50	102		508		5714
		MS6001B	MD	L/G		58,955.4	43,963	8140.4	11,517	12.30	322.30	146		546		5160
		MS7001EA	MD	L/G		121,362	90,500	7584.1	10,730	12.90	662.00	300		544		3600
		MS9001E	MD	L/G		175,272	130,700	7357.9	10,410	12.80	926.60	420		538		3000
		GE10	EG	L/G		15,086	11,250	8122.0	11,489	15.50	104.70	48		482		11,000
		GE10-1 STANDARD	EG	L/G		15,086.5	11,250	8114.9	11,481	15.50	104.70	48		482		11,000
		GE10-1 DLE	EG	L/G		15,089.2	11,252	8433.7	11,932	15.80	104.30	47		481		11,000

		GE10-2	MD	L/G		16,288.1	12,146	7620.9	10,782	15.60	103.60	47		483		7900
		GE10-2 DLE	MD	G		15,907.2	11,862	7762.2	10,982	15.80	103.60	47		489		7900
KAWASAKI HEAVY INDUSTRIES, LTD.	288	M1A-13A	EG	L/G		1991	1485	10,510.0	14,871	9.40	17.70	8		521		1800
		M1A-13D	EG	L/G		1991	1485	10,618.0	15,022	9.60	17.50	8		531		1800
		M1A-17D	EG	G		2260	1685	9576.0	13,548	10.50	17.70	8		521		1800
		M1T-13A	EG	L/G		3929	2930	10,662.0	15,085	9.40	35.40	16		521		1800
		M1T-13D	EG	G		3929	2930	10,771.0	15,239	9.60	35.10	16		531		1800
		M7A-01	EG	L/G		7410	5530	8590.0	12,150	13.00	48.00	22		545		1800
		M7A-02	EG	L/G		9120	6800	8390.0	11,870	16.00	59.50	27		516		1800
		M7A-01D	EG	L/G		7340	5470	8610.0	12,190	13.00	48.00	22		542		1800
		M7A-02D	EG	L/G		9040	6740	8410.0	11,900	16.00	59.50	27		513		1800
		M7A-03D	EG	L/G		10,460	7800	7580.0	10,730	16.00	59.50	27		523		1800
		L20A	EG	G		24,838	18,522	7418.0	10,496	18.60	131.80	60		541	1500	1800
		L30A	EG	G		40,391	30,120	6340.0	8970	24.90	195.60	89		470	1500	1800
		L30A	MD	G		41,546	30,980	6164.0	8720	24.90	195.30	89		470		5600
MAN DIESEL & TURBO SE	292	THM 1304-10N	MD	L/G		14,080	10,500	8370.0	11,840	10.00	102.50	47		490	3870	9450
		THM 1304-10N	EG	L/G		13,520	10,030	8720.0	12,330	10.00	102.50	47		490		9000
		THM 1304-12N	MD	L/G		16,760	12,500	8000.0	11,320	11.00	108.00	49		515	3870	9450
		THM 1304-12N	EG	L/G		16,090	12,000	8330.0	11,790	11.00	108.00	49		515		9000
		GT6	MD	G		9250	6900	7480.0	10,590	15.00	62.00	28		451	5400	12,600
		GT6	EG	G		8890	6630	7790.0	11,030	15.00	62.00	28		451		12,920

* This engine builder is not represented in this 2013 Edition of the Global Sourcing Guide with a section description of its products.

2013 Basic Specifications — GAS TURBINE ENGINES

Manufacturer	Page Reference	Model Number	Type (EG=Electric Generator, GG=Gas Generator, MD=Mechanical Drive, MN=Marine Propulsion)	Fuel (L=Liquid, G=Gaseous)	Gross Output MW	Continuous Output at ISO Conditions		Heat Rate		Pressure Ratio	Mass Flow		Turbine Inlet Temp (°C)	Exhaust Temp (°C)	Output Shaft Speed (r/min)	
						(bhp)	(kW)	(btu/hph)	(kJ/kWh)		(lb/s)	(kg/s)			min	max
MITSUBISHI HEAVY INDUSTRIES, LTD.	171	M701DA	EG	G			144,090		10,350	14.00	972.00	441		542	3000	
		M701F4	EG	G			324,300		9027	18.00	1570.00	712		592	3000	
		M701F5	EG	G			359,000		9000	21.00	1570.00	712		611	3000	
		M701G2	EG	G			334,000		9110	21.00	1625.00	737		587	3000	
		M701J	EG	G			470,000		8783	23.00	1900.00	861		638	3000	
		M701J	EG	G			470,000		8783	23.00	1900.00	861		638	3000	
		M501DA	EG	G			113,950		10,320	14.00	763.00	346		543	3600	
		M501F3	EG	G			185,400		9740	16.00	1011.00	458		613	3600	
		M501G1	EG	G			267,500		9211	20.00	1320.00	598		601	3600	
		M501GAC	EG	G			276,000		9046	20.00	1320.00	598		617	3600	
		M501J	EG	G			327,000		8783	23.00	1320.00	598		636	3600	
		M501JAC	EG	G			310,000		8783	23.00	1320.00	598		613	3600	
		MF-61	EG	G			5925		12,570	15.00	60.00	27		496	13800	
		MF-111	EG	G			14,570		11,630	15.00	121.00	55		530	9660	
		MF-221	EG	G			30,000		11,260	15.00	238.00	108		533	7200	
		MFT-8	EG	G			26,780		9310	21.00	190.00	86		464	5000	

MITSUI ENGINEERING & SHIPBUILDING CO., LTD.	209	MSC40	EG	L/G			3513		12,900	9.70		19		443		1800
		MSC50	EG	L/G			4600		12,266	10.30		19		509		1800
		MSC60	EG	L/G			5670		11,427	12.50		22		510		1800
		MSC70	EG	L/G			7964		10,503	16.00		27		508		1800
		MSC90	EG	L/G			9459		11,296	16.30		40		465		1800
		MSC100	EG	L/G			11,347		10,934	17.40		43		484		1800
		MSC130	EG	L/G			15,000		10,227	17.10		50		496		1800
		SB5	EG	L/G			1090		14,130	10.00		5		496		1800
MTU FRIEDRICHSHAFEN GMBH	197, 327	LM2500+G4	MN	L		47,370	35,320				205.00	93		549		
		LM2500	MN	L		33,600	25,060	6855.0	9700	19.30	155.00	70		566		3600
		LM2500+	MN	L		40,500	30,200	6608.0	9350	21.50	190.00	86		516		3600
		LM2500-PH STIG	EG	L/G		36,210	27,000	5986.0	8470	19.30	167.00	76		500		3600
		LM2500-PH STIG	MD	L/G		36,210	27,000	5986.0	8470	19.30	167.00	76		500		3600
		LM2500+(PK)	EG	L/G		41,840	31,200	6440.0	9114	22.20	192.00	87		515		3600
		LM2500+(PK)	MD	L/G		41,840	31,200	6440.0	9114	22.20	192.00	87		515		3600
		LM2500-PE	MD	L/G		30,180	22,500	6784.0	9600	17.90	152.00	69		524		3600
		LM2500-PE	EG	L/G		30,180	22,500	6784.0	9600	17.90	152.00	69		524		3600
		LM6000	MD	L/G		60,346	45,000									

* This engine builder is not represented in this 2013 Edition of the Global Sourcing Guide with a section description of its products.

2013 Basic Specifications — Gas Turbine Engines

Manufacturer	Page Reference	Model Number	Type: EG=Electric Generator, GG=Gas Generator, MD=Mechanical Drive, MN=Marine Propulsion	Fuel: L=Liquid, G=Gaseous	Gross Output MW	Continuous Output at ISO Conditions (bhp)	Continuous Output at ISO Conditions (kW)	Heat Rate (btu/hph)	Heat Rate (kJ/kWh)	Pressure Ratio	Mass Flow (lb/s)	Mass Flow (kg/s)	Turbine Inlet Temp (°C)	Exhaust Temp (°C)	Output Shaft Speed (r/min) min	Output Shaft Speed (r/min) max
NPO SATURN	*			G		15,3547			14,807	18.10	798.10	1362		559		14,200
OPRA TURBINES	286	OP16-3A	EG	L/G			1850		14,125	6.70	19.20	9		573	1500	1800
		OP16-3B(DLE)	EG	L/G			1840		14,175	6.70	19.20	9		573	1500	1800
PRATT & WHITNEY POWER SYSTEMS	*	ST6L-794	MN	L		810	604	10,514.0	14,876	7.00	6.90	3		561		33,000
		ST6L-795	EG/MD	L/G		909	678	10,301.0	14,575	7.40	7.10	3		589		33,000
		ST6L-812	MN	L		984	734	10,137.0	14,343	8.00	8.20	4		542		33,000
		ST6L-813	EG/MD	L/G		1137	848	9786.0	13,846	8.50	8.60	4		566		33,000
		ST18A	EG/MD/MN	L/G		2630	1961	8425.0	11,920	14.00	17.60	8		532		18,900
		ST40	EG/MD/MN	L/G		5416	4039	7688.0	10,878	16.90	30.70	14		544		14,875
		FT8	MN	L		33,340	24,862	6632.0	9384	18.80	183.30	83		462		3600/3000
		FT8-3	MN	L		36,825	27,481	6630.0	9381	19.70	189.80	86		482		3600/3000
		FT8	MD	L/G		34,685	25,865	6612.0	9356	19.50	188.50	86		458		5500
		FT8-3	MD	L/G		37,940	28,313	6580.0	9214	20.20	193.40	88		482		5500
		FT8 POWERPAC (DLN)	EG	G			25,495		9470	19.50	186.40	85		460		3600/3000
		FT8 POWERPAC	EG	L/G			27,970		9390	20.20	193.00	88		478		3600/3000

		FT8 SWIFTPAC (DLN)	EG	G			51,325		9410	19.50	186.40	85		460		3600/3000
		FT8 SWIFTPAC	EG	L/G			56,340		9326	20.20	386.00	175		478		3600/3000
ROLLS-ROYCE	257	501-KC5	MD	L/G		5500		8495.0		9.40	34.20	16		571	13,600	
		501-KC7	MD	L/G		7400		7902.0		13.50	46.20	21		520	13,600	
		Avon2648	MD	L/G		21,923		8323.0		9.60	179.00	81		426	5500	
		Avon2656	MD	L/G		22,807		8022.0		9.60	179.00	81		420	4950	
		RB211-G62 DLE	MD	L/G		37,465		6819.0		20.60	20[illegible].00	91		501	4800	
		RB211 - G62	MD	L/G		39,075		6743.0		21.30	206.00	93		500	4800	
		RB211- GT62 DLE	MD	L/G		41,084		6602.0		21.70	210.00	95		503	4800	
		RB211 - GT62	MD	L/G		41,495		6599.0		22.00	212.00	96		501	4800	
		RB211 - GT61 DLE	MD	L/G		44,230		6306.0		21.60	207.00	94		510	4850	
		RB211 - GT61	MD	L/G		45,316		6299.0		22.10	210.00	95		510	4850	
		RB211 - H63	MD	L/G		50,848		6134.0		23.00	235.00	107		486	6000	
		RB211 - H63 WLE	MD	L/G		59,005		6247.0		25.10	254.60	116		482	6000	
		Trent 60 DLE	MD	G		70,418		5939.0		34.00	337.80	153		440	3400	
		Trent 60 WLE	MD			79,120		6074.0		35.30	358.60	163				
		AG9140	MN	L		5300	3950	8648.0		11.30	34.40			553	14,340	
		RR 4500	MN	L		6030	4500	7894.0		14.30	46.[illegible]0			515	14,600	
		Spey	MN			26,150	19,500	6826.0		21.90	147.00			458	5500	
		WR-21	MN	L		33,850	25,242	6054.0		16.20	161.00			356	3600	
		MT30 (mech or gen set)	MN	L		48,275	36,000	6403.0		24.00	257.00			474	3600	
Continues		MT30 (mech only)	MN	L		53,640	40,000	6330.0		25.70	268.00			487	3600	

* This engine builder is not represented in this 2013 Edition of the Global Sourcing Guide with a section description of its products.

2013 Basic Specifications — Gas Turbine Engines

Manufacturer	Page Reference	Model Number	Type (EG=Electric Generator, GG=Gas Generator, MD=Mechanical Drive, MN=Marine Propulsion)	Fuel (L=Liquid, G=Gaseous)	Gross Output MW	Continuous Output at ISO Conditions		Heat Rate		Pressure Ratio	Mass Flow		Turbine Inlet Temp (°C)	Exhaust Temp (°C)	Output Shaft Speed (r/min)	
						(bhp)	(kW)	(btu/hph)	(kJ/kWh)		(lb/s)	(kg/s)			min	max
ROLLS-ROYCE Continued	257	501-KB5S	EG	L/G			3897	11,747.0		10.30	33.90	15		560	14,200	
		501-KB7S	EG	L/G			5245	10,848.0		13.90	46.60	21		498	14,600	
		501-KH5	EG	L/G			6447	8509.0	8971	12.50	40.60	18		530	14,600	
		RB211-G62 DLE	EG	L/G			27,216	9387.0	9904	20.70	201.20	91		501	4800	
		RB211-GT62 DLE	EG	L/G			29,845	9089.0	9589	21.60	210.20	95		503	4800	
		RB211-GT61 DLE	EG	L/G			32,130	8681.0	9159	21.60	206.90	94		510	4850	
		RB211-H63 WLE	EG	L/G			42,473	8679.0	9157	25.10	254.60	116				
		Trent 60 DLE 50Hz	EG				51,504	8104.0	8550	33.00	334.40	152		444	3000	
		Trent 60 DLE 60Hz	EG				51,685	8138.0	8586	34.00	340.80	155		440	3600	
		Trent 60 DLE ISI 50Hz	EG				58,000	8019.0	8461	37.10	369.30	168		423	3000	
		Trent 60 DLE ISI 60Hz	EG	L/G			58,000	8001.0	8441	36.70	366.20	166		422	3600	
		Trent 60 WLE	EG	L/G			60,480	8378.0	8839	37.10	376.00	171		426	3000	
		Trent 60 WLE	EG	L/G			61,210	8328.0	8787	36.60	370.00	168		433	3600	
		Trent 60 WLE ISI	EG	L/G			64,000	8273.0	8729	38.80	394.80	179		409	3000	
		Trent 60 WLE ISI	EG	L/G			64,000	8209.0	8661	37.70	382.80	174		416	3600	

SIEMENS AG ENERGY SECTOR	*	SGT-100	EG	L/G		6770	5050	8420.0	11,914	14.00	43.10	20		545	17,384	
		SGT-100	EG	L/G		7240	5400	8210.0	11,613	15.60	45.40	21		531	17,384	
		SGT-100	MD	L/G		7640	5700	7738.0	10,948	14.90	43.40	20		543	6500	13,650
		SGT-200	MD	L/G		10,300	7680	7616.0	10,776	12.60	64.90	29		493	5700	11,525
		SGT-200	EG	L/G		9050	6750	8122.0	11,492	12.30	64.50	29		466	11,053	
		SGT-300	EG	L/G		10,600	7900	8321.0	11,773	13.70	66.60	30		542	14,010	
		SGT-300	MD	L/G		11,000	8200	7350.0	10,400	13.30	63.90	29		498	5750	12,075
		SGT-400	EG	L/G		17,300	12,900	7319.0	10,355	16.80	86.80	39		555	9500	
		SGT-400	MD	L/G		18,000	13,400	7028.0	9943	16.80	86.80	39		555	4800	10,000
		SGT-400	EG	L/G		19,311	14,400	7128.0	10,084	18.90	97.70	44		546	9500	
		SGT-400	MD	L/G		20,115	15,000	6845.0	9684	18.90	97.70	44		546	4800	10,000
		SGT-500	EG	L/G		25,615	19,100	7540.0	10,664	13.00	215.90	98		369	3600	
		SGT-500	MD	L/G		26,177	19,520	7373.0	10,432	13.00	215.90	98		369	3450	
		SGT-500	MN	L/G		22,800	17,000	7950.0	11,250	12.00	206.00	93		379	800	3600
		SGT5-2000E	EG	L/G			172,000		10,190	12.10		531		537	3000	
		SGT5-4000F	EG	L/G			295,000		9053	18.80		692		586	3000	
		SGT5-8000H	EG	L/G			375,000		8999	19.20		829		627	3000	
		SGT-600	EG	L/G		33,215	24,770	7440.0	10,533	14.00	177.00	80		543	7700	
		SGT-600	MD	L/G		34,100	25,430	7250.0	10,256	14.00	177.00	80		543	3850	8085
		SGT6-2000E	EG	L/G			11,200		10,620	12.10		365		541	3600	
		SGT6-5000F	EG	L/G			232,000		9278	18.90		551		593	3600	
		SGT6-8000H	EG	L/G			274,000		8999	19.50		604		617	3600	
Continues		SGT-700	EG	L/G		44,012	32,820	6838.0	9675	18.70	209.00	95		533	6500	

* This engine builder is not represented in this 2013 Edition of the Global Sourcing Guide with a section description of its products.

2013 Basic Specifications — Gas Turbine Engines

Manufacturer	Page Reference	Model Number	Type (EG=Electric Generator, GG=Gas Generator, MD=Mechanical Drive, MN=Marine Propulsion)	Fuel (L=Liquid, G=Gaseous)	Gross Output MW	Continuous Output at ISO Conditions (bhp)	Continuous Output at ISO Conditions (kW)	Heat Rate (btu/hph)	Heat Rate (kJ/kWh)	Pressure Ratio	Mass Flow (lb/s)	Mass Flow (kg/s)	Turbine Inlet Temp (°C)	Exhaust Temp (°C)	Output Shaft Speed (r/min) min	Output Shaft Speed (r/min) max
SIEMENS AG ENERGY SECTOR Continued	*	SGT-700	MD	L/G		42,960	33,670	6661.0	9630	18.70	209.00	95		533	3250	6825
		SGT-750	EG	L/G		48,185	35,930	6570.0	9296	23.80	249.80	113		462	6100	
		SGT-750	MD	L/G		49,765	37,110	6362.0	9002	23.80	249.80	113		462	3050	6405
		SGT-800	EG	L/G		63,698	47,500	6755.0	9557	20.40	292.80	133		541	6608	
		SGT-800	EG	L/G		67,721	50,500	6649.0	9407	21.10	295.80	134		553		
SOLAR TURBINES INCORPORATED	Gas Turbine Tab	Taurus 70	MD	L/G		10,915	8140	7205.0	10,195	16.50	57.70	26		510		12,000
		Mars 90	EG	L/G			9450		11,300	16.30	88.50	40		465	1500	1800
		Mars 90	MD	L/G		13,220	9860	7655.0	10,830	16.30	88.50	40		465		9400
		Mars 100	EG	L/G			11,350		10,935	17.70	91.80	42		485	1500	1800
		Mars 100	MD	L/G		15,900	11,860	7395.0	10,465	17.10	91.80	42		485		9500
		Titan 130	EG	L/G			15,000		10,230	17.10	109.80	50		495	1500	1800
		Titan 130 Mobile Power Unit	EG	L/G			15,000		10,230	17.10	109.80	50		495	1500	1800
		Titan 130	MD	L/G		20,500	15,290	7025.0	9940	16.10	110.30	50		505		8500
		Titan 250	EG	L/G			21,745		9260	24.10	150.40	68		465	1500	1800
		Titan 250	MD	L/G		30,000	22,370	6360.0	9000	24.10	150.40	68		465		7000
		Saturn 20	MD	L/G		1590	1185	10,370.0	14,670	6.70	14.30	7		520		22,300

		Saturn 20	EG	L/G			1210		14,795	6.70	14.40	7		505	1500	1800
		Centaur 40	MD	L/G		4700	3500	9125.0	12,905	10.30	41.80	19		445		15,500
		Centaur 40	EG	L/G			3515		12,910	9.70	41.90	19		445	1500	1800
		Centaur 50	MD	L/G		6130	4570	8500.0	12,030	10.30	41.50	19		515		16,500
		Centaur 50	EG	L/G			4600		12,270	10.60	42.10	19		510	1500	1800
		Mercury 50	EG	G			4600		9351	9.90	39.00	18		377	1500	1800
		Taurus 60	EG	L/G			5670		11,425	12.50	48.00	22		510	1500	1800
		Taurus 60 Mobile Power Unit	EG	L/G			5670		11,425	12.50	48.00	22		510	1500	1800
		Taurus 60	MD	L/G		7700	5740	7965.0	11,265	11.50	47.70	22		510		13,950
		Taurus 65	EG	L/G			6300		10,945	15.00	46.50	21		549	1500	1800
		Taurus 70	EG	L/G			7965		10,505	17.60	53.50	27		505	1500	1800
TOSHIBA CORPORATION	*		EG	L/G	255,600			10,653.0	15.6		624.00		609	5231		
URAL TURBINE WORKS	*	GT-6	EG	G			6500		51,100.0	6.2		45.00	760	405	6075	
		GT-25U	EG	G			31,400		179,300.0	13.5		124.00	1060	471	5940	
		GT-16	EG	G						11.5		85.00	920	420	5100	
YANMAR CO. LTD.	*	AT900S	EG	L/G			699					4.30			1500	1800
		AT1200S	EG	L/G			883					5.70			1500	1800
		AT1800S	EG	L/G			1397					8.60			1500	1800
		AT2700S	EG	L/G			1936					13.30			1500	1800
		AT2900S	EG	L/G			2133					13.20			1500	1800
		AT360S	EG	L/G			268								1500	1800
		AT600S	EG	L/G			449					2.90			1500	1800

* This engine builder is not represented in this 2013 Edition of the Global Sourcing Guide with a section description of its products.

2013 Basic Specifications — Gas Turbine Engines

Manufacturer	Page Reference	Model Number	Type: EG=Electric Generator, GG=Gas Generator, MD=Mechanical Drive, MN=Marine Propulsion	Fuel: L=Liquid, G=Gaseous	Gross Output MW	Continuous Output at ISO Conditions (bhp)	Continuous Output at ISO Conditions (kW)	Heat Rate (btu/hph)	Heat Rate (kJ/kWh)	Pressure Ratio	Mass Flow (lb/s)	Mass Flow (kg/s)	Turbine Inlet Temp (°C)	Exhaust Temp (°C)	Output Shaft Speed (r/min) min	Output Shaft Speed (r/min) max
ZORYA-MASHPROEKT	294	UGT2500	EG	L/G			2850		12,631.0	12		16.50	951	460	1500	3600
		UGT3000	MN	L/G			3360	8100		13.5		15.50	1020	420	9700	
		UGT3000	MD	L/G			3360	8100		13.5		15.50	1020	420	9700	
		UGT5000	EG	L/G			5250			14.0		21.50		480	1500	3600
		UGT6000	EG	L/G			6700		11,730.0	16.6		31.00	1015	428	3000	9700
		UGT6000	MN	L/G			6700		11,730.0	16.6		31.00	1015	428	3000	9700
		UGT6000	MD	L/G			6700		11,730.0	16.6		31.00	1015	428	3000	9700
		UGT6000+	EG	L/G			8300		10,907.0	16.6		33.40	1100	442	3000	7900
		UGT6000+	MN	L/G			8300		10,907.0	16.6		33.40	1100	442	3000	7900
		UGT8000	MD	L/G			8300		10,907.0	16.6		33.40	1100	442	3000	7900
		UGT10000	EG	L/G			10,500		10,000.0	19.5		36.80	1200	490	3000	6500
		UGT10000	MN	L/G			10,500		10,000.0	19.5		36.80	1200	490	3000	6500
		UGT10000	MD	L/G			10,500		10,000.0	19.5		36.80	1200	490	3000	6500

		UGT16000	EG	L/G			16,300		11,613.0	12.8		98.50	865	354	3000	5300
		UGT16000	MN	L/G			16,300		11,613.0	12.8		98.50	865	354	3000	5300
		UGT16000	MD	L/G			16,300		11,613.0	12.8		98.50	865	354	3000	5300
		UGT15000	EG	L/G			17,500		10,284.0	19.6		72.20	1075	414	3000	5300
		UGT15000	MN	L/G			17,500		10,284.0	19.6		72.20	1075	414	3000	5300
		UGT15000	MD	L/G			17,500		10,284.0	19.6		72.20	1075	414	3000	5300
		UGT25000	EG	L/G			26,700		9725.0	21.6		89.00	1245	484	3000	5000
		UGT25000	MN	L/G			26,700		9725.0	21.6		89.00	1245	484	3000	5000
		UGT25000	MD	L/G			26,700		9725.0	21.6		89.00	1245	484	3000	5000
		UGT110000	EG	L/G			114,500		9862.0	14.7		365.00	1210	520	3000	
		GTE-45	EG	L/G			47,700			14		138.50		545	3000	
		GTE-60A	EG	L/G			63,500			18		174.50		510	3000	

2013 Basic Specifications — Combined-Cycle Engines

Manufacturer	Page Reference	Combined-Cycle Model Designation	Frequency (50 and/or 60 Hz)	Base Load Rating ISO Conditions, Gas Fuel Lower Heating Value (LHV) of Fuel			Number and Model of Gas Turbines	Gas Turbine Output (kW)	Steam Turbine Output (kW)
				Continuous Output (kW)	Heat Rate (kJ/kWh)	Efficiency (%)			
ALSTOM	290	KA24-2	60	664,000	6164	58.4	2 x GT24		
		KA11N2-2	60	349,000	6950	51.8	2 x GT11N2		
		KA26-2	50	935,000	6050	59.5	2 x GT26		
		KA26-1	50	467,000	6055	59.5	1 x GT26		
		KA13E2-3	50	850,000	6679	53.9	3 x GT13E2		
		KA13E2-2	50	565,000	6691	53.8	2 x GT13E2		
		KA13E2-1	50	281,000	6729	53.5	1 x GT13E2		
		KA11N2-2	50	345,000	7018	51.3	2 x GT11N2		
ANSALDO ENERGIA	300	1AE643-CC1M	50/60	111,700	6698	53.75	1 x AE64.3A	73,600	40,200
		1AE942-CC1M	50	258,400	6775	53.14	1 x AE94.2	165,800	96,600
		1AE943-CC1M	50	456,400	6116	58.86	1 x AE94.3A	306,000	157,900
		2AE643-CC1M	50/60	223,700	6689	53.82	2 x AE64.3A	147,100	80,600
		2AE942-CC1M	50	518,000	6761	53.25	2 x AE94.2	331,600	192,900
		2AE943-CC1M	50	913,200	6114	58.88	2 x AE94.3A	610,400	317,800
EBARA CORPORATION	*	FT8 Power Pac	50/60	32,280	7394	48.7	1 x FT8	24,700	7,580

		FT8 Twin Pac	50/60	65,310	7310	49.2	2 x FT8	49,660	15,650
MAN DIESEL & TURBO SE	292	2X THM 1304-12N	50/60	35,400	7550	47.7	2 x THM 1304-12N	24,000	11,400
MITSUBISHI HEAVY INDUSTRIES, LTD.	171	MPCP1 (M701)	50	212,500	7000	51.4	1 x M701DA	142,100	70,400
		MPCP2 (M701)	50	426,600	6974	51.6	2 x M701DA	284,200	142,400
		MPCP3 (M701)	50	645,000	6947	51.8	3 x M701DA	426,300	218,700
		MPCP1 (M701F4)	50	477,900	6000	60	1 x M701F4	319,900	158,000
		MPCP2 (M701F4)	50	958,800	5981	60.2	2 x M701F4	639,800	319,000
		MPCP1 (M701F5)	50	525,000	5902	61	1 x M701F5	354,000	171,000
		MPCP2 (M701F5)	50	1,053,300	5883	61.2	2 x M701F5	708,000	345,300
		MPCP1 (M701G)	50	498,000	6071	59.3	1 x M701G2	325,700	172,300
		MPCP2 (M701G)	50	999,400	6051	59.5	2 x M701G2	651,400	348,000
		MPCP1 (M701J)	50	680,000	5835	61.7	1 x M701J	463,000	217,000
		MPCP1 (M501)	60	167,400	7000	51.4	1 x M501DA	112,100	55,300
		MPCP2 (M501)	60	336,200	6974	51.6	2 x M50DA	224,200	112,000
		MPCP3 (M501)	60	506,200	6947	51.8	3 x M501DA	336,300	169,900
		MPCP1 (M501F)	60	285,100	6305	57.1	1 x M501F3	182,700	102,400
		MPCP2 (M501F)	60	572,200	6283	57.3	2 x M501F3	365,400	206,800
		MPCP1 (M501G)	60	398,900	6165	58.4	1 x M501G1	264,400	134,500
		MPCP2 (M501G)	60	800,500	6144	58.6	2 x M501G1	528,800	271,700
		MPCP1 (M501GAC)	60	412,400	6051	59.5	1 x M501GAC	273,600	138,800
Continues		MPCP2 (M501GAC)	60	826,100	6041	59.6	2 x M501GAC	547,200	278,900

* This engine builder is not represented in this 2013 Edition of the Global Sourcing Guide with a section description of its products.

2013 Basic Specifications — Combined-Cycle Engines

Manufacturer	Page Reference	Combined-Cycle Model Designation	Frequency (50 and/or 60 Hz)	Base Load Rating ISO Conditions, Gas Fuel Lower Heating Value (LHV) of Fuel			Number and Model of Gas Turbines	Gas Turbine Output (kW)	Steam Turbine Output (kW)
				Continuous Output (kW)	Heat Rate (kJ/kWh)	Efficiency (%)			
MITSUBISHI HEAVY INDUSTRIES, LTD. Continued	171	MPCP1(M501JAC)	60	450,000	< 5903	> 61.0	1 x M501JAC	310,000	140,000
		MPCP1 (M501J)	60	470,000	5854	61.5	1 x M501J	322,000	148,000
		MPCP2 (M501J)	60	942,900	5835	61.7	2 x M501J	644,000	298,900
MITSUI ENGINEERING & SHIPBUILDING CO., LTD.	209	MACS60	50/60	7720		41.88674843	1*MACS60	5670	2050
		MACS70	50/60	10,474		44.1372772	1*MACS70	7964	2510
		MACS90	50/60	12,579		41.30308457	1*MACS90	9459	3120
		MACS100	50/60	14,917		42.35019397	1*MACS100	11,347	3570
		MACS130	50/60	19,490		44.79172652	1*MACS130	15,000	4490
PRATT & WHITNEY POWER SYSTEMS	*	FT8 POWERPAC	50/60	32,910	7243	49.7	1 FT8	24,737	8755
		FT8 POWERPAC	50/60	36,570	7121	50.6	1 FT8-3	27,220	10,006
		FT8 SWIFTPAC	50/60	66,745	7143	50.4	2 FT8	49,828	18,020
		FT8 SWIFTPAC	50/60	74,185	7022	51.3	2 FT8-3	54,840	20,597
RABA ENGINES LTD.	*	D - CHP 160	50	160	215	85	x		
		G - CHP 130	50	130	180	85	x		
ROLLS-ROYCE * WATER INJECTED π SINGLE PRESSURE STEAM CYCLE, SUPPLEMENTARY FIRED TO 730C (1346F) τ SINGLE PRESSURE STEAM CYCLE, SUPPLEMENTARY FIRED TO	257	RB211-G62 DLE	50/60	37,725	7175	50.2	1 x RB211	26,716	# 12,045
		RB211-GT62 DLE	50/60	39,760	7005	51.4	1 x RB211	28,626	# 12,205
		RB211-GT61 DLE	50/60	42,640	6820	52.8	1 x RB211	31,171	# 12,593
		RB211-H63 WLE*	50/60	54,019	7085	50.8	1 x RB211	40,935	# 14,189

750C (1382F) # DUAL PRESSURE STEAM CYCLE		RB211-H63 WLE*	50/60	68,398	7428	48.5	1 x RB211	40,935	π 29,125
		Trent 60 DLE	50	64,232	6837	52.7	1 x Trent	50,068	# 15,261
		Trent 60 DLEISI*	50	73,109	6790	53	1 x Trent	58,000	# 16,289
		Trent 60 DLE	50	89,482	7172	50.2	1 x Trent	50,068	τ 41,348
		Trent 60 DLEISI*	50	101,693	7245	49.7	1 x Trent	58,000	τ 46,004
		Trent 60 DLE	60	64,601	6855	52.5	1 x Trent	50,492	# 15,211
		Trent 60 DLEISI*	60	72,921	6785	53.1	1 x Trent	58,000	# 16,097
		Trent 60 DLE	60	90,326	7191	50.1	1 x Trent	50,492	τ 41,791
		Trent 60 DLEISI*	60	100,955	7243	49.7	1 x Trent	58,000	τ 45,245
		Trent 60 WLE*	50	75,183	7109	50.6	1 x Trent	59,486	# 17,016
		Trent 60 WLEISI*	50	76,969	7200	50	1 x Trent	62,214	# 16,128
		Trent 60 WLE*	50	102,723	7389	48.7	1 x Trent	59,486	π 45,616
		Trent 60 WLEISI*	50	107,615	7481	48.1	1 x Trent	62,214	π 47,948
		Trent 60 WLE*	60	75,257	7056	51	1 x Trent	59,474	# 17,104
		Trent 60 WLEISI*	60	77,233	7126	50.5	1 x Trent	62,452	# 16,156
		Trent 60 WLE*	60	101,899	7340	49	1 x Trent	59,474	π 44,776
		Trent 60 WLEISI*	60	107,499	7432	48.4	1 x Trent	62,452	π 47,583
SIEMENS AG ENERGY SECTOR Continues	*	SCC-700 2X1	50/60	88,976	6857	52.7	2 x	62,240	26,740
		SCC-700 1X1 (dual pressure no reheat)	50/60	44,350	6884	52.3	1 x	31,210	13,140

* This engine builder is not represented in this 2013 Edition of the Global Sourcing Guide with a section description of its products.

2013 BASIC SPECIFICATIONS — COMBINED-CYCLE ENGINES

Manufacturer	Page Reference	Combined-Cycle Model Designation	Frequency (50 and/or 60 Hz)	Base Load Rating ISO Conditions, Gas Fuel Lower Heating Value (LHV) of Fuel			Number and Model of Gas Turbines	Gas Turbine Output (kW)	Steam Turbine Output (kW)
				Continuous Output (kW)	Heat Rate (kJ/kWh)	Efficiency (%)			
SIEMENS AG ENERGY SECTOR Continued	*	SCC-800 2X1 (dual pressure no reheat)	50/60	135,100	6618	54.4	2 x	92,100	44,400
		SCC-800 1X1 (dual pressure no reheat)	50/60	66,500	6703	53.7	1 x	46,000	21,400
		SCC5-2000E 1X1	50	253,000	6857	52.5	1 x	168,000	89,000
		SCC5-2000E 2X1	50	512,000	6780	53.1	2 x	336,000	184,000
		SCC5-4000F Single Shaft	50	431,000	6133	58.7	1 x		
		SCC5-4000F 2X1	50	862,000	6133	58.7	2 x	578,000	296,000
		SCC5-8000H Single Shaft	50	570,000	< 6000	> 60	1 x		
		SCC5-8000H 2x1	50	1,144,000	< 6000	> 60	2 x	75,000	390,000
		SCC6-2000E 1X1	60	171,000	7018	51.3	1 x	114,000	60,000
		SCC6-2000E 2X1	60	342,000	6923	52	2 x	228,000	12,000
		SCC6-5000F 1X1	60	345,000	6250	57.6	1 x	232,000	118,000
		SCC6-5000F 2X1	60	690,000	6205	58	2 x	464,000	236,000
		SCC6-8000H Single Shaft	60	410,000	< 6000	> 60	1 x		

		SCC6-8000H 2x1	60	820	< 6000	> 60	2x	584,000	270,000
		SCC-600 2X1 (dual pressure no reheat)	50/60	73,280	7071	50.9	2 x	47,780	26,450
		SCC-600 1X1 (dual pressure no reheat)	50/60	35,900	7220	49.9	1 x	23,880	12,600
TOSHIBA CORPORATION	*		50/60	117,700-480,000	6000-7175	50.2-60.0		75,100-254,100	44,000-148,500
URAL TURBINE WORKS	*	CCPP-450	50	450	7243		x CT-150-8	150,000	150,000
		CCPP-115	50	115			x P-38	77,000	38,000
		CCPP-230	50	230			T-53/67-8,0 (T-48/62-7,4); T-56/70-8,8; K-74-6,8; T-57/69-8,8, T-83/110-6,8	170,000	53,000
		CCPP-325	50	325			K-100-6,8/T-78/96-6,8	110,000	100,000
ZORYA-MASHPROEKT	294	UGT10000CC1	50	13,500		45.8	1 x UGT10000	10,000	3500
		UGT15000CC1	50	22,700		45.3	1 x UGT15000	16,000	6700
		UGT25000CC1	50	34,700		47.5	1 x UGT25000	25,000	9700
		UGT110000CC1	50	160,000		50.5	1 x UGT110000	110,000	50,000
		UGT10000CC20	50	27,500		46.5	2 x UGT10000	20,000	7500
		UGT15000CC2	50	45,800		45.9	2 x UGT15000	32,000	13,800
		UGT25000CC2	50	70,600		48.5	2 x UGT25000	50,000	20,600
		UGT110000CC2	50	325,000		52	2 x UG110000	220,000	105,000

* This engine builder is not represented in this 2013 Edition of the Global Sourcing Guide with a section description of its products.

Appendix A-2

MANUFACTURER	Elliott Energy Systems, Inc	Capstone Turbine Corporation	Capstone Turbine Corporation	Ingersoll-Rand Energy Systems[1]
MODEL	TA 100 CHP	C60	C30	PowerWorks250
TYPE	RECUP	RECUP	RECUP	RECUP
Power-kW	100	60	30	250
Power KVA	480 VAC	480 VAC	480 VAC	Note[2]
COMPR PRESS RATIO	4:1	Note[2]	Note[2]	4:1
Rotor Speed - rpm	Note[2]	96,000	96,000	45,000
Cooling				
Engine	Oil Cooled	Air Cooled	Air Cooled	Oil Cooled
Alternator	Oil Cooled	Air Cooled	Air Cooled	Oil Cooled
Heat Rate	12,355 Btu/kWhr	12,200 Btu/kWhr	13,100 Btu/kWhr	Note[2]
Exhaust Gas				
Temp	450°F	580°F	530°F	Note[2]
Flow	1.7 lbs/sec	1.07 lbs/sec	0.68 lbs/sec	1.84 kgs/sec
Total Weight w Encl				
Recuperated	1890 lbs	1671 lbs	1052 lbs	9000 lbs

[1]"PowerWorks 250-kWe Microturbines" Presented by Jim Kesseli at IGTI June 18, 2003
[2]Data not available at time of printing

Appendix B

Accessory Manufacturers

The following lists the manufacturers of some of the more common accessory and auxiliary equipment found in gas turbine packages. Also included in the list are sources for replacement gas turbine parts and components.

This list is separated into the following categories:

- Acoustic Enclosures
- Air/Oil Coolers
- Bearings
- Boroscopes
- Castings
- Coatings
- Component Parts
- Compressor Wash
- Control Instruments
- Control Systems
- Couplings
- Expansion Joints
- Filters (Fuel, Oil)
- Fuel Nozzles
- Fuel Treatment
- Inlet Air Cooling
- Inlet Air Filters
- Gearboxes
- Lubrication Systems
- Pumps
- Selective Catalytic Reduction
- Starters
- Switchgear
- Vibration Equipment

The list is offered only as a guide and should not be considered a complete list.

ACOUSTIC ENCLOSURES

Boet American Company
8895-T N. Military Trail
Suite 305 C
Palm Beach Gardens, FL 33410-6220
Phone: 407-622-7113
Fax: 407-622-7170

Braden Manufacturing
P.O. Box 1229
Tulsa, OK 74101
Phone: 918-272-5371
Fax: 918-272-7414

Central Metal Fabricators
900 S.W. 70 Avenue
Miami, FL 33144
Phone: 305-261-6262
Fax: 305-261-2156

Consolidated Fabricators, Inc.
17 St. Mark Street
Auburn, MA 01501
Phone: 508-832-9686
Fax: 508-832-7369

I.D.E. Processes Corp.,
Noise Control Div.
106-T 81st Ave.
Kew Gardens, NY 11415 1108
Phone: 718-54401177
Fax: 718-575-8050

Industrial Acoustics Company
1160 Commerce Avenue
Bronx, NY 10462
Phone: 718-931-8000
Fax: 718-863-1138

McGuffy Systems, Inc.
18635 Teige Road
Cypress, TX 77429
Phone: 281-255-6955
Fax: 281-351-8502

SPL Control, Inc.
Dave Potipcoe
1400 Bishop Street
P.O. Box 2252
Cambridge, Ontario,
Canada N1R 6W8
Phone: 519-623-6100
Fax: 519-623-7500

AIR/OIL COOLERS

Air-X-Changers
P.O. Box 1804
Tulsa, OK 74101
Phone: 918-266-1850
Fax: 918-266-1322

Alfa Laval Thermal Inc.
5400 International Trade Dr.
Richmond, VA 23231
Phone: 804-236-1361
Fax: 804-236-1360

GEA Rainey Corp.
5202 W. Channel Rd.
Catoosa, OK 74015
Phone: 918-266-3060
Fax: 918-266-2464

Hayden Industrial Division
1531 Pomona Road
P.O. Box 0848
Corona, CA 91716-0848
Phone: 909-736-2618
Fax: 909-736-2629

BEARINGS

Kaman Industrial
Technologies Corp.
1 Waterside Crossing
Windsor, CT 06095
Phone: 800-526-2626
Fax: 860-687-5170

Kingsbury Inc.
10385 Drummond Road
Philadelphia, PA 19154 3803
Phone: 215-824-4000
Fax: 215-824-4999

KMC Bearing Manufacturing & Repa...
8525 West Monroe Street
Houston, TX 77061
Phone: 713-944-1005
Fax: 713-944-3950

Magnetic Bearings Inc.
5241 Valleypark Drive
Roanoke, VA 24019
Phone: 703-563-4936
Fax: 703-563-4937

Magnolia Metal Corporation
One Magnolia Center,
P.O. Box 19110
Omaha, NE 68119
Phone: 800-852-9670
Fax: 402-455-8762

Milwaukee Bearing &
Machining, Inc.
W 134 N 5235 Campbell Dr.
Menomonee Falls, WI 53051
Phone: 262-783-1100
Fax: 262-783-1111

Odessa Babbitt Bearing Co.
6112 W. County Rd.
Odessa, TX 76764-3497
Phone: 915-366-2836
Fax: 915-366-4887

Orion Corporation
1111 Cedar Creek Road
Grafton, WI 53024
Phone: 262-377-2210
Fax: 262-377-0729

Pioneer Motor Bearing Company
129 Battleground Rd.
Kinds Mountain, NC 28086
Phone: 704-937-7000
Fax: 704-937-9429

Rexnord Corp.
P.O. Box 2022
Milwaukee, WI 53201
Phone: 800-821-1580
Fax: 414-643-3078

Waukesha Bearings Corp.
P.O. Box 1616
Waukesha, WI 53187-1 61 6
Phone: 262-506-3000
Fax: 262-506-3001

BOROSCOPES

Machida, Inc.
40 Ramland Road South
Orangeburg, NY 10962-2698
Phone: 800-431-5420
Fax: 845-365-0620

Olympus America Inc.
Two Corporate Center Drive
Melville, NY 11747
Phone: 800-455-8236
Fax: 631-844-5112

Videodoc Inc.
805 Center Street
Deer Park, TX 77536
Phone: 713-476-0470
Fax:

Lighting Products Div.,
Welch Allyn, Inc.
4619 Jordan Road
Skaneateles Falls, NY 13153 0187
Phone: 315-685-4347
Fax: 315-685-2854

CASTINGS

Charles W. Taylor & Son Ltd.
North Eastern Foundry
Templetown, South Shields
Tyne Wear NE33 5SE, UK
Phone: (4419)-14555507
Fax: (4419)-14560700

Chromalloy American Corp.,
Turbine Airfoil Div.
1400 N. Cameron St.
Harrisburg, PA 17103 1095
Phone: 717-255-3425
Fax: 717-255-3448

Chromally Connecticut
22 Barnes Industrial Rd.
P.O. Box 748
Wallingford, CT 06492 0748
Phone: 203-265-2811
Fax: 203-284-9825

Howmet Corporation
475 Steamboat Road
Greenwich, CT 06903
Phone: 203-661-4600
Fax: 203-625-8796

INCO Engineered Products, Inc.
Pond Meadow Rd.
Ivoryton, CT 06442
Phone: 860-767-0161
Fax: 860-767-3093

PCC Airfoils Inc.
25201 Chagrin Blvd., Suite 290
Beachwood, OH 44122 5633
Phone: 216-831-3590
Fax: 216-766-6217

Pennsylvania Precision
Cast Parts, Inc.
521 N. 3rd Ave.
P.O. Box 1429
Lebanon, PA 17042
Phone: 717-273-3338
Fax: 717-273-2662

Utica Corp.
2 Halsey Rd.
Whitesboro, NY 13492
Phone: 315-768-8070
Fax: 315-768-8005

COATINGS

Aerobraze Corporation
940 Redna Terrace
Cincinnati, Oh 45215
Phone: 513-772-1461
Fax: 513-772-0149

Englehard Corporation Surface Te…
12 Thompson Road
East Windsor, CT 06088
Phone: 860-623-9901
Fax: 860-634-657

Fusion, Inc.
6911 Fulton
Houston, TX 77022
Phone: 713-691-6547
Fax: 713-691-1903

General Plasma
Derlan Industries Ltd.
12-A Thompson Rd.
East Windsor, CT 06088
Phone: 860-623-9910
Fax: 860-623-4657

G.T.C. Gas Turbines Ltd.
Dundee, DD4 ONZ
Scotland
Phone: 441-382-73931 1
Fax: 441-382-736019

Howmet Corp.,
Quality & Technical
Services Laboratories (QTS)HIP
1600-T South Warner St.
Whitehall, MI 49461
Phone: 231-894-7475
Fax: 231-894-7499

Ivar Rivenaes A/S
Damsgardsvel 35
Bergen, 5000, Norway
Phone: 475-290-230
Fax: 475-296-325

Liburdi Engineering Ltd.
400 Highway 6 N.
Dundas, ON L9H 7K4, CAN
Phone: 905-689-0734
Fax: 905-689-0739

Progressive Technologies, inc.
4201 Patterson S.E.
Grand Rapids, MI 49512-4015
Phone: 616-957-0871
Fax: 616-957-3484

Sermatech Technical Services
155 S. Limerick Rd.
Limerick, PA 19468
Phone: 610-948-5100
Fax: 610-948-1729

Treffers Precision, Inc.
(A Praxair Surface
Technologies Co.)
1021-T N. 22nd Ave.
Phoenix, AZ 85009 3717
Phone: 602-744-2600
Fax: 602-744-2660

Turbine Metal Technology
7237 Elmo Street
Tujunga, CA 91042
Phone: 818-352-8721
Fax: 818-352-8726

Walbar Metals Inc.
Peabody Division
Peabody Industrial Center
Box T
Peabody, MA 01961-3369
Phone: 978-532-2350
Fax: 978-532-6867

COMPONENT PARTS

Bet-Shemesh Engines Ltd.
West Industrial Zone
Bet-Shemesh 99000
Israel
Phone: 9722909206
Fax: 9722911970

Birken Turbine Controls, Inc.
5 Old Windson Road
Bloomfield, CT 06002
Phone: 860-242-0448
Fax: 860-726-1981

Enginetics Brake Corp.
7700 New Carlisle Pike
Huber Heights, OH 45424
Phone: 937-754-3260
Fax: 937-754-6505

H&E Machinery Inc.
334 Comfort Road
Ithaca, NY 14850
Phone: 607-277-4968
Fax: 607-277-1193

INCO Engineered Products Inc.
Turbo Products International Div.
110 Pond Meadow Road
Ivoryton, CT 06442
Phone: 860-767-0161
Fax: 860-767-3093

Sabo International Inc.
1010 Farmington Avenue
P.O. Box 270403 West
Hartford, CT 06127-0403
Phone: 203-231-7800
Fax: 203-231-7778

Turbo Parts
P.O. Box 1257
606 Pierce Road
Clifton Park, NY 12065
Phone: 518-877-3056
Fax: 518-877-0921

COMPRESSOR WASH

Brent America, Inc.
921 Sherwood Drive
Lake Bluff, IL 60044
Phone: 847-295-1660
Fax: 847-295-8748

Brent Industrial PLC
Industrial Division
Ridgeway, Iver
Buckinghamshire SLO 9JJ, UK
Phone: (4475)-3651812
Fax: (4475)-3630314

Rochem Industrial
Rochem Technical Services, Ltd.
2185 South Dixie Avenue
Suite 214
Kettering, OH 45409
Phone: 513-298-6841
Fax: 513-298-8390

ROTRING Engineering GmbH
Markstr. 2
P.O. Box 1511
Buxtehude, 21614, Germany
Phone: 494-16-174090
Fax: 494-16-1740999

Spraytec Inc.
P.O. Box 676
Brookfield, CT 06804
Phone: 203-775-2802
Fax: 203-775-9339

Turbotect Ltd
P.O. Box, CH-5401
Baden, Switzerland
Phone: 41-562-005020
Fax: 41-562-005022

CONTROL INSTRUMENTS

Action Instruments
8601 Aero Drive
San Diego, CA 92123
Phone: 858-279-5726
Fax: 858-279-6290

Amot Controls Corporation
401 First Street
Richmond, CA 94801
Phone: 510-236-8300
Fax: 510-223-4950

Dwyer Instruments Inc.
P.O. Box 373
Michigan City, IN 46361
Phone: 219-879-8000
Fax: 219-872-9057

Fisher Rosemount
8301 Cameron Road
Austin, TX 78754
Phone: 512-832-2190
Fax: 512-418-7505

Moore Products Company
Sumneytown Pike
Springhouse, PA 19477
Phone: 215-646-7400
Fax: 215-283-6358

National Instruments
11500 N. Mopac Expwy., Bldg. B
Austin, TX 78759
Phone: 800-433-3488
Fax: 512-794-8411

Newport Electronics Inc.
2229 S. Yale Street
Santa Ana, CA 92704-4426
Phone: 714-540-4914
Fax: 714-546-3022

CONTROL SYSTEMS

Bailey Controls
29801 Euclid Avenue
Wickliffe, OH 44092
Phone: 440-585-8500
Fax: 440-585-8756

Continental Controls Corp
8845 Rehco Rd.
San Diego, CA 92121
Phone: 619-453-9880
Fax: 858-453-5078

Fisher-Rosemount Systems, Inc.
8301 Cameron Rd.
Austin, TX 78754 3827
Phone: 512-835-2190
Fax: 512-834-7313

Foxboro Measurement & Instruments Div.
33 Commercial St.
Foxboro, MA 02035
Phone: 888-369-2676
Fax: 508-549-6750

Honeywell
Industrial Automation and Control
16404 North Black Canyon Hwy.
Phoenix, AZ 85023
Phone: 602-863-5000
Fax: 602-313-5121

HSDE, Inc.
A Subsidiary of
VO/PER THORNYCRO...
14900 Woodham Drive
Suite A-1 30
Houston, TX 77073
Phone: 713-821-8448
Fax: 713-821-8442

Innovative Control Systems, Inc.
Chaucer Square
Route 9
Clifton Park, NY 12065
Phone: 518-383-8078
Fax: 518-383-8082

Yokogawa Industrial Automation
4 Dart Road
Newnan, GA 30265
Phone: 770-254-0400
Fax: 770-251-6427

Moog Controls
300 Jamison Road
East Aurora, NY 14052
Phone: 716-655-2000
Fax: 716-655-1803

Petrotech
108 Jarrell Dr.
P.O. Box 503
Belle Chasse, Louisiana 70037
Phone: 504-394-5500
Fax: 504-394-6117

Precision Engine Controls Corpor…
11661 Sorrento Valley Road
San Diego, CA 92121
Phone: 858-792-3217
Fax: 858-792-3200

Power & Compression Systems
P.O. Box 3028
Mission Viejo, CA 92691
Phone: 949-582-8545
Fax: 949-582-8992

Rockwell Automation
1201 South Second Street
Milwaukee, Wisconsin 53204-2496
Phone: 414-382-2000
Fax. 414-382-4444

Rosemount Inc.
1201 North Main Street
Orrville, OH 44667-0901
Phone: 216-682-9010

Woodward Governor Company
Engine & Turbine Controls division
5001 N. Second St., P.O. Box 7001
Rockford, IL 61125 7001
Phone: 815-877-7441
Fax: 815-639-6033

COUPLINGS

American VULKAN Corp.
P.O. Drawer 673
2525 Dundee Rd.
Winter Haven, FL 33882-0673
Phone: 863-324-2424
Fax: 863-324-4008

Coupling Corporation of America
2422 S. Queen Street
York, PA 17402-4995
Phone: 800-394-3466
Fax: 717-741-0886

Euroflex Transmissions
India Pvt…
Plot No. 99, C.I.E., Phase - II
Gandhi Nagar, Bala Nagar
Hyderabad, 500 037, India
Phone: 914-027-9775
Fax: 914-027-9523

Flexibox International Ltd.
Flexibox Ltd.
Nash Road Trafford Park
Manchester, M17 1SS, England
Phone: 446-187-22482
Fax: 446-187-21654

Flexibox, Inc.
2407 Albright
Houston, TX 77017
Phone: 713-944-6690
Fax: 713-946-8252

Kop-Flex Emerson Power Transmission Corp.
P.O. Box 1696
Baltimore, MD 21203
Phone: 410-768-2000
Fax: 410-787-8424

Lucas Aerospace Power Transmission
211 Seward Avenue
P.O. Box 457
Utica, NY 13503-0457
Phone: 315-793-1419
Fax: 315-793-1415

Marland Clutch
Division of Zurn Industries
650 E. Elm Street
La Grange, IL 60525-0308
Phone: 708-352-3330
Fax: 708-352-1403

N.B. ESCO Transmissions S.A.
Culliganlaan, 3
Diegen, B-1831 Belgium
Phone: (32)272-04-880
Fax: (32)272-12-827

Ovado Couplings USA
1447-T New Litchfield St.
Torrington, CT 06790
Phone: 860-489-1817
Fax: 860-489-3867

Rexnord Corp.
P.O. Box 2022
Milwaukee, WI 53201
Phone: 800-821-1580
Fax: 414-643-3078

Shackelford Wattner/Texas Custom...
5824 Waltrip Street
Houston, TX 77087
Phone: 713-644-5595
Fax: 713-644-6334

SSS Clutch Company
610 West Basin Road
New Castle, DE 19720
Phone: 302-322-8080
Fax: 302-322-8548

Voith Transmissions Inc.
25 Winship Road
York, PA 17402
Phone: 717-767-3200
Fax: 717-767-3210

Voith Turbo GmbH
Voithstresse 1
P.O. Box 1555
74564 Crailsheim, Germany
Phone: 079-51-7320
Fax: 079-51-732500

Zurn Industries, Inc.
1801 Pittsburgh Avenue
Erie, PA 16514
Phone: 814-480-5100
Fax: 814-453-5891

EXPANSION JOINTS

Bachmann Industries, Inc.
416 Lewiston Junction Rd.
Auburn, ME 04211-2150
Phone: 207-784-1903
Fax: 207-784-1904

Badger Industries
101 Badger Dr.
Zelienople, PA 16063
Phone: 724-452-4500
Fax: 724-452-0802

Croll-Reynolds Engineering Co.
2400 Reservoir Avenue
Trumbull, CT 06611
Phone: 203-371-1983
Fax: 203-371-0615

EFFOX, Inc.
9759 Inter Ocean Drive
Cincinnati, OH 45246
Phone: 513-874-8915
Fax: 513-874-1343

Expansion Joint Systems, Inc.
10035-TR Prospect Avenue
Suite 202
Santee, CA 92071
Phone: 800-482-2808
Fax: 619-562-0636

Flextech Engineering Inc.
1142 E., Route 66. P.O. Box 725
Glendora, CA 91740
Phone: 626-914-2494
Fax: 626-914-1404

J.M. Clipper Corporation
P.O. Drawer 2340
Nacogdoches, TX 75963-2340
Phone: 409-560-8900
Fax: 409-560-8998

KE Burgamnn A/S
Expansion Joint Divson
Park Alle 34
6600 Vejen
Denmark
Phone: 457-536-181 1
Fax: 457-536-1532

McGuffy Systems, Inc.
18635 Teige Road
Cypress, TX 77429
Phone: 281-255-6955
Fax: 281-351-8502

Pathway Bellows Inc.
115 Franklin Road
P.O. Box 3027
Oak ridge, TN 37831
Phone: 423-483-7444
Fax: 423-482-5600

Senior Flexonics, Inc.
Powerflex Expansion Joint Division
Hertford Road
Waltham Cross
Hertfordshire EN8 7Yd, England
Phone: (0199)-2788557
Fax: (0199)-2788554

Senior Flexonics, Inc.
Expansion Joint Division
2400 Longhorn Industrial Dr.
New Braunfels, TX 78130 2530
Phone: 800-332-0080
Fax: 830-629-6899

Wahlco Metroflex
29 Lexington St.
Lewiston, ME 04240
Phone: 800-272-6652
Fax: 207-784-1338

FILTERS (FUEL, OIL)

Alfa Laval Inc.
4405 Cox Rd., Ste. 130
Glen Allen, VA 23060
Phone: 866-253-2528

Allen Filters
P.O. Box 747
Springfield, MO 65801
Phone: 800-865-3208
Fax: 417-865-2469

Westfalia Separator, Inc.
100 Fairway Court
Northvale, NJ 07647
Phone: 201-767-3900
Fax: 201-767-3416

Contect Inc.
332 Federal Road
Brookfield, CT 06804
Phone: 203-775-8445
Fax: 203-775-9339

Ecolochem, Inc.
4545 Patent Road
Norfolk, VA 23502
Phone: 804-959-0640
Fax: 757-855-1478

Industrial Filter Manufacturers
10244 Hedden Road
Evansville, IN 47725
Phone: 812-867-4730
Fax: 812-867-4744

North American Filter Corp. (NAFCO)
200 Westshore Blvd.
Newark, NY 14513
Phone: 800-265-8943
Fax: 315-331-3730

Pall Process Filtration Co.
2200 Northern Blvd.
East Hills, NY 11548-1289
Phone: 516-484-5400
Fax: 516-484-5228

Southeastern Metal Products Div., Pneumafil Corp.
P.O. Box 16348, 1420 Metals Dr.
Charlotte, NC 28297 8804
Phone: 704-596-4017
Fax: 704-596-3844

FUEL NOZZLES

Fuel Systems Textron
700 North Centennial Street
Zeeland, MI 49464
Phone: 616-772-9171
Fax: 616-772-7322

Gas Turbines Fuel Systems Ltd.
Unit E22
Wellheads Industrial Centre
Dyce, Aberdeen, AB2 OGA Scotland
Phone: 012-247-771133
Fax: 012-247-725275

Natole Turbine Enterprises
P.O. Box 1167
La Porte, TX 77572
Phone: 713-470-9226
Fax: 713-470-9676

Parker Hannifin Corp.
6035 Parkland Blvd.
Cleveland, OH 44124
Phone: 800-272-7537
Fax: 440-266-7400

P.T. H&E Turbo Indonesia
JI Raya Narogong Km, 15
Ciketing Udik, Bantar Gebang
Bekasi 17310, West Java
Indonesia
Phone: 021-823-0718
Fax: 021-823-0131

FUEL TREATMENT

ALFA LAVAL Power Business Unit
955 Mearns Rd.
Warminster, PA 18974-0556
Phone: 800-975-2532
Fax: 215-957-4859

Fuel Systems Textron
700-T N. Centennial
Zeeland, MI 49464
Phone: 616-772-9171
Fax: 616-772-7322

Howmar International Ltd.
Albany Park Estate, Frimley Rd.
Camberley
Surrey, GU15 2QQ
England
Phone: 012-766-81101
Fax: 012-766-81107

Petrolite Corporation
5455 Old Spanish Trail
Houston, TX 77023
Phone: 713-926-1162

ROTRING Engineering GmbH
Markstr. 2
P.O. Box 1511
Buxtehude, 21614, Germany
Phone: 494-16-174090
Fax: 494-16-1740999

GEARBOXES

BHS Sonthofen
Hans-Boeckler-Str. 7
Sonthofen, Allgaeu 87527, Germany
Phone: 083-21-8020
Fax: 083-21-802689

Cincinnati Gear Company
5657 Wooster Pike
Cincinnati, OH 45227
Phone: 513-271-7700
Fax: 513-271-0049

Flender-Graffenstaden
Rue du Vieux Moulin
BP 84 -1
F67402 Illkrich Cedex
France
Phone: 338-867-6000
Fax: 338-867-0617

Kreiter Geartech
2530 Garrow Street
Houston, TX 77003
Phone: 713-237-9793
Fax: 713-237-1207

Lufkin Industries, Inc.
407 Kiln Street
P.O. Box 849
Lufkin, TX 75902-0849
Phone: 409-637-5612
Fax: 409-637-5883

MAAG Gear Company, Ltd.
P.O. Box CH-8023
Zurich, Switzerland
Phone: 41-012-787878
Fax: 41-012-787880

Philadelphia Gear corporation
181 South Gulph Road
King of Prussia, PA 19406
Phone: 610-265-3000
Fax: 610-337-5637

Renk AG
Gogginger Strasse 73
D-86159, Augsburg
Phone: 082-157-00534
Fax: 082-157-00460

Westech Gear Corp.
Division of Philadelphia Gear Corp.
2600 East Imperial Highway
Lynwood, CA 90262
Phone: 310-605-2600
Fax: 310-898-3592

INLET AIR COOLING

Atomizing Systems Inc.
One Hollywood Ave.
Ho Ho Kus, NJ 07423 1433
Tel: 888-265-3364
Fax: 201-447-6932

Baltimore Aircoil Co.
7595 Montivideo Rd.
Jessup, MD 20794
Tel: 410-799-6200
Fax: 410-799-6416

Fogco Systems, Inc.
1051 N. Fiesta Blvd.
Gilbert, AZ 85233
Tel: 480-507-6478
Fax: 480-838-2232

HRT - Power L.P.
P.O. Box 2728
Conroe, TX 77305
Tel: 713-827-0001
Fax: 713-827-0002

Mee Industries
204 West Pomona Ave.
Monrovia, CA 91016
Phone 800-732-5364
Fax 626-359-4660

Munters Corp.
Marketing Dept.,
79 Monroe St.,
P.O. Box 640
Amesbury, MA 01913 3204
Tel: 978-241-1100
Fax: 978-241-1215

Nortec Industries Inc.
826 Proctor Ave., P.O. Box 698
Ogdensburg, NY 13669
Tel: 315-425-1255
Fax: 613-822-7964

Premier Industries, Inc.
P.O. Box 37136
Phoenix, AZ 85069
Tel: 800-254-8989
Fax: 602-997-5998

Refrigeration Resources Co.,
Div. of Mechanical Resources, Inc.
360 Crystal Run Rd.
Middletown, NY 10941
Phone: 800-448-7716
Fax: 845-692-9032

INLET AIR FILTERS

AAF International
P.O. Box 35690
10300 Ormsby Park Place
Suite 600
Louisville, KY 40223-6169
Phone: 502-637-0011
Fax: 502-637-0321

Atomizing Systems Inc.
One Hollywood Ave.
Ho Ho Kus, NJ 07423 1433
Phone: 888-265-3364
Fax: 201-447-6932

Baltimore Aircoil Co.
7595 Montivideo Rd.
Jessup, MD 20794
Phone: 410-799-6200
Fax: 410-799-6416

Dollinger Corporation
PO Box 537
Stanley, NC 28164
Phone: 540-726-2500
Fax: 540-726-2577

Donaldson Company Inc., Gas Turbine Systems
1400 W. 94th St., P.O. Box 1299
Minneapolis, MN 55431 2370
Phone: 952-887-3543
Fax: 952-887-3843

Farr Company
2221 Park Place,
El Segundo, CA 90245
Phone: 310-727-6300
Fax: 800-643-9086.

Filteco International N.V.
Steenweg op Brussels 365
B3090 Overijse, Belgium
Phone: (32)026-68-0520
Fax: (32)026-68-0801

Flanders Filters inc.
531 Flanders Filters Rd.
Washington, NC 27889 1708
Phone: 252-946-8081
Fax: 252-946-3425

Fogco Systems, Inc.
1051 N. Fiesta Blvd.
Gilbert, AZ 85233
Phone: 480-507-6478
Fax: 480-838-2232

Freudenberg Faservliesstoffe KG
GB Gasfiltration
Weinheim, 69465 Germany
Phone: 496-201-805537
Fax: 496-201-886299

Freudenberg Nonwovens L.P.
Viledon Filter Division
2975 Pembroke Road
Hopkinsville, KY 42240
Phone: 502-885-9420
Fax: 502-887-5187

HRT - Power L.P.
P.O. Box 2728
Conroe, TX 77305
Phone: 713-827-0001
Fax: 713-827-0002

Locker Air Maze Limited
P.O. Box 17
Folly Lane
Warringtin, WA5 5NP, England
Phone: 019-256-54321
Fax: 019-254-14588

Mee Industries
204 West Pomona Ave.
Monrovia, CA 91016
Phone 800-732-5364
Fax 626-359-4660

Munters Corp.
Marketing Dept., 79 Monroe St., P.O. Box 640
Amesbury, MA 01913 3204
Phone: 978-241-1100
Fax: 978-241-1215

Nortec Industries Inc.
826 Proctor Ave., P.O. Box 698
Ogdensburg, NY 13669
Phone: 315-425-1255
Fax: 613-822-7964

Pneumafil Gas Turbine Div,
P.O. Box 16348, Charlotte, NC 28297
Phone # 704-399-7441,
Fax # 704-398-7507

Premier Industries, Inc.
P.O. Box 37136
Phoenix, AZ 85069
Phone: 800-254-8989
Fax: 602-997-5998

Refrigeration Resources Co., Div. of Mechanical Resources, Inc.
360 Crystal Run Rd.
Middletown, NY 10941
Phone: 800-448-7716
Fax: 845-692-9032

TDC Filter Mfg., Inc.
1331 S. 55th Court
Cicero, IL 60804-1211
Phone: 800-424-1910
Fax: 708-863-4472

Trox (UK) Ltd.
Caxton Way
Theford, Norfolk iP24 35Q
United Kingdom
Phone: (0184)-2754545
Fax: (0184)-2763051

LUBRICATION SYSTEMS

Buffalo Pumps Inc.
874 Oliver Street
North Tonawanda, NY 14120-3298
Phone: 716-693-1850
Fax: 716-693-6303

CECO Filters Inc.
1029 Conshohocken Road
P.O. Box 683
Conshohocken, PA 19428-0683
Phone: 800-220-1127
Fax: 610-825-3108

Kim Hotstart Mfg. Co.
P.O. Box 11245
E. 5723 Alki 42
Spokane, WA 9921-0245
Phone: 509-534-6717
Fax: 800-224-5550

Lubrecon Systems Inc.
11215 Jones Road W.
Suite H
Houston, TX 77065
Phone: 713-890-6400

Lubrication Systems Company
1740 Stebbins Dr.
Houston, TX 77043
Phone: 713-464-6266
Fax: 713-464-9871

SESCO (Systems Engineering & Sales Co.)
3805 E. Pontiac St.
Fort Wayne, IN 46803
Phone: 800-422-2028
Fax: 260-424-0607

PUMPS

ALSTHOM USA Inc.
4 Skyline Drive
Hawthorne, NY 10532-2160
Phone: 914-347-5149
Fax: 914-347-5432

BW/IP International Inc.
Pump Division
200 Oceangate Blvd., Suite 900
Long Beach, CA 90802
Phone: 562-436-0500
Fax: 562-436-7203

Demag Delaval
Turbomachinery Corp.
840 Nottingham Way
P.O. Box 8788
Trenton, NJ 08650-0788
Phone: 609-890-5112
Fax: 609-587-7790

Dresser Roots - Roots Division
Dresser Industries
900 W. Mount Street
Connersville, IN 47331
Phone: 317-827-9200
Fax: 317-825-7669

Gorman-Rupp Co.
270 Bowman St., P.O. Box 1217
Mansfield, OH 44901 1217
Phone: 419-755-1011
Fax: 419-755-1404

IMO Industries Inc.
IMO Pump Division
1009 Lenox Dr., Building Four West
Lawrenceville, NJ 08648
Phone: 609-896-7600

Ingersoll-Dresser Pump Co.
150 Allen Rd.
Liberty Corner, NJ 07938
Phone: 908-647-6800
Fax: 908-689-3095

Johnson Pumps of America
9950 W. Lawrence Ave., Ste. 300
Schiller Park, IL 60176-1216
Phone: 847-671-7867
Fax: 847-671-7909

Magnatex Pumps, Inc.
8730-T Westpark, P.O. Box 770845
Houston, TX 77063
Phone: 713-972-8666
Fax: 713-972-8680

Peerless Pump
A Member Of The Sterling Fluid
Systems Group
P.O. Box 7026
Indianapolis, IN 46207 7026
Phone: 317-925-9661
Fax: 317-924-7388

Roper Pump Co.
P.O. Box 269
Commerce, GA 30529
Phone: 800-876-9160
Fax: 706-335-5505

Roto-Jet Pump
EnviroTech Pumpsystems
P.O. Box 15619
721 North B Street
Sacramento, CA 95852
Phone: 916-444-7195
Fax: 916-441-6550

Sulzer Pumps
Divisional Headquarters
Hegifeldstrasse 10
P.O. Box 65
CH-8404 Winterthur, Switzerland
Phone: 41-522-621155
Fax: 41-522-620040

Sulzer Bingham Pumps, Inc.
2800 N.W. Front Ave.
Portland, OR 97210
Phone: 503-226-5200
Fax: 503-226-5286

SELECTIVE CATALYTIC REDUCTION (SCR)

Bachmann Industries, Inc.
416 Lewiston Junc. Rd., PO Box 2150
Auburn, ME 04211-2150
Phone: 207-784-1903
Fax: 207-784-1904

Engelhard Corp.
Environmental Catalyst Group
101-T Wood Ave.
Iselin, NJ 08830
Phone: 732-205-5000
Fax: 732-632-9253

Foster Wheeler Power Systems, Inc.
Perryville Corporate Park
Clinton, NJ 08809 4000
Phone: 908-730-4000
Fax: 908-730-5315

Johnson Matthey
Catalytic Systems Division
434 Devon Park Dr.
Wayne, PA 19087 1889
Phone: 610-971-3100
Fax: 610-971-3116

STARTERS

Gali Internacional, S.A.
Division Air Starting System
Apartado 11- 08181
Sentment
Barcelona, Spain
Phone: 9-37.15.31.11
Fax: 9-37.26.43.85

Hilliard Corporation
Hilco Division
102 W. Forth Street
Elmira, NY 14902-1504
Phone: 607-733-7121
Fax: 607-737-1108

Ingersoll-Rand
Engine Starting Systems
P.O. Box 8000
1725 U.S. Highway 1 North
Southern Pines, NC 28387
Phone: 910-692-8700
Fax: 910-692-7822

Koenig Engineering, Inc.
410 Eagleview Blvd.
Suite 104
Exton, PA 19341
Phone: 610-458-0153
Fax: 610-458-0404

Lucas Electrical Systems - Heavy...
Larden Road
Action
London, UK W3 7RP
Phone: (4401)-81743311 1
Fax: (4401)-817438276

Magnetech Specialty Company Inc.
550 Interstate 10 South
Beaumont, TX 77707
Phone: 409-842-4145
Fax: 409-842-4297
Fax: 607-737-1108

POW-R-QUIK, Ltd.
5518 Mitchelldale
Houston, TX 77092-7218
Phone: 713-682-0077
Fax: 713-682-8980

SWITCHGEAR

ABB
501 Merrit 7
P.O. Box 5308
Norwalk, CT 06856
Phone: 203-750-7648
Fax: 203-750-7788

Eaton Corp.
Cutler-Hammer Group
1000 Cherrington Pkwy.
Moon Township, PA 15108
Phone: 412-893-3300

Siemans Energy & Automation, Inc.
Automation Division
100 Technology Drive
Alpharetta, GA 36202
Phone: 404-740-3000

VIBRATION EQUIPMENT

Bently Nevada Corp.
1631 Bently Parkway Ave.
Minden, NV 89423
Phone: 775-782-3611
Fax: 775-782-9337

Bruel & Kjaer Instruments, Inc.
2364 Park Central Blvd.
Decatur, GA 30035
Phone: 800-332-2040
Fax: 404-981-8370

Computational Systems, Inc.
835 Innovation Dr.
Knoxville, TN 37932
Phone: 865-675-2400
Fax: 865-675-3100

Dantec Measurement
Technology, Inc.
777 Corporate Drive
Mahwah, NJ 07430
Phone: 201-512-0037
Fax: 201-512-0120

Diagnostic Systems Corp.
3995 Via Oro Avenue
Long Beach, CA 990810
Phone: 310-522-0474
Fax: 310-522-0274

Dynalco Controls
3690 N.W. 53rd St.
Fort Lauderdale, FL 33309 2452
Phone: 954-739-4300, Ext. 219
Fax: 954-484-3376

Endevco Corp.
30700 Rancho Viejo Road
San Juan Capistrano, CA 92675
Phone: 800-982-6732
Fax: 949-661-7231

Entek
1700 Edison Dr.
Cincinnati, OH 45150
Phone: 513-784-2766
Fax: 513-563-7831

epro KANIS Elektronik Gmbh
Jobkesweg 3
Gronau, D-48599, Germany
Phone: 025-627-09242
Fax: 025-627-09255

Fabreeka International Inc.
Engineered Systems
1023 Turnpike Street
Stoughton, MA 02138
Phone: 800-322-7352
Fax: 781-341-3983

Metrix Instrument Co.
1711 Townhurst Dr.
Houston, TX 77043-2899
Phone: 713-461-2131
Fax: 713-461-8223

PMC/BETA
1711 Townhurst
Houston, TX 77043-2810
Phone: 800-638-7494
Fax: 713-461-8223

Radian Corp.
8501 N. Mo-Pac Blvd.
Austin, TX 78759
Phone: 512-454-4797
Fax: 512-454-7129

Schenck Trebel
535 Acorn Street
Deer Park, NY 11729
Phone: 800-877-2357
Fax: 631-242-4147

Scientific-Atlantic, Inc.
13122 Evening Creek Dr.
South San Diego, CA 92128
Phone: 619-679-6000
Fax: 619-679-6400

Shinkawa Electric Co., Ltd.
International Trade Dept.
4-3-3 Shinkoujimachi Bldg 3F
Koujimachi
Chiyoda-ku, Tokyo, 102, Japan
Phone: (03)326-234417
Fax: (03)326-22171

SKF Condition Monitoring
4141 Ruffin Rd.
San Diego, CA 92123
Phone: 800-959-1366
Fax: 858-496-3531

Solartron Inc.
Industrial Data Monitoring
964 Marcon Blvd. #200
Allentown, PA 18103
Phone: 610-264-5034
Fax: 610-264-5329

The Balancing Company
898 Center Drive
Vandalia, OH 45377
Phone: 937-898-9111
Fax: 937-898-6145

Vibration Specialty Corp.
100 Geiger Road
Philadelphia, PA 19115-1090
Phone: 215-698-0800
Fax: 215-677-8874

Vibro-Meter Corp.
3995 Via Oro Avenue
Long Beach, CA 90810
Phone: 310-522-0474
Fax: 310-522-0274

Vitec Inc.
24755 Highpoint Rd.
Cleveland, OH 44122
Phone: 800-321-6384
Fax: 216-464-5324

Wireless Data Corporation
620 Clyde Avenue
Mountain View, CA 94043
Phone: 650-967-9100
Fax: 650-967-7727

Zonic Corporation
50 West Technecenter Drive
Milford, OH 45150
Phone: 513-248-1911
Fax: 513-248-1589

Appendix C-1

Characteristics of Particles And Particle Dispersoids

This chart, shown on the following pages, is provided courtesy of SRI International.

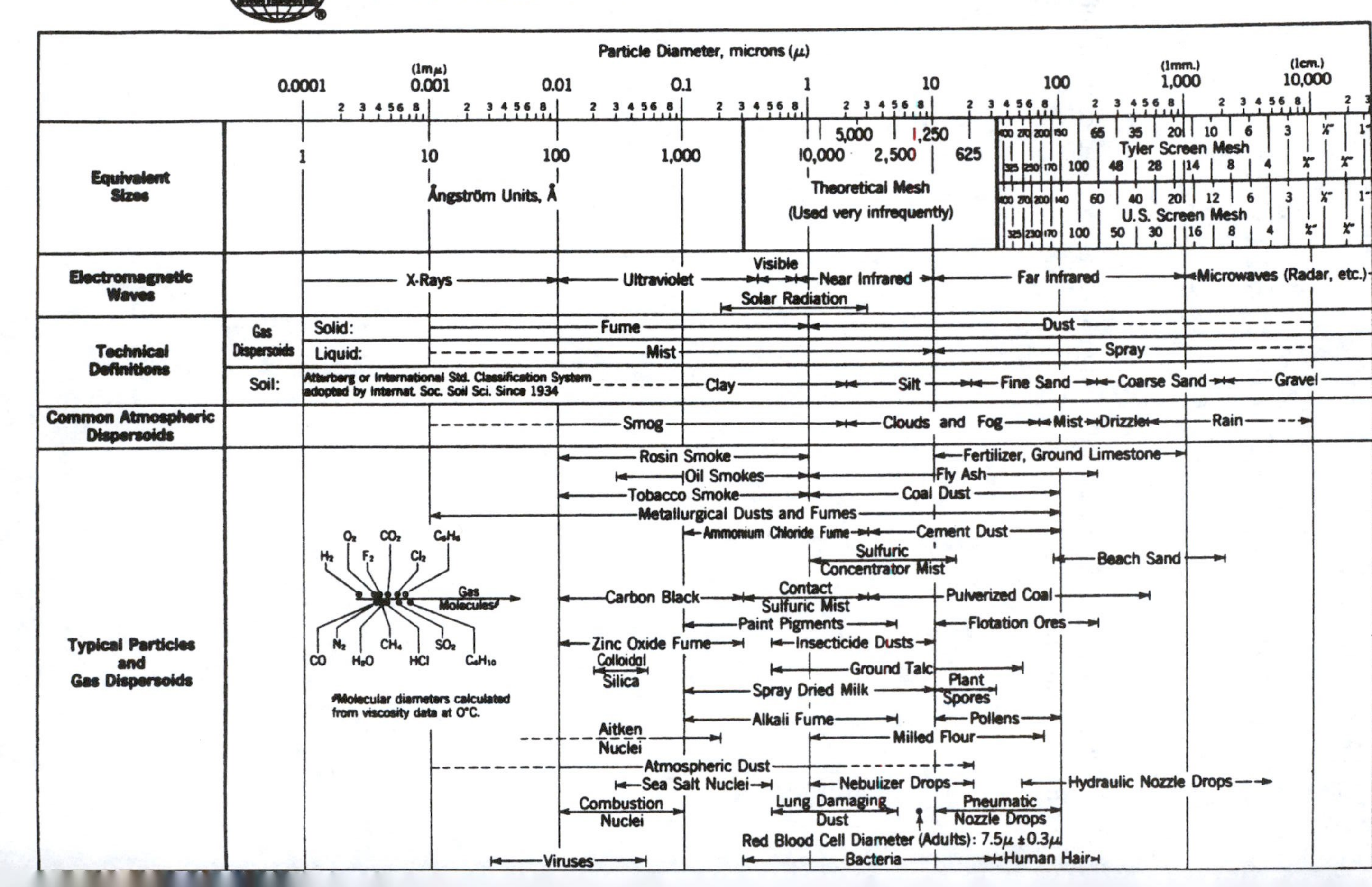
CHARACTERISTICS OF PARTICLES AND PARTICLE DISPERSOIDS
SRI International
Particle Diameter, microns (μ)
0.0001 0.001 (1mμ) 0.01 0.1 1 10 100 1,000 (1mm.) 10,000 (1cm.)
Equivalent Sizes
1 10 100 1,000 10,000
Ångström Units, Å
5,000 1,250 2,500 625
Theoretical Mesh
(Used very infrequently)
Tyler Screen Mesh
U.S. Screen Mesh
Electromagnetic Waves
X-Rays
Ultraviolet
Visible
Near Infrared
Far Infrared
Microwaves (Radar, etc.)
Solar Radiation
Technical Definitions
Gas Dispersoids
Solid:
Liquid:
Fume
Dust
Mist
Spray
Soil:
Atterberg or International Std. Classification System adopted by Internat. Soc. Soil Sci. Since 1934
Clay
Silt
Fine Sand
Coarse Sand
Gravel
Common Atmospheric Dispersoids
Smog
Clouds and Fog
Mist
Drizzle
Rain
Typical Particles and Gas Dispersoids
Rosin Smoke
Fertilizer, Ground Limestone
Oil Smokes
Fly Ash
Tobacco Smoke
Coal Dust
Metallurgical Dusts and Fumes
Ammonium Chloride Fume
Cement Dust
Sulfuric Concentrator Mist
Beach Sand
Carbon Black
Contact Sulfuric Mist
Pulverized Coal
Paint Pigments
Flotation Ores
Zinc Oxide Fume
Insecticide Dusts
Colloidal Silica
Ground Talc
Spray Dried Milk
Plant Spores
Alkali Fume
Pollens
Aitken Nuclei
Milled Flour
Atmospheric Dust
Sea Salt Nuclei
Nebulizer Drops
Hydraulic Nozzle Drops
Combustion Nuclei
Lung Damaging Dust
Pneumatic Nozzle Drops
Red Blood Cell Diameter (Adults): 7.5μ ±0.3μ
Viruses
Bacteria
Human Hair
Gas Molecules*
*Molecular diameters calculated from viscosity data at 0°C.

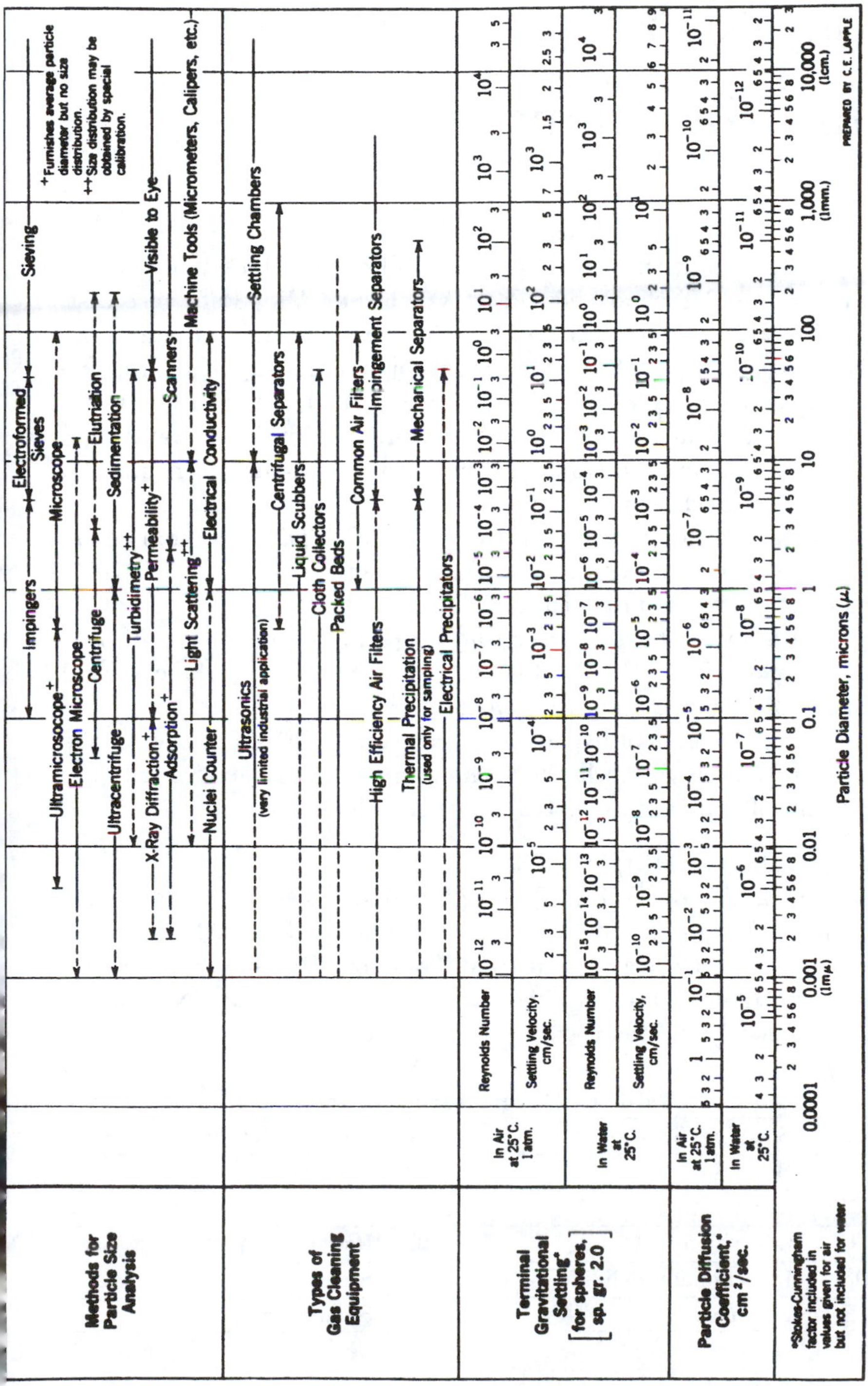

Methods for Particle Size Analysis

Impingers — Electroformed Sieves — Sieving

Ultramicroscope+ — Microscope

Electron Microscope

Centrifuge — Elutriation

Ultracentrifuge — Sedimentation

Turbidimetry++

X-Ray Diffraction+ — Permeability+ — Visible to Eye

Adsorption+ — Scanners

Light Scattering++ — Machine Tools (Micrometers, Calipers, etc.)

Nuclei Counter — Electrical Conductivity

+Furnishes average particle diameter but no size distribution.

++Size distribution may be obtained by special calibration.

Types of Gas Cleaning Equipment

Ultrasonics (very limited industrial application) — Settling Chambers

Centrifugal Separators

Liquid Scubbers

Cloth Collectors

Packed Beds

Common Air Filters

High Efficiency Air Filters — Impingement Separators

Thermal Precipitation (used only for sampling) — Mechanical Separators

Electrical Precipitators

Terminal Gravitational Settling* [for spheres, sp. gr. 2.0]

In Air at 25°C. 1 atm.

Reynolds Number: 10^{-12}, 10^{-11}, 10^{-10}, 10^{-9}, 10^{-8}, 10^{-7}, 10^{-6}, 10^{-5}, 10^{-4}, 10^{-3}, 10^{-2}, 10^{-1}, 10^{0}, 10^{1}, 10^{2}, 10^{3}, 10^{4}

Settling Velocity, cm/sec.: 10^{-5}, 10^{-4}, 10^{-3}, 10^{-2}, 10^{-1}, 10^{0}, 10^{1}, 10^{2}, 10^{3}

In Water at 25°C.

Reynolds Number: 10^{-15}, 10^{-14}, 10^{-13}, 10^{-12}, 10^{-11}, 10^{-10}, 10^{-9}, 10^{-8}, 10^{-7}, 10^{-6}, 10^{-5}, 10^{-4}, 10^{-3}, 10^{-2}, 10^{-1}, 10^{0}, 10^{1}, 10^{2}, 10^{3}, 10^{4}

Settling Velocity, cm/sec.: 10^{-10}, 10^{-9}, 10^{-8}, 10^{-7}, 10^{-6}, 10^{-5}, 10^{-4}, 10^{-3}, 10^{-2}, 10^{-1}, 10^{0}, 10^{1}

Particle Diffusion Coefficient,* cm²/sec.

In Air at 25°C. 1 atm.: 1, 10^{-1}, 10^{-2}, 10^{-3}, 10^{-4}, 10^{-5}, 10^{-6}, 10^{-7}, 10^{-8}, 10^{-9}, 10^{-10}, 10^{-11}

In Water at 25°C.: 10^{-5}, 10^{-6}, 10^{-7}, 10^{-8}, 10^{-9}, 10^{-10}, 10^{-11}, 10^{-12}

*Stokes-Cunningham factor included in values given for air but not included for water

Particle Diameter, microns (μ): 0.0001, 0.001 (1mμ), 0.01, 0.1, 1, 10, 100, 1,000 (1mm.), 10,000 (1cm.)

PREPARED BY C. E. LAPPLE

Reprinted from Stanford Research Institute Journal, Third Quarter, 1961. Single copies 8-1/2 by 11 inches free or $15 per hundred. 20 by 26 inch wall chart $15 each. Both charts available from Dept. 380.

SRI International • 333 RAVENSWOOD AVE. • MENLO PARK, CALIFORNIA 94025 • (415) 326-6200

Appendix C-2

Material Chart

The information provided in the following tables is excerpted from the United Technologies, Pratt & Whitney Alloy Reference List.

This list is a compilation of the specification condition and form, nominal chemistry, coefficient of expansion and density of gas turbine alloys listed by AMS, PWA specification and AISI, AA or brand name.

Each AMS and PWA specification number is immediately followed by the form(s) in which the material is supplied (a nomenclature key is provided in Table C-2.1).

The specification condition in the SPEC. COND. column refers to the condition in which the material is received by the purchaser (a nomenclature key is provided in Table C-2.2).

The alloy chemistry percentages are listed on top in the row for each alloy. The "R" indicates that the remainder of the chemical composition is made up of that element.

The coefficients of expansion (micro in./in./°F) are listed on the bottom in the row for each alloy. The temperatures pertaining to the coefficient of expansion is listed below the element name.

Example: Incoloy 901 (page 303)
AMS 5660 and 5661 (B=Bar, F= Forging)//PWA Spec 1003 (Bar, Forging), 1022 (Forging), 1024 (Forging)//SPEC COND.: SSP=Solution, Stabilization and Precipitation Heat Treated; SP= Solution and Precipitation Heat Treated//Considering the column under Ni=Nickel: 42.5% is nickel, and the coefficient of expansion at 200°F is 7.75.

Table C-2.1. Specification Form

AT	Seamless Tubing	PMET	Powder Metallurgy
B	Bar	PR	Preform
BI	Billet	R	Rings
CC	Centrifugally Cast	RO	Rod
DS	Directionally Solidified	S	Sheet
E	Extrusion	SC	Sand Cast

Table C-2.1. Specification Form (*Continued*)

F	Forging	ST	Strip
HIP	Hot Isostatically Pressed	SXL	Single Crystal
MT	Mechanical Tubing	W	Wire
P	Plate	WDT	Welded and Drawn Tubing
PC	Investment Cast	WT	Welded Tubing
PM	Permanent Mold	WW	Weldwire

Table C-2.2. Specification Condition

AC	As Cast	SCT	Subzero Cooled and Tempered
AHCF	Annealed and Hot or Cold Finished	SCW	Solutioned and Cold Worked
ANN	Annealed	SEO	Solutioned, Equalized and Overtempered
AO	As Ordered	SO	Solution Heat Treated and Overaged
AP	Annealed and Precipitation Heat Treated	SOL	Solution Heat Treated
CD	Cold Drawn	SP	Solution and Precipitation Heat Treated
CT	Cold Worked and Tempered	SR	Stress Relieved
ELI	Extra Low Interstitial	SSP	Solution, Stabilization and Precipitation Heat Treated
EO	Equalized and Overtempered	SSR	Solution and Stress Relieved
EQ	Equalized	SST	Solutioned and Stabilized
H	Hardened	STP	Stabilized and Precipitation Heat Treated
HCF	Hot or Cold Finished	STS	Stabilized and Solutioned
HCFT	Hot or Cold Finished or Tempered	SWP	Solutioned, Cold Worked, and Precipitation Heat Treated
HS	Homogenized and Solutioned	TEM	Tempered
OT	Overtempered	VAR	Various

WROUGHT NON-HARDENABLE NICKEL BASE ALLOYS

AMS	PWA SPEC.	AISI or BRAND	SPEC. COND.	-300	Ni 200	Cr 600	Co 800	C 1000	Fe 1200	Ti 1400	Al 1600	Mo 1800	2000	OTHERS	DENSITY
5553 S,ST *1*	6000 B,F 6005 S,ST,P 6015 W, WW	NICKEL 200, 201 (A NICKEL)	ANN ANN ANN	5.8	R 7.0	8.25	8.6	8.8	8.9	9.0	9.05	9.1			0.321
5540 S,ST,P 5580 AT 7232 RIVET 5665 B,F,R 5687 W 5683 WW *2*	1060 AT 1062 WDT 1070 AT	INCONEL 600	ANN ANN ANN CD	5.6	R 7.0	15.5 7.8	8.0	8.2	8.0 8.4	8.8	9.1	9.6			0.305
5715 B,F,R 5870 S,ST,P *3*		INCONEL 601	ANN ANN		R 7.6	22.5 8.15	8.3	0.05 8.5	14.0 8.9	9.2	1.35 9.5	9.8	10.2		0.291
4	1165S 1168 WW	INCONEL 617	SOL CD		R 6.4	22.0 7.4	12.5 7.6	0.07 7.7	8.0	8.4	1.0 8.7	9.0 9.0	9.2		0.302
5599 S,ST,P 5581 AT,WT 5666 B,F,R 5837 WW *5*	1069 AT 1072 WDT	INCONEL 625	ANN ANN SOL SOL	5.1	R 7.1	21.5 7.45	7.6	7.8	8.2	8.5	8.8	9.0		Cb/Ta 3.65	0.305
5530 S,ST,P *6*		HASTELLOY C	SOL		R 6.25	15.5 7.0	7.35	7.6	5.5 7.9	8.2	8.45	16.0 8.7		W 3.8	0.323
5607 S,ST,P 5771 B,F,R *7*		HASTELLOY N	SOL SOL	5.3	R 6.4	7.0 6.9	7.1	0.06 7.3	7.45	7.75	8.1	16.5 8.45	8.7	B 0.01	0.317
5711 B,F,R 5873 S,ST,P 5838 WW *8*		HASTELLOY S	SOL CD		R 6.7	15.8 7.0	7.15	7.35	7.6	7.95	0.3 8.3	15.2 8.65	8.9	La 0.05 Mn 0.7 Si 0.5	0.316
5755 B,F,R 5786 WW *9*		HASTELLOY W	SOL SOL	4.9	R 6.2	5.0 6.95	7.25	7.35	5.5 7.4	7.7	8.1	24.5 8.4			0.325
5536 S,ST,P 5754 B,F,R 5587 AT 5588 WDT 7237 RIVET 5798 WW *10*	1038 S,ST	HASTELLOY X	SOL SOL SOL CD	5.7	R 7.7	22.0 7.9	1.5 8.1	0.10 8.3	18.5 8.55	8.8	9.0	9.0 9.2	9.4	W 0.6	0.299
5890 B,F,E 5865 S,ST *11*		TD NICKEL	SR SR		R 5.8	7.4	7.75	7.85	8.0	8.25	8.6	8.8	9.0	ThO_2 2.2	0.322
12															

WROUGHT PRECIPITATION HARDENABLE NICKEL BASE ALLOYS

AMS	PWA SPEC.	AISI or BRAND	SPEC. COND.	−300	Ni 200	Cr 600	Co 800	C 1000	Fe 1200	Ti 1400	Al 1600	Mo 1800	2000	OTHERS	DENSITY
5702 B,F,R 5703 B,F,R 5606 S,ST,P 1	1025 B,F,R 1026 B,F,R	INCONEL 706	SOL SSP SOL		41.5 7.4	16.0 8.4	8.6	8.75	37.9 8.95	1.75 9.5				Cb/Ta 2.9	0.292
5596 S,ST,P 5589 AT 5662 B,F,R 5663 B,F,R 5832 WW 7466/7467 BOLT 2	1009 B,F 1010 B,F,R 1065 WDT 1085 B,F,R 96 BOLT	INCONEL 718	SOL SOL SP ANN SP SR SP SOL	6.2	R 7.1	19.0 7.9	8.1	0.05 8.25	18.0 8.45	0.9 8.8	0.6 9.4	3.0 9.75	9.9	Cb/Ta 5.2	0.297
5542 S,ST,P 5582 AT 5598 S,ST,P 5667 B,F,R 5668 B,F,R 5670 B,F,R 5671 B,F,R 5698 W 5699 W 3		INCONEL X-750	SOL SOL EQ SP SOL CD CD	5.7	R 6.7	15.5 7.5	7.75	0.08 8.0	7.0 8.3	2.5 8.7	0.7 9.2	9.7		Cb/Ta 1.0	0.298
5660 B,F 5661 B,F 4	1003 B,F 1022 F 1024 F	INCOLOY 901	SSP SP SP		42.5 7.75	12.5 8.05	8.25	0.05 8.5	36.3 8.8	2.9 9.2	9.65	5.8		B 0.015	0.294
5872 S,ST,P 5	1087 WW	C-263	SOL CD		R 6.3	20.0 7.0	20.0 7.4	0.06 7.6	7.9	2.15 8.3	0.45 9.0	5.85 9.6			0.302
5856 BI/PR (PMET) 6	1102 E (PMET) 299 BOLT (PMET)	AF2-1DA	AO SSP		R	12.0	10.0	0.33 6.9	7.1	3.0 7.4	4.5 7.8	3.0 8.5		B 0.015 Ta 1.5 W 6.0 Zr 0.10	0.299
7															
8															

WROUGHT PRECIPITATION HARDENABLE NICKEL BASE ALLOYS (CONT'D)

AMS		PWA SPEC.	AISI or BRAND	SPEC. COND.	−300	Ni 200	Cr 600	Co 800	C 1000	Fe 1200	Ti 1400	Al 1600	Mo 1800	2000	OTHERS	DENSITY
7253 NUT 7471 BOLT 5708 B,F 5709 B,F 5706 B,F,R 5828 WW 5707 B,F,R 5704 F 5544 S,ST,P 5586 WT	1	92 BOLT 686 B,F 1005 F 1007 F 1016 F 1027 F 1057 F 1164 S	WASPALOY	SSP SOL SSP SOL SSP SSP SSP SSP SSP ANN	6.0	R 7.0	19.5 7.5	13.5 7.7	0.05 7.9	8.1	3.0 8.4	1.4 8.9	4.2 9.6	10.4	B 0.007 Zr 0.06	0.298
	2	1021 B 1013 F	ASTROLOY	SOL SSP		R 7.4	15.0 7.55	17.0 7.6	0.06 7.75	8.0	3.5 8.35	4.0 9.0	5.0 9.65	10.3	B 0.025	0.290
	3	689 B,F	UDIMET 700	SSP		R 7.4	15.0 7.55	18.5 7.6	0.07 7.75	8.0	3.3 8.35	4.3 9.0	5.0 9.65	10.3	B 0.03	0.288
	4	1073 E+F (PMET) 1074 E+F (PMET) 1106 E+F (PMET)	MODIFIED IN-100	SSP SSP SSP		R 6.9	12.4 7.2	18.5 7.3	0.07 7.5	7.8	4.3 8.1	5.0 8.5	3.2 9.1	9.8	B 0.02 V 0.8 Zr 0.06	0.284
	5	1099 HIP (PMET) 1100 HIP (PMET)	IMPROVED IN-100	SSP SSP		R 6.9	12.4 7.2	18.5 7.3	0.025 7.5	7.8	4.3 8.1	5.0 8.5	3.2 9.1	9.8	B 0.02 Cb 1.4 Hf 0.4	0.286
	6															
	7															
	8															

CAST NICKEL BASE ALLOYS

AMS	PWA SPEC.	AISI or BRAND	SPEC. COND.	Ni 200	Cr 600	Co 800	C 1000	Fe 1200	Ti 1400	Al 1600	Mo 1800	W 2000	OTHERS	DENSITY
5405 PC *1*	663 PC	B-1900	AC	R 6.4	8.0 7.0	10.0 7.2	0.11 7.4	7.6	1.0 7.8	6.0 8.2	6.0 8.7	9.3	B 0.015 Ta 4.3 Zr 0.07	0.296
5406 PC *2*	1455 PC	B-1900+Hf	AC	R 6.4	8.0 7.0	10.0 7.2	0.11 7.4	7.6	1.0 8.2	6.0 8.7	6.0 9.3		B 0.015 Hf 1.15 Ta 4.25 Zr 0.08	0.296
5408 CC *3*		HASTELLOY S	SOL	R 6.7	7.0	7.15	7.35	7.6	7.95	0.3 8.3	15.2 8.65	8.9	La 0.05 Mn 0.7 Si 0.5	0.316
5390 PC *4*		HASTELLOY X	AC	R 7.7	22.0 7.9	1.5 8.1	8.3	18.5 8.55	8.8	9.0	9.0 9.2	0.6 9.4		0.299
5397 PC *5*	658 PC	IN-100	AC	R 6.9	9.5 7.2	15.0 7.3	0.17 7.5	7.8	4.75 8.1	5.5 8.5	3.0 9.1	9.8	B 0.015 V 1.0 Zr 0.06	0.280
6	661 PC	INCONEL 600	AC	R 7.0	15.5 7.8	8.0	0.15 8.2	8.5 8.4	8.8	9.1	9.6		Cb/Ta 2.3	0.305
5401 PC 5402 PC *7*	1468 PC	INCONEL 625	AC SOL	R 7.1	21.5 7.4	7.8	7.8	8.2	8.5	8.8	9.0		Cb/Ta 3.65	0.305
5391 PC *8*	655 PC	INCONEL 713C	AC	R 6.6	13.5 7.3	7.5	0.14 7.75	8.0	0.9 8.3	6.0 8.65	4.5 9.2	9.8	B 0.010 Cb/Ta 2.1 Zr 0.08	0.286
5377 PC *9*		INCONEL 713 LC	AC	R 6.6	12.0 7.3	7.5	0.05 7.75	8.0	0.70 8.3	6.0 8.65	4.5 9.2	9.8	B 0.010 Cb 2.0 Zr 0.10	0.289
5383 PC *10*	649 PC 1469 PC+HIP	INCONEL 718	VAR VAR	R 7.1	19.0 7.9	8.1	8.25	18.0 8.45	0.9 8.8	0.6 9.4	3.0 9.75	9.9	Cb/Ta 5.1	0.297
11	1451 PC	INCONEL 738	SOL	R 6.1	16.0 6.9	8.5 7.2	0.13 7.5	7.7	3.4 8.0	3.4 8.4	1.75 8.8	2.6 9.4	B 0.012 Cb 0.85 Ta 1.75 Zr 0.12	0.294
12														

CAST NICKEL BASE ALLOYS (CONT'D)

AMS	PWA SPEC.	AISI or BRAND	SPEC. COND.	Ni 200	Cr 600	Co 800	C 1000	Fe 1200	Ti 1400	Al 1600	Mo 1800	W 2000	OTHERS	DENSITY
1	1456 PC 1467 PC+HIP	MODIFIED INCONEL 792	AC SOL	R 5.9	12.2 6.7	9.0 6.9	0.12 7.1	7.3	4.1 7.5	3.5 7.85	1.9 8.3	3.8 8.9	B 0.015 Hf 0.5 Ta 3.9 Zr 0.1	0.299
2	1422 DS	MAR-M-200 + Hf	SOL	R 6.4	9.0 6.8	10.0 7.0	0.14 7.2	7.4	2.0 7.7	5.0 8.1	8.6	12.5 9.2	B 0.015 Cb 1.0 Hf 2.0	0.311
3	1447 PC	MAR-M-247	AC	R 6.7	8.4 7.1	10.0 7.2	0.15 7.4	7.6	1.1 7.8	5.5 8.2	0.65 8.7	10.0 9.3	B 0.015 Hf 1.4 Ta 3.1 Zr 0.055	0.311
4	1480 SXL		SOL	R 6.4	10.0 6.8	5.0 7.0	7.2	7.4	1.5 7.7	5.0 8.1	8.6	4.0 9.2	Ta 12.0	0.312
5384 PC 5		UDIMET 500	SP	R 6.7	18.0 7.4	18.0 7.6	7.8	8.0	2.9 8.4	2.9 8.9	4.0 9.6	10.3	B 0.006	0.290
6	656 PC	UDIMET 700	SOL	R 7.4	15.0 7.5	15.3 7.6	0.07 7.7	8.0	3.4 8.4	4.3 8.9	4.4 9.7	10.3	B 0.016	0.288
7	652 PC 1471 PC	WASPALOY	SSP SOL	R 7.0	19.5 7.5	13.5 7.7	0.07 7.9	8.1	3.0 8.4	1.25 8.9	4.25 9.55	10.4	B 0.005	0.298
8														
9														
10														
11														
12														

WROUGHT COBALT BASE ALLOYS

AMS	PWA SPEC.	AISI or BRAND	SPEC. COND.	Co 200	Cr 600	Ni 800	C 1000	Fe 1200	Mo 1400	W 1600	Ti 1800	Al 2000	OTHERS	DENSITY
5537 S 5759 B,F,R 5796 WW 7236 RIVET *1*	1064 WDT	HAYNES 25 (L-605)	SOL ANN SOL	R 6.8	20.0 7.6	10.0 7.8	0.10 8.0	8.3	8.6	15.0 9.0	9.4	9.8	Mn 1.5	0.329
5772 B,F,R 5608 S,ST,P 5801 WW *2*	1042 S 1088 WDT	HAYNES 188	SOL CD SOL	41.3 6.7	22.0 7.4	22.0 7.8	0.1 8.2	8.7	9.2	14.5 9.6	9.9	10.2	La 0.075 Si 0.35	0.324
5758 B 5844 B 5845 B 7468 BOLT *3*	115 BOLT	MP35N	SOL SCW SWP	35.3 7.1	10.0 8.2	35.0 8.3	8.7		9.75					0.304
5841 B 5842 B 5843 B 7475 BOLT *4*		MP159	SOL SCW SWP	36.0 7.9	19.0 7.9	25.0 8.1	8.4	9.0 8.7	7.0		2.9	0.2	Cb 0.5	0.302
5789 WW *5*	744 (PMET)	STELLITE 31 (X-40)	AO CD	R 7.1	25.5 7.7	10.5 8.0	0.5 8.2	8.5	8.8	7.5 9.05	9.3	9.4		0.311
6														

CAST COBALT BASE ALLOYS

AMS	PWA SPEC.	AISI or BRAND	SPEC. COND.	Co 200	Cr 600	Ni 800	C 1000	Fe 1200	Mo 1400	W 1600	Ti 1800	Al 2000	OTHERS	DENSITY
5385 PC *7*		STELLITE 21 (MOD VITALLIUM)	AC	R	27.0 7.65	2.75 7.8	0.25 8.0	8.2	5.5 8.5					0.300
5382 PC *8*		STELLITE 31 (X-40)	AC	R 7.1	25.5 7.7	10.5 8.0	0.5 8.2	8.5	8.8	7.5 9.05	9.3	9.4		0.311
9	653 PC 654 PC	WI-52	AC AC	R 7.0	21.0 7.5	7.8	0.45 8.0	1.75 8.25	8.55	11.0 8.85	9.2	9.5	Cb/Ta 2.0	0.322
10	657 PC	MAR-M-302	AC	R 6.8	21.5 7.1	7.4	0.85 7.6	1.0 7.8	8.05	10.0 8.3	8.7	9.2	Ta 9.0 Zr 0.25	0.333
11	647 PC	MAR-M-509	AC	R 6.7	23.4 7.4	10.0 7.65	0.6 7.9	8.2	8.45	7.0 8.75	0.2 9.1		Ta 3.5 Zr 0.5	0.317
12														

WROUGHT AND CAST CORROSION AND HEAT RESISTANT AUSTENITIC IRON BASE ALLOYS

AMS	PWA SPEC.	AISI or BRAND	SPEC. COND.	Fe −300	C −100	Cr 200	Ni 400	Mo 600	Mn 800	Ti 1000	Cb/Ta 1200	OTHERS	DENSITY
5688 W 5517 S,ST 5518 S,ST 5519 S,ST 5636 B,W 5637 B,W 5515 S,ST,P 5516 S,ST,P 1		301, 302	CW CW SOL	R 7.7	0.15 8.7	18.0 9.0	8.5 9.4	9.7	9.9	10.1	10.3		0.286
5560 AT 5565 WT 5513 S,ST,P 5697 W 7228 RIVET 5639 B,F,MT,R 5647 B,F,MT,R 2		304	SOL SOL SOL	R 7.7	8.7	19.0 9.0	9.5 9.4	9.7	9.9	10.1	10.3		0.286
5521 S,ST,P 5572 AT 5577 WT 5651 B,F,MT,R 5694 WW 5365 SC 5366 PC 3		310	SOL SOL SOL AC	R 7.4	8.2	25.0 8.4	20.5 8.7	9.0	9.25	9.5	9.8		0.286
5524 S,ST,P 5573 AT 5648 B,F,MT,R 5690 W 5360 PC 5361 SC,CC 5692 WW 4		316	SOL SOL SOL CD	R 7.7	8.7	18.0 8.9	13.0 9.3	2.25 9.6	9.85	10.1	10.3		0.286
5510 S,ST,P 5570 AT 5576 WT 5645 B,F,MT,R 5689 W 5557 AT,WT 5559 WT 5		321	SOL SOL SOL SOL	R 7.7	8.7	18.0 9.05	10.0 9.4	9.7	9.9	0.5 10.1	10.3		0.286
5512 S,ST,P 5558 WT 5571 AT 5575 WT 5646 B,F,MT,R 5680 WW 5362 PC 5363 SC,CC 5674 W 5556 AT,WT 5654 B,F,MT,R 7229 RIVET 6	770 WDT 767 ST	347	SOL. SOL SOL SOL SOL	R 7.5	8.5	18.0 9.0	10.5 9.4	9.7	9.95	10.1	0.8 10.35		0.286
7	772 B,F	INVAR 36 FM	ANN	R 1.9	2.3	0.9	36.0 1.7	3.8	0.5 5.2	6.2	6.9	Se 2.0	0.294
8	773 B,F	INVAR 42	ANN	R		2.5	42.0 2.6	2.7	0.5 3.5	4.8	5.8		0.293
5595 S,ST,P 5656 B,F,R 5562 AT 5561 WDT 9		NITRONIC 40 (21-6-9)	SOL SOL CD	R		20.0 9.3	8.5 9.6	10.1	9.0 10.4	10.6	10.85	N 0.27	0.283
10	1084 B	NITRONIC 60	SOL	R		17.0 8.8	8.5 9.2	9.6	8.0 9.8	10.0	10.3	N 0.13 Si 4.0	0.275

WROUGHT AND CAST CORROSION AND HEAT RESISTANT MARTENSITIC AND FERRITIC IRON BASE ALLOYS

AMS	PWA SPEC.	AISI or BRAND	SPEC. COND.	Fe −100	C 200	Cr 400	Ni 600	Mo 800	Mn 1000	V 1200	Cb/Ta	OTHERS	DENSITY
5504 S,ST,P 5591 AT 5613 B,F,MT,R 5776 WW 5350 PC 5351 SC 1	1089 F	410,410 MOD	ANN ANN TEM	R 5.1	0.12 5.6	12.5 5.9	6.2	6.4	6.6	6.7			0.279
5616 B,F,MT,R 5508 S,ST,P 5817 WW 5354 PC 7470 BOLT 2	1464 (PC+HIP)	GREEK ASCOLOY	ANN ANN TEM	R	0.17 5.3	13.0 5.6	2.0 5.9	6.1	6.3	6.5		W 3.0	0.285
5627 B,F,MT,R 5503 S,ST,P 3		430	ANN ANN	R	5.5	17.0 5.8	6.1	6.3	6.5	6.7			0.275
5630 B,F,W 5618 B,F 4		440C	HCF ANN	R 5.2	1.0 5.8	17.0 6.1	6.25	0.5 6.3	6.45				0.280
5	765 B,F	440C MOD	ANN	R 5.1	1.1 5.8	14.0 5.8	6.0	4.0 6.1	6.2	6.3			0.280
6	771 B	NM-100 (B-100)	ANN	R	1.2 5.3	17.5	5.8	6.2	6.3	0.75 6.5		B 0.007 Co 9.5 W 10.5	0.303
5548 S,ST 5554 AT 5745 B,F 5546 S,ST 7		AM-350	ANN EO CT	R	0.10 6.3	16.5 6.5	4.5 6.8	2.9 7.1	0.85 7.2	7.05		N 0.10 (SC 850)	0 282
5780 WW 5743 B,F 5547 S,ST 5549 P 5744 B,F		AM-355	ANN SEO SOL TEM	R	0.12	15.5	4.5	2.9	0.85			N 0.09	0.282
5359 SC 5368 PC 8			SOL	R	0.12 6.9	15.0 7.1	4.0 7.4	2.3 7.65	0.75 7.75	7.6		N 0.08 (SCT 1000)	0.281
5749 B,F,MT,W 9		BG-42	AHCF	R	1.15 5.95	14.75 6.25	6.45	4.0 6.65	0.45 6.85	1.2		Si 0.3	0.280
10	779 B	WD-65	ANN	R	1.15 6.1	14.1 6.1	6.1	4.0 6.3	6.8	2.75		Co 5.75 W 2.25	0.282
11													
12													

WROUGHT AND CAST CORROSION AND HEAT RESISTANT MARTENSITIC AND FERRITIC IRON BASE ALLOYS (CONT'D)

AMS	PWA SPEC.	AISI or BRAND	SPEC. COND.	Fe −100	C 200	Cr 400	Ni 600	Mo 800	Mn 1000	V 1200	Cb/Ta	OTHERS	DENSITY
5604 S,ST,P 5643 B,F,MT,R 5622 B,F,MT,R 7474 BOLT		17-4PH	SOL SOL SP	R		16.5	4.0				0.3	Cu 4.0	0.283
5398 SC,CC 5355 PC 5342 PC 5343 PC 5344 PC 1			SOL SP SP	R	6.3	16.1 6.5	4.1 6.6	6.8			0.28	Cu 3.15 (H 1075)	0.282
5528 S,ST,P 5568 WT 5529 S,ST 5678 W 2		17-7PH	SOL SCW CD	R	5.6	17.0 6.1	7.0 6.3	6.6				Al 1.0 (H 1050)	0.276
5520 S,ST,P 3		PH 15-7 Mo	ANN	R	6.1	15.0 6.1	7.1 6.1	2.5 6.3	6.6			Al 1.1 (H 1050)	0.277
5346 PC 5347 PC 5356 PC 5400 PC 5357 PC 5659 B,F,R,W 5862 S,ST,P 4		15-5PH	SP SP SOL SOL	R	6.3	14.8 6.5	4.5 6.6	6.8			0.3	Cu 3.5 (H 1025)	0.283
5617 B,F,W 5860 S,ST,P 5	36150 BOLT	CUSTOM 455	SOL SP	R	5.9	12.0 6.1	8.5 6.3	6.6	6.85		0.3	Cu 2.0 Ti 1.1	0.282
5718 B,F,R,MT,W,E 5719 B,F,R,MT,W,E 6		JETHETE M-152	ANN ANN	R	0.11 6.0	11.8 6.1	2.5 6.3	1.8 6.45	0.7 6.6	0.33 6.8		N 0.03	0.280
7	791 WW 798 S 1079 B,F	AM-363	ANN ANN ANN	R	0.03 5.95	11.5 6.15	4.5 6.3	6.45	6.55			Ti 0.5	0.283
5655 B,F 8		C-422	TEM	R	0.22	12.5	0.75	1.0		0.22		Si 0.4 W 1.0	0.282
9													
10													
11													
12													

WROUGHT HEAT RESISTANT PRECIPITATION HARDENABLE IRON BASE ALLOYS

AMS	PWA SPEC.	AISI or BRAND	SPEC. COND.	Fe −300	C −100	Cr 200	Ni 400	Mo 600	Mn 800	V 1000	Ti 1200	OTHERS	DENSITY
5525 S,ST,P 5805 WW 7235 RIVET 5895 B,F,MT,R 5731 B,F,MT,R 5858 S,ST,P 5737 B,F,MT,R 7250 NUT 5732 B,F,MT,R,W 7251 NUT 7477 BOLT 7478 BOLT 7482 STUD *1*	1029 F	A-286 (MOD TINIDUR)	SOL CD SP SOL SOL SP SP SP SP	R 7.3	7.9	15.0 9.3	25.5 9.35	1.25 9.5	9.7	0.3 9.85	2.1 9.95	B 0.006	0.287
2	1191 B,F 1192 B,F	INCOLOY 909	SOL SOL	40.4		4.5	37.5 4.3	4.2	4.3	5.0	1.55 5.8	Cb/Ta 4.75 Co 14.0 Si 0.38	0.296
5768 B,F,R 5769 B,F,R 5585 WT 5532 S,ST,P 5794 WW *3*		N-155 (MULTIMET)	SP SOL ANN	30.9 6.6	0.12 7.1	21.0 8.0	20.0 8.3	3.0 8.6	1.5 8.85	9.1	9.4	Cb/Ta 1.0 Co 20.0 W 2.5	0.296

WROUGHT ALLOY STEELS

AMS	PWA SPEC.	AISI or BRAND	SPEC. COND.	Fe −300	C −100	Cr 200	Ni 400	Mo 600	Mn 800	V 1000	Si 1200	OTHERS	DENSITY
6260 B,F,MT 6265 B,F,MT 6267 B,F,MT *4*		9310	HCF HCF	R 4.9	0.1 5.1	1.2 6.5	3.25 6.8	0.12 7.05	0.55 7.3	7.5			0.283
6263 B,F,MT *5*		9315	HCF	R 4.9	0.14 5.1	1.2 6.5	3.25 6.8	0.12 7.3	0.55 7.7	8.0			0.283
6	724 B,F 742 B,F,MT	BOWER 315	ANN ANN	R 4.8	0.13 5.4	1.55 6.45	2.8 6.7	5.0 6.95	0.55 7.1	7.2	7.4		0.286
6322 B,F,R 6323 MT 6358 S,ST,P 6325 B,F 6327 B,F 7252 NUT 7452 BOLT *7*		8740	HCF HCF TEM TEM	R 4.5	0.41 5.5	0.5 6.5	0.55 6.8	0.25 7.05	0.88 7.3	7.5	0.25 7.7		0.283
8													

WROUGHT ALLOY STEELS (CONT'D)

AMS	PWA SPEC.	AISI or BRAND	SPEC. COND.	Fe −300	C −100	Cr 200	Ni 400	Mo 600	Mn 800	V 1000	Si 1200	OTHERS	DENSITY
6370 B,F,R 6371 MT 6351 S,ST,P *1*		4130	AO HCF ANN	R 5.0	0.3 5.9	0.95 6.8	7.4	0.2 7.6	0.5 7.9	8.0	0.25 8.3		0.283
6414 B,F,MT 6415 B,F,MT 6359 S,ST,P 6454 S,ST,P *2*		4340	HCFT ANN	R 4.7	0.4 5.5	0.8 6.5	1.8 6.8	0.25 7.1	0.75 7.5	7.9	0.28 8.2		0.283
6304 B,F,MT 6305 B,F,MT 7454 BOLT 7459 BOLT *3*	733 F 768 F	17-22A	ANN TEM TEM	R	0.45	0.95 6.8	7.0	0.55 7.3	0.65 7.5	0.30 7.75	0.28 8.0		0.283
6440 B,F,MT 6444 B,F,MT *4*	723 B,F,MT	52100	HCF ANN	R 5.0	1.04 5.2	1.45 6.3	6.8	7.2	0.35 7.5	7.8	0.25		0.283
6491 B,F,MT 6490 B,F,MT *5*	725 B,F 793 B,F,MT	M-50	AO HCF ANN	R 5.5	0.81	4.0 6.2	6.6	4.25 6.85	7.05	1.0 7.4	7.8		0.283
6522 P 6527 B,F *6*		AF-1410	OT	R	0.15	2.0 8.0	10.0 8.0	1.0 7.7	7.3	7.3	7.8	Co 14.0	0.283
6437 S,ST,P 6488 F 6485 F 6487 F 7464 BOLT 6485 B 6487 B 6488 B *7*		H-11	ANN AO TEM HCF	R	0.4	5.0 6 1	1.3 6.5	0.3 6.8	0.5 7.1	0.9 7.25			0.283
8													

WROUGHT COPPER AND COPPER-BERYLLIUM ALLOYS

AMS	PWA SPEC.	AISI or BRAND	SPEC. COND.	Cu −300	Be −100	200	400	600	800	1000		OTHERS	DENSITY
4501 S,ST,P 4602 B,RO *9*		COPPER	CW CW	R 7.9	8.75	9.4	9.6	9.85	10.1	10.25			0.324
4530 S,ST,P 4650 B,F,RO 4725 W 4532 S,ST *10*		BERYLCO 25	SOL SCW	R 7.8	1.9 8.75	9.3	9.4	9.9	10.4				0.298
11													

WROUGHT AND CAST TITANIUM BASE ALLOYS

AMS	PWA SPEC.	AISI or BRAND	SPEC. COND.	Ti −300	Al −100	Sn 200	V 400	Zr 600	Mo 800	1000	1200	OTHERS	DENSITY
4902 S,ST,P 4942 AT 4941 WT 1		Ti COMM PURE 40K YS	ANN ANN	99.0 3.0		4.8	5.2	5.3	5.4	5.5	5.6		0.163
4900 S,ST,P 4951 WW 2		Ti COMM PURE 55K YS	ANN ANN	99.0 3.0		4.8	5.2	5.3	5.4	5.5	5.6		0.163
4901 S,ST,P 4921 B,F,R 3		Ti COMM PURE 70K YS	ANN ANN	99.0 3.0		4.8	5.2	5.3	5.4	5.5	5.6		0.163
4910 S,ST,P 4926 B,R 4966 F 4953 WW 4909 S,ST,P(ELI) 4924 B,F,R(ELI) 4	1261 PC	A-110AT Ti-5Al-2.5Sn	ANN ANN ANN	R 4.3	5.0 4.6	2.5 5.2	5.2	5.25	5.3	5.4	5.5		0.161
4943 AT 5	1260 AT	Ti-3Al-2.5V	ANN	R	3.0	2.5 5.35	5.4	5.45	5.5	5.55			0.162
4911 S,ST,P 4967 B,F,R 4935 E,R 4928 B,F 4930 B,F,R(ELI) 4956 WW(ELI) 4965 B,F,R 7460 BOLT 7461 BOLT 4954 WW 6	1215 F 1262 PC 1264 (PC+HIP) 1213 E 1228 B,F,R	Ti-6Al-4V	ANN ANN SSR ANN ANN SP CD	R 4.3	6.0 4.6	4.8	4.0 5.0	5.1	5.2	5.3	5.5		0.160
4976 F 4975 B,R 4919 S,ST,P 7	1209 B,F 1210 B 1214 F 1222 WW 1223 B,F 1224 B,F 1225 B,F 1226 B,F 1231 S,ST,P 1265 (PC+HIP)	Ti-6-2-4-2	SP SP SP ANN ANN SP SP AP ANN SP/SSR	R	6.0	2.0 5.0	5.25	4.0 5.4	2.0 5.55	5.65		Si 0.08 (PWA 1209 0.10 MAX)	0.164
4972 B,R 4973 F 4933 E,R 4955 WW 4915 S,ST,P 4916 S,ST,P 8	91 BOLT 1202 B,F	Ti-8-1-1	SST SST CD ANN	R	8.0	4.7	1.0 4.85	5.0	1.0 5.2	5.8			0.158

WROUGHT AND CAST TITANIUM BASE ALLOYS (CONT'D)

AMS	PWA SPEC.	AISI or BRAND	SPEC. COND.	Ti −300	Al −100	Sn 200	V 400	Zr 600	Mo 800	1000	1200	OTHERS	DENSITY
1	1212 B,F 1216 B,F 1220 B,F 1227 F	Ti-6-2-4-6	SOL SP SSP STP	R	6.0	2.0 5.0	5.2	4.0 5.35	6.0 5.5	5.55			0.168
2													
3													
4													
5													
6													
7													
8													
9													
10													
11													
12													
13													
14													
15													

WROUGHT ALUMINUM BASE ALLOYS

AMS		AA or BRAND	SPEC. COND.	Al −300	Mg −100	Mn 200	Cu 400	Si 600	Ti	Ni	Cr	OTHERS	DENSITY
4121 B,W 4133 F 4135 F 4153 E 4029 S 4029 P 4314 R	1	2014	T6 T6 T651	R 10.1	0.5 11.3	0.8 12.25	4.5 13.15	0.9 13.65					0.101
4035 S,P 4087 AT 4037 S 4086 AT 4088 AT 4152 E 4120 B,RO,W 4037 P 4119 B,RO,W	2	2024	O T3 T4, 351 T351	R 10.4	1.5 11.8	0.6 12.65	4.4 13.15	13.75					0.101
4115 B,R 4160 E 4025 S,P 4079 AT 4080 AT 4116 B,RO,W 4146 F 4161 E 4026 S,P 4081 AT 4117 B,R,RO,W 4127 F 4150 E,R 4027 S 4082 AT 4083 AT 4027 P 4117 B,RO 4312 R 4312 R	3	6061	O T4 T6 T6 T651 T652	R 10.3	1.0 11.8	13.0	0.25 13.6	0.6 14.2			0.25		0.098
4044 S,P 4122 B,R,RO,W 4126 F 4154 E 4045 S,P 4141 F 4123 B,RO	4	7075	O T6 T73 T651	R 10.5	2.5 12.0	13.2	1.6 13.7				0.3	Zn 5.6	0.101
4031 S,P 4143 F 4313 R 4068 AT 4066 AT 4144 F,R 4162 E 4163 E	5	2219	O T6 T351 T851 T852 T8511 T3511	R 10.0	11.3	0.3 12.4	6.3 13.1	13.7	0.06			V 0.1 Zr 0.8	0.102
4130 F	6	2025	T6	R	12.05	0.8 12.6	4.5 13.1	0.85 13.7					0.101
4132 F	7	2618	T61	R	1.6	12.3	2.3 12.8	0.18 13.45	0.07	1.1		Fe 1.1	0.100
	8												
	9												

CAST ALUMINUM BASE ALLOYS

AMS		AA or BRAND	SPEC. COND.	Al −300	Mg −100	Mn 200	Cu 400	Si 600	Ti	Ni	Cr	OTHERS	DENSITY
4227 SC	1	A240.0 (A140)	F	R	6.0	0.5 12.25	8.0 13.0	13.6		0.5			0.101
4222 SC	2	242.0P (142)	T77	R	1.5	12.6	4.0 13.2	13.6	0.12	2.0			0.101
4212 SC 4281 PM 4214 SC 4280 PM	3	355.0	T6 T71	R 9.9	0.5 11.2	12.4	1.2 13.0	5.0 13.8					0.098
4215 (PREMIUM)	4	C355.0	T6P	R 9.9	0.5 11.2	12.4	1.2 13.0	5.0 13.8					0.098
4217 SC 4260 PC 4284 PM 4285 CC	5	356.0	T6	R	0.3	11.85	12.5	7.0 13.0					0.097
4218 (PREMIUM)	6	A356.0	T61P	R	0.3	11.85	12.5	7.0 13.0					0.097
4219 (PREMIUM)	7	A357.0	T61	R	0.58	11.95	12.6	7.0	0.15			Be 0.055	0.096
4226 SC	8	224.0	T7	R 10.0	11.3	0.35 12.4	5.0 13.1	13.7				V 0.1 Zr 0.18	0.102
4225 SC	9	HIDUMINIUM RR-350 (203.0P)	SO	R		0.25 12.4	5.0 13.1	13.75		1.5			0.102
4223 PM,SC 4229 PM,SC	10	A201.0	T4 T7	R	0.25	0.3 10.6	4.5 12.5	13.9	0.25			Ag 0.7	0.101
	11												
	12												
	13												
	14												
	15												
	16												

CAST MAGNESIUM BASE ALLOYS

AMS		BRAND	SPEC. COND.	Mg −300	Al −100	Mn 200	Zn 400	Zr 603	Ag			OTHERS	DENSITY
4418 SC	1	QE22A	T6	R		14.4	15.1	0.7 15.7	2.5			Di 2.1	0.066
4434 SC 4453 PC 4484 PM	2	AZ92A	T6	R	9.0	0.1 14.5	2.0 15.2	15.7					0.066
4442 SC	3	EZ33A	T5	R		14.5	2.5 15.0	0.75 15.5				Ce 3.25	0.066
4439 SC	4	ZE41A	T5	R		14.4	4.25 15.1	0.7 15.7				Ce 1.25	0.066
	5												
	6												
	7												
	8												
	9												
	10												
	11												
	12												
	13												
	14												
	15												
	16												

Appendix C-3

Interrelationship of Gas Turbine Engine Parameters

The chart contains all the gas turbine engine parameter interrelationships in a general matrix. These equation sets provide a means of estimating steady state and transient variations in engine performance parameters for any gas turbine under most conceivable sets of input conditions. Solutions require only that sufficient parameters are available to generate at least as many equations as there are unknowns. This matrix was published in the book *Gas Turbine Engine Parameter Interrelationships*, written by Louis A. Urban, and published by Hamilton Standard Division in 1969.

This chart is reprinted courtesy of Hamilton Standard Division of United Technologies Corporation.

	TURBINE INLET TEMP. VARIATION $\frac{\partial T_{t4}}{T_{t4}}$	ENGINE SPEED VARIATION $\frac{\partial N}{N}$	SHAFT POWER EXTRACTION $\partial \frac{SHP}{W_a}$
$\partial T_{t4}/T_{t4}$	1	0	0
$\partial N/N$	0	1	0
$\partial W_a/W_a$ INLET	0	μ	0
$\partial T_{t3}/T_{t3}$	$\frac{k-1}{2k\eta_c}$	$\mu\frac{k-1}{k\eta_c}$	0
$\partial P_{t3}/P_{t3}$	$\frac{1}{2}$	μ	0
$\frac{\partial W_f}{W_f}$	$\frac{T_{t4}}{\Delta T_b}-\frac{k-1}{2k\eta_c}\frac{T_{t3}}{\Delta T_b}$	$\mu\left(1-\frac{k-1}{k\eta_c}\frac{T_{t3}}{\Delta T_b}\right)$	0
$\frac{\partial W_f/P_{t3}}{W_f/P_{t3}}$	$\frac{T_{t4}}{\Delta T_b}-\frac{k-1}{2k\eta_c}\frac{T_{t3}}{\Delta T_b}-\frac{1}{2}$	$-\mu\frac{k-1}{k\eta_c}\frac{T_{t3}}{\Delta T_b}$	0
$\partial Z_2/Z_2$	1	μ	0
$\partial Z_3/Z_3$	$\frac{1}{2}$	0	0
$\partial W_g/W_g$	0	μ	0
$\frac{\partial T_{t5}}{T_{t5}}$	$1+\frac{k-1}{k}\left(\frac{2-\alpha}{2\phi_t}\right)$	$-\mu\frac{k-1}{k}\frac{\alpha}{\phi_t}$	$\frac{-550}{Jc_p T_{t4}\eta_t}\left[\frac{\frac{k-1}{k}+\phi_t}{\phi_t}\right]$
$\frac{\partial P_{t5}}{P_{t5}}$	$\frac{2+\phi_t-\alpha}{2\phi_t}$	$\mu\frac{\phi_t-\alpha}{\phi_t}$	$\frac{-550}{RT_{t4}\eta_t}\left[\frac{\frac{k-1}{k}+\phi_t}{\phi_t}\right]$
$\frac{\partial A_n}{A_n}$	$-\frac{k+1}{2k}\left(\frac{2-\alpha}{2\phi_t}\right)-\epsilon_n\left(\frac{2+\phi_t-\alpha}{2\phi_t}\right)$	$\mu\left[\frac{k+1}{2k}\frac{\alpha}{\phi_t}-\epsilon_n\left(\frac{\phi_t-\alpha}{\phi_t}\right)\right]$	$\frac{550}{RT_{t4}\eta_t}\left[\frac{\frac{k-1}{k}+\phi_t}{\phi_t}\right]\cdot\left[\frac{k+1}{2k}+\epsilon_n\right]$
$\frac{\partial F_n}{F_n}$	$\frac{F_g}{F_n}\left[\frac{k-1}{2k}\left(\frac{2-\alpha}{2\phi_t}\right)+\frac{\phi_n}{2}\left(\frac{2+\phi_t-\alpha}{2\phi_t}\right)+\frac{1}{2}\right]$	$\mu\left\{1+\frac{F_g}{F_n}\left[\frac{\phi_n}{2}-\frac{\alpha}{\phi_t}\left(\frac{k-1}{2k}+\frac{\phi_n}{2}\right)\right]\right\}$	$-\frac{550}{RT_{t4}\eta_t}\left[\frac{\frac{k-1}{k}+\phi_t}{\phi_t}\right]\cdot\frac{F_g}{F_n}\left[\frac{k-1}{2k}+\frac{\phi_n}{2}\right]$
$\frac{\partial HP_{PT}}{HP_{PT}}$	$1+\frac{\phi_{PT}}{2}+\left(\frac{2-\alpha}{2\phi_t}\right)\left[\frac{k-1}{k}+\phi_{PT}\right]$	$\mu\left[1+\phi_{PT}-\frac{\alpha}{\phi_t}\left(\frac{k-1}{k}+\phi_{PT}\right)\right]$	$\frac{-550}{RT_{t4}\eta_t}\left[\frac{\frac{k-1}{k}+\phi_t}{\phi_t}\right]\cdot\left[\frac{k-1}{k}+\phi_{PT}\right]$

	TURBINE INLET TEMP. VARIATION $\frac{\partial T_4}{T_4}$	ENGINE SPEED VARIATION $\frac{\partial N}{N}$	SHAFT POWER EXTRACTION $\partial\frac{SHP}{W_a}$
$\frac{\partial T_{t5}}{T_{t5}}$	$1 - \eta_{PT}\frac{k-1}{2k}$	$-\mu\eta_{PT}\frac{k-1}{k}$	0
$\partial T_{t2}/T_{t2}$	0	0	0
$\partial P_{t2}/P_{t2}$	0	0	0
$\partial P_{am}/P_{am}$	0	0	0
$\partial\eta_c/\eta_c$	0	0	0
$\frac{\partial\Delta P_b/P_{t3}}{\Delta P_b/P_{t3}}$	0	0	0
$\partial\eta_b/\eta_b$	0	0	0
$\partial A_4/A_4$	0	0	0
$\partial\eta_t/\eta_t$	0	0	0
$\partial\eta\ /\eta$	0	0	0

	ENGINE INLET TEMP. VARIATION $\frac{\partial T_{t2}}{T_{t2}}$	ENG. IN. PRESS. VAR. $\frac{\partial P_{t2}}{P_{t2}}$	AMB. PRESS. VAR. $\frac{\partial P_{am}}{P_{am}}$	COMP. DISCHARGE AIR BLEED (For 1% Bleed, $\frac{\partial W_a}{W_a} = -.01$) $\frac{\partial W_a}{W_a}$
$\partial T_{t4}/T_{t4}$	0	0	0	0
$\partial N/N$	0	0	0	0
$\partial W_a/W_a$ INLET	$-\frac{1+\mu}{2}$	1	1	0
$\partial T_{t3}/T_{t3}$	$\frac{k+1}{2k} - \frac{k-1}{k\eta_c}\frac{\mu}{2}$	0	0	$\frac{k-1}{k\eta_c}$
$\partial P_{t3}/P_{t3}$	$-\frac{1+\mu}{2}$	1	1	1
$\frac{\partial W_f}{W_f}$	$-\left[\frac{1}{2} + \frac{k+1}{2k}\frac{T_{t3}}{\Delta T_b}\right]$ $-\frac{\mu}{2}\left[1 - \frac{k-1}{k\eta_c}\frac{T_{t3}}{\Delta T_b}\right]$	1	1	$1 - \frac{k-1}{k\eta_c}\frac{T_{t3}}{\Delta T_b}$
$\frac{\partial W_f/P_{t3}}{W_f/P_{t3}}$	$-\frac{T_{t3}}{\Delta T_b}\left(\frac{k+1}{2k} - \frac{k-1}{k\eta_c}\frac{\mu}{2}\right)$	0	0	$-\frac{k-1}{k\eta_c}\frac{T_{t3}}{\Delta T_b}$
$\partial Z_2/Z_2$	$-\frac{1+\mu}{2}$	0	0	1

	ENGINE INLET TEMP. VARIATION $\frac{\partial T_{t2}}{T_{t2}}$	ENG. IN. PRESS VAR. $\frac{\partial P_{t2}}{P_{t2}}$	AMB. PRESS. VAR. $\frac{\partial P_{am}}{P_{am}}$	COMP. DISCHARGE AIR BLEED (For 1% Bleed, $\frac{\partial W_a}{W_a} = -.01$) $\frac{\partial W_a}{W_a}$
$\partial Z_3/Z_3$	0	0	0	0
$\partial W_g/W_g$	$-\frac{1+\mu}{2}$	1	1	1
$\frac{\partial T_{t5}}{T_{t5}}$	$\frac{k-1}{k}\left[\frac{1+\mu}{2}\frac{\alpha}{\phi_t}-\frac{1}{\phi_t}\right]$	0	0	$\frac{k-1}{k}\left(\frac{1-\alpha}{\phi_t}\right)$
$\frac{\partial P_{t5}}{P_{t5}}$	$-\left(\frac{1+\mu}{2}\right)\left(\frac{\phi_t-\alpha}{\phi_t}\right)-\frac{1}{\phi_t}$	1	1	$\frac{1+\phi_t-\alpha}{\phi_t}$
$\frac{\partial A_n}{A_n}$	$\frac{k+1}{2k}\left(\frac{1}{\phi_t}-\frac{1+\mu}{2}\frac{\alpha}{\phi_t}\right)+\epsilon_n\left[\frac{1+\mu}{2}\left(\frac{\phi_t-\alpha}{\phi_t}\right)+\frac{1}{\phi_t}\right]$	$-\epsilon_n$	0	$-\frac{k+1}{2k}\left(\frac{1-\alpha}{\phi_t}\right)-\epsilon_n\left(\frac{1+\phi_t-\alpha}{\phi_t}\right)$
$\frac{\partial F_n}{F_n}$	$-\left(\frac{1+\mu}{2}\right)\left\{1+\frac{F_g}{F_n}\left[\frac{\phi_n}{2}+\frac{1}{\phi_t}\left(\frac{2}{1+\mu}-\alpha\right)\left(\frac{k-1}{2k}+\frac{\phi_n}{2}\right)\right]\right\}$	$1+\frac{F_g}{F_n}\frac{\phi_n}{2}$	1	$1+\frac{F_g}{F_n}\left[\frac{\phi_n}{2}+\left(\frac{1-\alpha}{\phi_t}\right)\left(\frac{k-1}{2k}+\frac{\phi_n}{2}\right)\right]$
$\frac{\partial HP_{PT}}{HP_{PT}}$	$-\left(\frac{1+\mu}{2}\right)\left[1+\phi_{PT}+\frac{1}{\phi_t}\left(\frac{2}{1+\mu}-\alpha\right)\left(\frac{k-1}{k}+\phi_{PT}\right)\right]$	$1+\phi_{PT}$	1	$1+\phi_{PT}+\left(\frac{1-\alpha}{\phi_t}\right)\left(\frac{k-1}{k}+\phi_{PT}\right)$
$\frac{\partial T_{t6}}{T_{t6}}$	$\left(\frac{1+\mu}{2}\right)\eta_{PT}\frac{k-1}{k}$	$-\eta_{PT}\frac{k-1}{k}$	0	$-\eta_{PT}\frac{k-1}{k}$
$\partial T_{t2}/T_{t2}$	1	0	0	0
$\partial P_{t2}/P_{t2}$	0	1	1	0
$\partial P_{am}/P_{am}$	0	0	1	0
$\partial \eta_c/\eta_c$	0	0	0	0
$\frac{\partial \Delta P_b P_{t3}}{\Delta P_b P_{t3}}$	0	0	0	0
$\partial \eta_b/\eta_b$	0	0	0	0
$\partial A_4/A_4$	0	0	0	0
$\partial \eta_t/\eta_t$	0	0	0	0
$\partial \eta/\eta$	0	0	0	0

	COMP. INLET AIRFLOW VAR. $\frac{\partial W_a}{W_a}$	COMPRESSOR EFFICIENCY VARIATION $\frac{\partial \eta_c}{\eta_c}$	BURNER PRESS. LOSS VARIATION $\partial \frac{\Delta P_b}{P_{t3}}$	BURNER EFF. VAR. $\frac{\partial \eta_b}{\eta_b}$
$\partial T_{t4}/T_{t4}$	0	0	0	0
$\partial N/N$	0	0	0	0
$\partial W_a/W_a$ INLET	1	0	0	0
$\partial T_{t3}/T_{t3}$	$\frac{k-1}{k\eta_c}$	$-\frac{k-1}{k\eta_c}\frac{1}{\alpha}$	$\frac{k-1}{k\eta_c}\left(\frac{1}{1-\frac{\Delta P_b}{P_{t3}}}\right)$	0
$\partial P_{t3}/P_{t3}$	1	0	$\frac{1}{1-\Delta P_b/P_{t3}}$	0
$\frac{\partial W_f}{W_f}$	$1-\frac{k-1}{k\eta_c}\frac{T_{t3}}{\Delta T_b}$	$\frac{k-1}{k\eta_c}\frac{T_{t3}}{\Delta T_b}\frac{1}{\alpha}$	$-\frac{k-1}{k\eta_c}\frac{T_{t3}}{\Delta T_b}\left(\frac{1}{1-\frac{\Delta P_b}{P_{t3}}}\right)$	−1
$\frac{\partial W_f/P_{t3}}{W_f/P_{t3}}$	$-\frac{k-1}{k\eta_c}\frac{T_{t3}}{\Delta T_b}$	$\frac{k-1}{k\eta_c}\frac{T_{t3}}{\Delta T_b}\frac{1}{\alpha}$	$\frac{-1}{1-\frac{\Delta P_b}{P_{t3}}}\left(1+\frac{k-1}{k\eta_c}\frac{T_{t3}}{\Delta T_b}\right)$	−1
$\partial Z_2/Z_2$	1	0	0	−1
$\partial Z_3/Z_3$	0	0	$\frac{-1}{1-\Delta P_b/P_{t3}}$	−1
$\partial W_g/W_g$	1	0	0	0
$\frac{\partial T_{t5}}{T_{t5}}$	$-\frac{k-1}{k}\frac{\alpha}{\phi_t}$	$\frac{k-1}{k}\frac{1}{\phi_t}$	$-\frac{k-1}{k}\frac{\alpha}{\phi_t}\left(\frac{1}{1-\frac{\Delta P_b}{P_{t3}}}\right)$	0
$\frac{\partial P_{t5}}{P_{t5}}$	$\frac{\phi_t-\alpha}{\phi_t}$	$\frac{1}{\phi_t}$	$-\frac{\alpha}{\phi_t}\left(\frac{1}{1-\frac{\Delta P_b}{P_{t3}}}\right)$	0
$\frac{\partial A_n}{A_n}$	$\frac{k+1}{2k}\frac{\alpha}{\phi_t}-\epsilon_n\left(\frac{\phi_t-\alpha}{\phi_t}\right)$	$-\frac{1}{\phi_t}\left[\frac{k+1}{2k}+\epsilon_n\right]$	$\frac{\alpha}{\phi_t}\left(\frac{1}{1-\frac{\Delta P_b}{P_{t3}}}\right)\left[\frac{k+1}{2k}+\epsilon_n\right]$	0
$\frac{\partial F_n}{F_n}$	$1+\frac{F_g}{F_n}\left[\frac{\phi_n}{2}-\frac{\alpha}{\phi_t}\left(\frac{k-1}{2k}+\frac{\phi_n}{2}\right)\right]$	$\frac{F_g}{F_n}\frac{1}{\phi_t}\left[\frac{k-1}{2k}+\frac{\phi_n}{2}\right]$	$-\frac{F_g}{F_n}\frac{\alpha}{\phi_t}\left(\frac{1}{1-\frac{\Delta P_b}{P_{t3}}}\right)\left(\frac{k-1}{2k}+\frac{\phi_n}{2}\right)$	0
$\frac{\partial HP_{PT}}{HP_{PT}}$	$1+\phi_{PT}-\frac{\alpha}{\phi_t}\left(\frac{k-1}{k}+\phi_{PT}\right)$	$\frac{1}{\phi_t}\left[\frac{k-1}{k}+\phi_{PT}\right]$	$-\frac{\alpha}{\phi_t}\left(\frac{1}{1-\frac{\Delta P_b}{P_{t3}}}\right)\left(\frac{k-1}{k}+\phi_{PT}\right)$	0

	COMP. INLET AIRFLOW VAR. $\frac{\partial W_a}{W_a}$	COMPRESSOR EFFICIENCY VARIATION $\frac{\partial \eta_c}{\eta_c}$	BURNER PRESS. LOSS VARIATION $\partial \frac{\Delta P_b}{P_{t3}}$	BURNER EFF. VAR. $\frac{\partial \eta_b}{\eta_b}$
$\frac{\partial T_{t6}}{T_{t6}}$	$-\eta_{rt}\frac{k-1}{k}$	0	0	0
$\partial T_{t2}/T_{t2}$	0	0	0	0
$\partial P_{t2}/P_{t2}$	0	0	0	0
$\partial P_{am}/P_{am}$	0	0	0	0
$\partial \eta_c/\eta_c$	0	1	0	0
$\frac{\partial \Delta P_b P_{t3}}{\Delta P_b P_{t3}}$	0	0	$\frac{1}{\Delta P_b/P_{t3}}$	0
$\partial \eta_b/\eta_b$	0	0	0	1
$\partial A_4/A_4$	0	0	0	0
$\partial \eta_t/\eta_t$	0	0	0	0
$\partial \eta/\eta$	0	0	0	0

	TURBINE AREA VARIATION $\frac{\partial A_4}{A_4}$	TURBINE EFF. VAR. $\frac{\partial \eta_t}{\eta_t}$	GAS GEN. EXH. PRESS. VAR. $\frac{\partial P_{t5}}{P_{t5}}$	EXH. NOZ. OR POWER TURB EFF. VAR. $\frac{\partial \eta}{\eta}$
$\partial T_{t4}/T_{t4}$	0	0	0	0
$\partial N/N$	0	0	0	0
$\partial W_a/W_a$ INLET	0	0	0	0
$\partial T_{t3}/T_{t3}$	$-\frac{k-1}{k\eta_c}$	0	0	0
$\partial P_{t3}/P_{t3}$	-1	0	0	0
$\frac{\partial W_f}{W_f}$	$\frac{k-1}{k\eta_c}\frac{T_{t3}}{\Delta T_b}$	0	0	0
$\frac{\partial W_f/P_{t3}}{W_f/P_{t3}}$	$1+\frac{k-1}{k\eta_c}\frac{T_{t3}}{\Delta T_b}$	0	0	0
$\partial Z_2/Z_2$	0	0	0	0

	TURBINE AREA VARIATION $\frac{\partial A_4}{A_4}$	TURBINE EFF. VAR. $\frac{\partial \eta_t}{\eta_t}$	GAS GEN. EXH. PRESS VAR. $\frac{\partial P_{t5}}{P_{t5}}$	EXH. NOZ. or POWER TURB EFF. VAR. $\frac{\partial \eta}{\eta}$
$\partial Z_3/Z_3$	1	0	0	0
$\partial W_g/W_g$	0	0	0	0
$\frac{\partial T_{t5}}{T_{t5}}$	$\frac{k-1}{k}\frac{\alpha}{\phi t}$	0	0	0
$\frac{\partial P_{t5}}{P_{t5}}$	$-\frac{\phi t-\alpha}{\phi t}$	$\frac{1}{\phi t}$	1	0
$\frac{\partial A_n}{A_n}$	$\frac{k-1}{2k}\frac{\alpha}{\phi t} + (1+\epsilon_n)\left(\frac{\phi t-\alpha}{\phi t}\right)$	$\frac{-1}{\phi t}(1+\epsilon_n)$	$-(1+\epsilon_n)$	0
$\frac{\partial F_n}{F_n}$	$\frac{F_g}{F_n}\left[-\frac{\phi_n}{2} + \frac{\alpha}{\phi t}\left(\frac{k-1}{2k}+\frac{\phi_n}{2}\right)\right]$	$\frac{F_g}{F_n}\frac{\phi_n}{2\phi t}$	$\frac{F_g}{F_n}\frac{\phi_n}{2}$	$\frac{1}{2}\frac{F_g}{F_n}$
$\frac{\partial HP_{PT}}{HP_{PT}}$	$\frac{\alpha}{\phi t}\left(\frac{k-1}{k}+\phi_{PT}\right)-\phi_{PT}$	$\frac{\phi_{PT}}{\phi t}$	ϕ_{PT}	1
$\frac{\partial T_{t6}}{T_{t6}}$	$\eta_{PT}\frac{k-1}{k}$	$-\eta_{PT}\frac{k-1}{k}\frac{1}{\phi t}$	$-\eta_{PT}\frac{k-1}{k}$	$-\frac{k-1}{k}\frac{1}{\phi_{PT}}$
$\partial T_{t2}/T_{t2}$	0	0	0	0
$\partial P_{t2}/P_{t2}$	0	0	0	0
$\partial P_{am}/P_{am}$	0	0	0	0
$\frac{\partial \eta_c/\eta_c}{\partial \Delta P_b P_{t3}}$	0	0	0	0
$\Delta P_b P_{t3}$	0	0	0	0
$\partial \eta_b/\eta_b$	0	0	0	0
$\partial A_4/A_4$	1	0	0	0
$\partial \eta_t/\eta_t$	0	1	0	0
$\partial \eta/\eta$	0	0	0	1

$$Z = \frac{w_f}{P\left(1 - \frac{T_{t3}}{T_{t4}}\right)} \qquad \mu = \left.\frac{\partial W_a/W_a}{\partial N/N}\right|_{T_{t4} = \not{c}} \qquad \alpha = \frac{\frac{k-1}{k}\left(\frac{P_{t3}}{P_{t2}}\right)^{\frac{k-1}{k}}}{\left(\frac{P_{t3}}{P_{t2}}\right)^{\frac{k-1}{k}} - 1} \qquad \phi_t = \frac{\frac{k-1}{k}}{\left(\frac{P_{t4}}{P_{t5}}\right)^{\frac{k-1}{k}} - 1}$$

$$\phi_n = \phi_{PT} = \frac{\frac{k-1}{k}}{\left(\frac{P_{t5}}{P_{am}}\right)^{\frac{k-1}{k}} - 1} \qquad \epsilon_n = \frac{\frac{k+1}{2} - \left(\frac{P_{t5}}{P_{am}}\right)^{\frac{k-1}{k}}}{k\left[\left(\frac{P_{t5}}{P_{am}}\right)^{\frac{k-1}{k}} - 1\right]} = 0 \text{ Beyond Choking}$$

$$\frac{F_g}{F_n} = \frac{1}{1 - M_n\sqrt{\frac{k\,\dot{T}_{am}}{2\,\eta_n\,T_{t4}}\left[\frac{\frac{k-1}{k} + \phi_t}{\phi_t}\right]\left[\frac{k-1}{k} + \phi_n\right]}}$$

Hamilton Standard

DIVISION OF UNITED AIRCRAFT CORP

UA®

Appendix C-4

Classification of Hazardous Atmospheres

Area classification are divided into CLASS, DIVISION, and GROUPS. As shown in the following table:

CLASS		DIVISION		GROUP	
I	Gases, Vapors	1	Normally Hazardous	A B C D	Acetylene (581°F) Hydrogen (968°F) Ethylene (842°F) Fuels such as: Methane (999°F) Propane (842°F) Butane (550°F) Gasoline (536-880°F) Naphtha (550°F)
I	Gases, Vapors	2	Not Normally Hazardous	A B C D	Same as Div. 1 Same as Div. 1 Same as Div. 1 Same as Div. 1
II	Combust- ible Dust	1	Normally Hazardous	E F G	Carbonaceous metal dust Carbonaceous dust Resistivity $>10^2$ and $<10^8$ ohm-centimeter Carbonaceous dust Resistivity $> 10^8$ ohm-centimeter
		2	Not Normal- ly Hazardous	F	Carbonaceous dust Resistivity $>10^5$ ohm-centimeter Same as Div. 1
III	Easily Ignitable Fibers and Flyings				

The above information has been extracted from NFPA 70-1990, National Electric Code. National Electric Code and NEC are Registered Trademarks of the National Fire Protection Association, Inc. Quincy, MA.

Appendix C-5

Guidelines for Selecting Gas Turbine Inlet Air Filtration

Project Name ____________________ Date ____________________

Gas Turbine Make ________ Model ________ Max Inlet Flow ____________CFM

Total Inlet ΔP ____________in. W.G. (Clean) Min Inlet Flow ____________CFM

Inlet ΔP Per Stage ________in. W.G. (Clean Electrical Criteria ____________

Geographic Location ______________

1. Configuration
   Number of Levels __________, Side Inlets Per Level __________
   Outlet: Bottom ____________, Rear ____________

2. Size Limitations
   Length __________ Ft, Width __________ Ft, Height __________ Ft.

3. Trash Screen (Hinged)
   Mesh: 1" ______________ Material: Stainless Steel ________
   Other ______________ Galv. Steel ________
   ΔP ____________in. W.G. Aluminum ________
   Other ________

4. Bug Screen
   Material ________________ Velocity ____________ FPM
   Dirt Holding Capacity_____#/Sq. In. Other ______________
   Failure ΔP ____________in. W.G. ΔP ____________in. W.G. (Clean)

5. Weather Hood	Material ______________	Rain ______ Snow _______
6. Inertial Separator	Material: Mild Steel ________	Efficiency ______________
	Stainless Steel ______	ΔP _________in. W.G.(Clean)
	Area Class: C1 ___, Div ___, Gp ___	Motors Qty _____
	Paint ________________	Motor HP ______________
	Fans: Explosion Proof ______	Non-Sparking ___
7. Pre-Filter	Roll Type ______________	Self-Contained ___________
	Pad Type _______________	Other ________________
	Efficiency ________	Dirt Holding Capacity___ #/Sq. In.
8. Intermediate Filter	Roll Type ______________	Self-Contained___________
	Pad Type _______________	Other ________________
	Efficiency ________	Dirt Holding Capacity___ #/Sq. In.
9. High Efficiency Filter	Roll Type ______________	Self-Contained ___________
	Pad Type _______________	Other ________________
	Efficiency ________	Dirt Holding Capacity___ #/Sq. In.
10. Self Clean Filter	Cartridge Type ___________	Efficiency _________
	Pad Type _______________	Dirt Holding Capacity___ #/Sq. In.
11. Blow-In Doors	If blow-in doors are required,than limit switches should be specified to alarm when doors open.	
12. Internal Lights	Explosion Proof ___ Yes ___ No	Quantity _______
	Lumens _____________	Voltage ______________

13. Platforms and Ladders — Required to Service
    Motors ______
    Junction Boxes ______
    Other ______
    Other Requirements ______

14. Access Doors — Requirements ______ Location ______

15. Support Steel — Required — Not Required

16. Pressure Switches

Quantity	______		Alarm	Shutdown
Location	______	Setting	______	______ in. W.G.
	______		______	______ in. W.G.
	______		______	______ in. W.G.

17. Enclosure Finish — Manufacturer's Std ______ Other ______

18. General
    a) Collapsing Pressure Of Filter house (>21" Std) ______ In. W.G.
    b) No Field Welding
    c) No Corteen
    d) All Electrical Components Will Be Wired To Junction Box(s) Which Will Be Provided At Location ______
    e) Bolts And Gaskets To Be Furnished In Vendor Plant.
    f) Fit-Up Of Components For Air Filter In Vendor Plant.

19. Anti-Ice — Piccolo Tubes: Required ______ Not Required ______
    Materiel: Stainless Steel ______ Other ______

Appendix C-6

Air/Oil Cooler Specifications Check List

1. Site Conditions
   - A. Location ____________________
   - B. Altitude ____________________
   - C. Ambient Temperature Range
     Max __________ Min ___________

2. Type
   - A. Vertical ____ Horizontal ____
   - B. Indoor _____ Outdoor _____
   - C. Louvers: With _____ Without _____
     Drives: Electric _____ Hydraulic _____

3. Number of Cooling Sections ____________

4. Section _____ Gas Generator
   - A. Oil Flow ______________gpm
   - B. Heat Load_____________Btu/hr
   - C. Design Temp 300°F (150°C)
     Other ____________
   - E. Oil Outlet Temp 140°F (60°C)
     Other ___________
   - F. Allowable ΔP __5 psig (Oil Side)

D. Design Press.__________psig

5. Section _____ Power Turbine

A. Oil Flow_____________gpm

B. Heat Load____________Btu/hr

C. Design Temp 300°F (150°C)

Other ____________

D. Design Press.__________psig

6. Codes

A. Design to ASME

Stamp ______Yes _____No

B. Section VIII Boiler Code for

Location _______________

7. Electrical Requirements

A. Codes UL_____ CSA _____

B. Hazardous Area Classification

Class ___, Div ___ Group ___

8. Fan Requirements

A. Non-Sparking ___ Yes ___ No

G. Type Oils: _____________________

Specific Gravity ___________________

E. Oil Outlet Temp 140°F (60°C)

Other ___________

F. Allowable ΔP __5 psig (Oil Side)

G. Type Oils: ___________________

Specific Gravity ___________________

C. API Std __661__

D. Customer Spec _____________________

C. Available Power

AC_____, Ph _____, Hz ____

DC _____

Other ____________________

B. Manually Adjustable Pitch Blades

____ Yes ____ No

9. Drive Requirements

A. Hydraulic Drives

__________ Direct Connected

B. Electric Drive Connection

____ Direct _____ Indirect

Indirect Electric Drive ______belt _____gear

10. Vibration Switch - Remote Resetable

SPDT _______ (for hydraulic drives)

DPDT _______ (for electric drives)

11. Special Conditions

A. Physical Location

Wall Mounted _____ Floor _____

B. Louvers Required

Inside _____ Outside _____

C. Louver Actuators - When Required To Be Fail-safe, Closing Outside Louvers

D. Actuator Temperature Control Switches _____

E. Outside Louvers To Be Weather Tight _____

F. Air Actuators To Be Tubed To Coupling On Radiator ____________

12. Paint

A. Per Customer Spec __________

B. Per Manufacturer Spec ________

C. Vendor Standard _________

D. Galvanized ______ Stainless Steel ______

13. Electrical Conditions

A. Actuators, temperature controls, vibration switches to be wired to external junction boxes to meet electrical codes.

B. Electric motor drives (and motor heater) to include conduit wiring to external surface of radiator enclosure.

14. Noise
    A. Noise levels to meet customer spec for field requirement ______
    B. Outside source noise to be provided by Purchaser _______

15. Hinged Access Door To Inlet Side Of Radiator Is Required Unless Otherwise Specified

16. Provision for floor drain of radiator enclosure box _______ Yes _______ No

17. Support Legs _______ Yes _______ No

18. Tube Bundles To Be
    Cleanable _________
    Turbulators Removable _________
    Type Of Fin Construction:
        Rolled In ________
        Tension Wound ________
        Welded "L" ________
        Other ________

    Material: Steel with Aluminum Fins,
        Other_______________
    Fouling Factor <u>0.0005</u>
        Std/Customers Spec ____________

19. Hydraulic Pressure Available ____ psig
        Other _____ psig

20. Glycol/Water Heating Coil By Spec ______ (if required)

21. Insulation By Spec _________ (if required)

22. Packaging For Shipment Per Spec ________________

23. Spare Parts List Required _____ Yes ______ No

24. Commercial Considerations
    - A. Delivery Commitment _______
    - B. Progress Reports __ Yes __ No
      Monthly _____ Other _____
    - C. Warrantees Std ____ Other ____
    - D. Sourcing ____________________

25 Drawings
    - A. Proposal Outline ______________________
    - B. Schedule For Approval Drawings (from date of order) _____________________
    - C. Schedule For Final Certified Drawings (from date of order) _________________

26 Quality and Inspection
    - A. Witness and Inspection Notice ______________________________________
    - B. Shipping Release ____
    - C. Material Certification ____
    - D. Radiographic Requirements ____

27. Installation and Maintenance Manuals
    - A. Number Required __________________
    - B. Delivery Required __________________

Appendix C-7

Gaseous Fuel Properties

This information is reprinted from an article "Gas Turbine Fuels," by W.S.Y Hung, Ph.D., courtesy of Solar Turbines Incorporated.

FUEL GAS PROPERTIES

The physical, thermodynamic and chemical fuel-gas properties relevant to gas turbine operations are given here for single-component pure gases (Table C-7-1), for binary fuel-gas mixtures of methane/nitrogen (Table C-7-2), for methane/carbon-dioxide (Table C-7-3), and methane/hydrogen (Table C-7-4). Finally, fuel gas properties from a selection of typical fuel gases, derived from different coal gasification processes, are summarized for medium-Btu gas (Table C-7-5) and low-Btu gas (Table C-7-6).

Parameters	Units	NH_3 100%	CO 100%	H_2 100%	CH_4 100%	C_2H_6 100%	C_3H_8 100%	C_4H_{10} 100%	C_5H_{12} 100%	C_6H_{14} 100%
Molecular Weigh	$\frac{lb}{mol}$	17.03	28.01	2	16.04	30.07	44.10	58.12	72.15	86.18
Lower Heating Value	$\frac{Btu}{lb}$	7410	4343	51,579	21,504	20,416	19,929	19,665	19,501	19,394
Lower Heating Value	$\frac{Btu}{scf}$	347	321	274	909	1620	2320	3018	3714	4413
Adiabatic Temperature Rise, $\Delta T_{ab_{stoich}}$	°C	1700	2232	2103	1924	1980	1994	2000	2006	2008
Stoich Air/Fuel Ratio	$\frac{lb}{lb}$	6.06	2.48	34.3	17.23	16.17	15.75	15.53	15.4	15.31
Fuel Flow Mass Ratio	$\frac{(Btu/lb)_{CH_4}}{Btu/lb}$	2.9	4.95	0.42	1	1.05	1.08	1.09	1.1	1.11
Wobbe Index $\frac{LHV}{\sqrt{\frac{M_{Gas}}{M_{Air}}}}$	$\frac{Btu}{scf}$	452	326	1039	1222	1590	1880	2130	2353	2558
Fuel Flow Volume Ratio	$\frac{(Wobbe)_{CH_4}}{Wobbe}$	2.7	3.75	1.18	1	0.77	0.65	0.57	0.52	0.48
Limit of Flammability Lower, L_{25}	% Vol.	15	12.5	4	5	3	2.1	1.8	1.4	1.2
Limit of Flammability Higher, U_{25}	% Vol.	28	74	76	15	12.4	9.5	8.4	7.8	7.4
Limit of Flammability Ratio- U_{25}/L_{25}		1.87	5.92	18.75	3	4.13	4.52	4.67	5.57	6.17

Table C-7-1. Properties of Pure Gases.

Parameters	Units	1/0	.88/.12	.66/.34	.55/.45	.44/.56	.33/.67	.22/.78	.11./89	/-55 .945
Molecular Weight	$\frac{lb}{mol}$	16.04	17.48	20.11	21.43	22.75	24.07	25.38	26.7	27.36
Lower Heating Value	$\frac{Btu}{lb}$	21,506	17,369	11,321	8854	6673	4731	2990	1421	694
Lower Heating Value	$\frac{Btu}{scf}$	909	800	600	500	400	300	200	100	50
Adiabatic Temperature Rise, $\Delta T_{ab_{stoich}}$	°C	1924	1887	1841	1795	1729	1629	1457	1090	686
Stoich Air/Fuel Ratio	$\frac{lb}{lb}$	17.24	13.92	9.07	7.1	5.35	3.79	2.4	1.14	0.556
Fuel Flow Mass Ratio	$\frac{(Btu/lb)_{CH_4}}{Btu/lb}$	1	1.24	1.9	2.43	3.22	4.56	7.19	16.13	30.99
Wobbe Index $\frac{LHV}{\sqrt{\frac{M_{Gas}}{M_{Air}}}}$	$\frac{Btu}{scf}$	1222	1030	720	581	451	329	214	104	51.4
Fuel Flow Volume Ratio	$\frac{(Wobbe)_{CH_4}}{Wobbe}$	1	1.19	1.7	2.1	2.71	3.71	5.71	11.76	23.77
Limit of Flammability Lower, L_{25}	% Vol.	5	5.68	7.58	9.18	11.86	16.15	24.23	NF	NF
Limit of Flammability Higher, U_{25}	% Vol.	15.5	15.36	19.15	21-59	24.27	29.15	35.23	NF	NF
Limit of Flammability Ratio- U_{25}/L_{25}		3.1	2.7	2.53	2.35	2.05	1.805	1.45	—	—

Table C-7-2. Properties of CH_4/N_2 Gases.

Parameters	Units	1/0	.88/.12	.66/.34	.55/.45	.44/.56	.33/.67	.22/.78	.11/.89	.055 .945
Molecular Weight	$\frac{lb}{mol}$	16.04	19.4	25.55	28.63	31.7	34.78	37.88	40.93	42.47
Lower Heating Value	$\frac{Btu}{lb}$	21,505	15,650	8911	6628	4788	3273	2005	927	447
Lower Heating Value	$\frac{Btu}{scf}$	909	800	600	500	400	300	200	100	50
Adiabatic Temperature Rise, $\Delta T_{ab_{stoich}}$	°C	1924	1887	1791	1720	1623	1482	1256	837	456
Stoich Air/Fuel Ratio	$\frac{lb}{lb}$	17.24	12.55	7.14	5.31	3.84	2.62	1.61	0.74	0.36
Fuel Flow Mass Ratio	$\frac{(Btu/lb)\ CH_4}{Btu/lb}$	1	1.37	2.41	3.24	4.49	6.67	10.73	23.20	48.11
Wobbe Index $\frac{LHV}{\sqrt{\frac{M_{Gas}}{M_{Air}}}}$	$\frac{Btu}{scf}$	1222	978	639	503	382	274	175	84	41
Fuel Flow Volume Ratio	$\frac{(Wobbe)\ CH_4}{Wobbe}$	1	1.25	1.91	2.43	3.20	4.46	6.96	14.55	29.8
Limit of Flammability Lower, L_{25}	% Vol.	5	5.68	8.075	10.1	12.56	17.3	NF	NF	NF
Limit of Flammability Higher, U_{25}	% Vol.	15.1	15.36	18.575	20.59	23.25	27.15	NF	NF	NF
Limit of Flammability Ratio- U_{25}/L_{25}		3.1	2.7	2.3	2.04	1.88	1.57	—	—	—

Table C-7.3 Properties of CH_4/CO_2 Gases

Parameters	Units	1/0	.9/.1	.8/.2	.7/.3	.6/.4	.5/.5	.4/.6	.3/.7	.2/.8	.1/.9	0/1
Molecular Weight	$\frac{lb}{mol}$	2.02	3.02	4.82	6.22	7.63	9.03	10.43	11.83	13.24	14.64	16.04
Lower Heating Value	$\frac{Btu}{lb}$	51,579	37,466	31,565	28,324	26,275	24,862	23,830	23,042	22,421	21,919	21,504
Lower Heating Value	$\frac{Btu}{scf}$	274	338	401	464	528	592	655	719	782	846	$09
Adiabatic Temperature Rise, $\Delta T_{ab_{stoich}}$	°C	2103	2052	2018	1994	1977	1963	1952	1943	1936	1929	1924
Stoich Air/Fuel Ratio	$\frac{lb}{lb}$	34.3	26.3	22.9	21.1	19.9	19.1	18.6	18.1	17.76	17.5	17.23
Fuel Flow Mass Ratio	$\frac{(Btu/lb)_{CH_4}}{Btu/lb}$	0.42	0.57	0.68	0.76	0.82	0.86	0.90	0.93	0.96	0.98	1
Wobbe Index $\frac{LHV}{\sqrt{\frac{M_{Gas}}{M_{Air}}}}$	$\frac{Btu}{scf}$	1039	982	982.8	1002	1029	1059	1091	1124	1157	1189	1221
Fuel Flow Volume Ratio	$\frac{(Wobbe)_{CH_4}}{Wobbe}$	1.18	1.24	1.24	1.22	1.19	1.15	1.12	1.09	1.06	1.03	1
Limit of Flammability Lower, L_{25}	% Vol.	4	4.08	4.17	4.26	4.35	4.44	4.55	4.65	4.76	4.88	5
Limit of Flammability Higher, U_{25}	% Vol.	75	53.57	41.67	34.09	28.85	25	22.06	19.74	17.86	16.3	15
Limit of Flammability Ratio, U_{25}/L_{25}		18.75	13.13	9.99	8	6.63	5.63	4.35	4.25	3.75	3.34	3

Table C-7-4. Properties of H_2/CH_4 Gases.

Process	Blue Water Gas	Coal Gas Vertical Retorting	Winkler Fluidized Bed	Lurgi Fixed Bed	Koppers Totzek Entrained Bed	Texaco	Brit. Gas Moving Bed Stagging	Foster Wheeler Fluidized Bed	Coke Oven Gas
Air									
Oxygen									
Steam									
Analysis: %Mol									
O_2	—	0.4		0.4					
N_2	4.5	6.2	1.0	0.32	1.13	1.13	0.45	0.67	2.8
CO_2	4.7	4.0	19	28.87	6.27	9.22	1.07	11.80	4.0
H_2O	—	—	—	0.26	0.37	0.36	0.36	0.36	—
CO	41	18	38	19.52	55.11	53.06	58.73	43.46	8.0
H_2	49	49.4	40	38.80	3718	35.97	30.87	36.46	51.2
CH_4	0.8	20	2	11.20	—	0.26	7.78	7.35	30.6
C_2H_6	—	—	—	—	—	—	0.54	—	3.0
H_2S	—	—	—	1.0	—	0.20	—	—	
Molecular Weight	15.9	13.1	20.4	21.1	19.3	20.1	19.21	19.51	11.76
LHV (Btu/scf)	273	383	250	274	278	271	354	306	993
LHV (Btu/lb)	6504	11,051	4640	4918	5469	5125	6982	5954	15,890
Wobbe Index Btu/scf	368	568	298	320	341	326	434	373	773
Adiabatic Temp. Rise									
ΔT stoich. (OC)	2073	1978	1924	1785	2100	2063	2108	1996	1965
A/F stoich. by wt.	4.03	7.87	2.91	3.4	3.34	3.10	4.48	3.86	11.9
Fuel Flow Mass Ratio	3.31	1.95	4.63	4.37	3.93	3.75	3.08	3.61	1.35
Fuel Flow Vol. Ratio	3.32	2.15	4.11	3.82	3.58	4.20	2.81	3.28	1.58
Limits of Flammability									
Lower U_{25}	6.38	5.6	7.48	7.50	7.56	7.86	7.04	7.23	4.86
Upper U_{25}	70.0	38.8	66.7	45.9	72.6	71.6	55.4	55.5	30.2
Ratio	11.0	6.91	8.92	6.12	9.60	9.10	7.86	7.68	6.21
H—/CO (by Vol.)	1.195	2.744	1.053	1.99	0.675	0.678 1	0.526	0.839	6.4

Table C-7-5. Properties of Medium-Btu Gas.

Process Air	Fire Flood in Situ	Producer Gas from Coal	Blast Furnace Gas	Lurgi Static Bed	Lurgi Static Bed + Steam	Foster Wheeler Fluidized Bed	Bureau of Mines 7644	Winkler Fluidized Bed	Wellman Galusha Fixed Bed	IGT U-Gas Fluidized Bed	Shale Oil Retorting + in Situ
Oxygen											
Steam											
Analysis % Mat											
O_2	0.12	—	—	—	—	—	—	—			
N_2	80	52.4	60	40.26	30.2	50.4	54.5	55.3	50	52	52
CO_2	17	4	11	14.18	10.7	0.5	7.2	10	4	10	10
H_2O	—	—	——	0.257	27.8	0.5	—	—	—	—	—
CO	OAS	29	27	16.71	10.7	31.8	20	22	29	19	20
H_2	—	12	2	22.98	15.7	15.6	15.5	12	15	14	15
CH_4	1	2.6	—	4.97	4.4	0.5	2.8	0.7	2	5	3
C_2H_6 +	1.25	—	—	—	—	—	—	—	—	—	—
H_2S	0.48	—	—	0.63	0.5	0.7	—	—	—	—	—
Molecular Weight	31	25.2	29.25	23.7	22.4	23.97	24.8	26.4	24.5	25.4	25.4
LHV (Btu/scf)	50	150	92.3	165	118	150	132	110	152	145	133
LHV (Btu/lb)	612	2250	1194	2646	2006	2386	2021	1578	2357	2164	1983
Wobbe Index	48	161	91.8	183	134	165	143	115	166	155	142
ΔT_{stoich} (°C)	628	1574	1204	1548	1258	1646	1453	1338	1603	1467	1439
(A/F) stoich by wt.	2.08	1.41	1.462	1.81	1.40	1.48	1.30	1.040	1.5	1.4	1.28
Fuel Flow Mass Ratio	35.1	9.56	18	8.13	10.7	9.01	10.6	13.6	9.1	9.9	10.8
Fuel Flow Vol Ratio	25.3	7.59	13.3	6.67	9.12	7.41	8.62	10.6	7.4	7.9	8.6
Limits of Flammability											
Lower U_{25}	—	17.9	37.59	12.6	18	15.06	16.6	20.5	16.54	17.7	12.5
Upper U_{25}	—	66.1	73.06	61.8	60.57	69.8	64.95	71.17	66.88	59.0	38.1
Ratio	—	3.7	1.94	4.11	3.36	4.64	3.91	3.47	4.04	3.33	2.88
(H_2/CO) by vol	—	0.414	0.074	1.375	1.467	0.491	7.75	0.545	0.517	0.737	0.75

Table C-7-6. Properties of Low-Btu Gas.

Appendix C-8

Liquid Fuel Properties

This information is reprinted from an article "The Use of Liquid Fuel in Industrial Gas Turbines," by M.A. Galica and K.H. Maden, courtesy of Solar Turbines Incorporated.

Typical Properties of Liquid Fuels

	True Distillates		Ash-bearing Fuels	
Fuel Type	Kerosene	No. 2 Distillate	Blended Residuals and Crudes	Heavy Residuals
Specific Gravity, 60°F/60°F	0.78-0.83	0.82-0.86	0.80-0.92	0.92-1.05
Viscosity, cSt, 100°F	1.4-2.2	2.0-4.0	2-100	100-1800
Flash Point, °F	130-160	150-200	50-200	175-265
Pour Point, °F	–50	(–10)–30	15-110	15-95
Lower Heating Value Btu/lb	18,480-18,700	18,330-18,530	18,000-18,620	17,200-18,000
MJ/kg	42.9-43.4	42.6-43.0	41.8-43.2	39.9-41.8
Filterable Dirt, % Max	0.002	0.005	0.05	0.2
Carbon Residue (10% Bottoms), %	0.01-0.1	0.03-0.3	—	—
Carbon Residue (100% Sample), %	—	—	0.3-3	2-10
Sulfur, %	0.01-0.1	0.1-0.8	0.2-3	0.5-4
Nitrogen, %	0.002-0.01	0.005-0.06	0.06-0.2	0.05-0.9
Hydrogen, %	12.8-14.5	12.2-13.2	12.0-13.2	10-12.5
Ash, ppm	1-5	2-50	25-200	100-1000
Sodium Plus Potassium, ppm	0-0.05	0-1	1-100	1-350
Vanadium, ppm	0-0.1	0-0.1	0.1-80	5-400
Lead, ppm	0-0.05	0-1	0-1	0-25
Calcium, ppm	0-1	0-2	0-10	0-50

Typical Properties of Alcohols

Parameters	Methanol	Ethanol	Isopropanol
Formula	CH_3OH	CH_3CH_2OH	$(CH_3)_2\ CHOH$
Molecular Weight	32.04	46.07	60.09
Specific Gravity (60°F.60°F)	0.796	0.794	0.789
Boiling Temperature (°F)	149	172	180
Flash Point (°F)	52	55	53
Lower Heating Value (Btu/lbm)	8,570	11,500	19,070
Autoignition Temperature (°F)	867	793	750
Flammability Limits (Vol. %)			
Lower	6.7	4.3	2.0
Higher	36.0	19.0	12.0

Appendix C-9

List of Symbols

A	Area, Ft^2
Btu	British thermal unit
C	Velocity of the air entering the compressor or air and combustion products leaving the turbine, *ft per sec*
CDP	Compressor discharge pressure, *psia*
CDT	Compressor discharge temperature, °R
c_p	Specific heat at constant pressure, Btu/lb_m °*F*
c_v	Specific heat at constant volume, Btu/lb_m °*F*
°E	Engler, degrees
EGT	Exhaust gas temperature, °*R*
g_c	Acceleration due to gravity, ft/sec^2
h_1	Enthalpy of the fluid entering, Btu/lb_m
h_2	Enthalpy of the fluid leaving, Btu/lb_m
h_i	Enthalpy of the fluid entering, Btu/lb_m
h_o	Enthalpy of the fluid leaving, Btu/lb_m
HP	Horsepower
Hr	Hours
J	Ratio of work unit to heat unit, 778.2 $ft\ lb_f/Btu$
k	Ratio of specific heats, c_p/c_v
KE_1	kinetic energy of the fluid entering, Btu/lb_m
KE_2	Kinetic energy of the fluid leaving, Btu/lb_m
kJ	Kilojoules
kW	Kilowatts
Mo	Mass of a given volume of oil
Mw	Mass of a given volume of water
MW	Power output (electric generator)
MW	Molecular weight
N_1	Single spool or low pressure compressor-turbine rotor speed, rpm
N_2	Dual spool high pressure compressor-turbine rotor speed, rpm
N_3	Power turbine rotor speed, rpm
NO_x	Oxides of nitrogen

P_2	Compressor inlet pressure, *psia*
P_3	Single spool compressor discharge pressure, single spool low pressure compressor discharge pressure, *psia*
P_4	Dual spool low pressure compressor discharge pressure, *psia*
P_5	Turbine inlet pressure, *psia*
P_3/P_2	Single spool compressor pressure ratio, dual spool low pressure compressor pressure ratio
P_4/P_2	Dual spool compressor pressure ratio
P_4/P_3	Dual spool high pressure compressor pressure ratio
P_{AMB}	Atmospheric total pressure, *psia*
PE_1	Potential energy of the fluid entering, Btu/lb_m
PE_2	Potential energy of the fluid leaving, Btu/lb_m
P_i	Inlet pressure, *psia*
P_o	Discharge pressure, *psia*
cP	Absolute or Dynamic Viscosity, Centipoises
P_{PT}	Power input to the load device, Btu/sec
Ps_3	Low pressure compressor out pressure (static), psig
Ps_4	High pressure compressor out pressure (static), psig
P_{t7}	Exhaust total pressure, *psia*
Q	Volume flow cubic feet per second (CFS)
${}_1Q_2$	Heat transferred to or from the system, Btu/lb_m
Q_m	Mechanical bearing losses of the driven equipment, Btu/sec
Q_m	Mechanical bearing losses of the power turbine, Btu/sec
Q_r	Radiation and convection heat loss, Btu/sec
R_c	Compressor pressure ratio
R_t	Turbine pressure ratio
secs R.I.	Redwood No 1 (UK), seconds
cST	Kinematic viscosity = Absolute viscosity/density. centistokes
SUS	Saybolt Universal (USA), seconds
T_{1DB}	Dry bulb temperature upstream of the cooler, $°R$
T_{2DB}	Dry bulb temperature downstream of the cooler, $°R$
T_{2WB}	Wet bulb temperature downstream of the cooler, $°R$
T_2	Compressor inlet temperature, $°R$
T_3	Single spool compressor discharge temperature, dual spool low pressure compressor discharge temperature, $°R$
T_4	Dual spool low pressure compressor discharge temperature, $°R$

T_5	Turbine inlet temperature, °*R*
T_7	Exhaust gas temperature, °*R*
T_{AMB}	Ambient air temperature, °*R*
T_{EXH}	Turbine total exhaust temperature, °*R*
Ti	Inlet temperature, °*R*
TIT	Turbine total inlet temperature, °*R*
T_o	Discharge temperature, °*R*
U	Velocity, *feet per sec*
D	Diameter, in^2, ft^2
V	Vibration
W	Mass flow rate, lb_m/sec
W_2	Relative velocity
W_a	Air flow entering the compressor, *lb/sec*
W_f	Fuel flow, lb_m/sec
W_f/CDP	Gas turbine burner ratio units, $lb_m/hr\text{-}psia$
W_g	Turbine inlet gas flow, lb_m/sec
${}_1W_2$	Work per unit mass on or by the system, $ft\text{-}lb_f/lbm$
${}_1W_2/J$	Work, Btu/lb_m
Z_{ave}	Average compressibility factor of gas
β_2	Relative direction of air leaving the stator vanes
η_c	Compressor efficiency
η_t	Gas generator efficiency
η_{PT}	Power extraction turbine efficiency
θ_1	Direction of air leaving the stator vanes

Appendix C-10

Conversion Factors

TO OBTAIN	MULTIPLY	BY
bars	atmospheres	1.0133
cm Hg @ 32°F (0°C)	atmospheres	76.00
ft of water @ 40°F (4°C)	atmospheres	33.90
in. Hg @ 32°F (0°C)	atmospheres	29.92
in Hg @ 60°F (15°C)	atmospheres	30.00
lb/sq. ft	atmospheres	2,116
lb/sq. in.	atmospheres	14.696
cm Hg @ 32°F (0°C)	bars	75.01
in. Hg @ 32°F (0°C)	bars	29.53
lb/sq ft.	bars	2088
lb/sq in.	bars	14.50
ft-lb	Btu	778.26
Horsepower-hr	Btu	3.930×10^{-4}
joules	Btu	1055
kilogram calories	Btu	2.520×10^{-1}
kilowatt-hr	Btu	2.930×10^{-4}
grams	carats	2.0×10^{-1}
ounces	carats	7.055×10^{-3}
ft.	centimeters	3.281×10^{-2}
in.	centimeters	$3.937\ 10^{-1}$
kilogram/hr m	centipoise	3.60
lb/sec ft	centipoise	6.72×10^{-4}
Poise	centipoise	1×10^{-2}
ft of water @ 40°F (4°C)	cm Hg	4.460×10^{-1}
in. H_2O @ 40°F (4°C)	cm Hg	5.354
kilogram/sq m.	cm Hg	135.95
lb/sq. in.	cm Hg	1.934×10^{-1}
lb/sq./ft	cm Hg	27.85
ft/sec	cm/sec	3.281×10^{-2}
mph	cm/sec	2.237×10^{-2}
cu in.	cu cm (dry)	6.102×10^{-2}

gal U.S.	cu cm (liquid)	2.642×10^{-4}
liters	cu cm (liquid)	1×10^{-3}
cu cm	cu ft (dry	2.832×10^{4}
cu in.	cu ft (dry)	1728
liters	cu ft (liquid)	28.32
lb	cu ft H_2O	64.428
gal U.S.	cu ft (liquid)	7.481
cu m/min	cu ft/min	2.832×10^{-2}
gallons/hr	cu ft/min	4.488×10^{2}
liters/sec	cu ft/min	4.719×10^{-1}
cu cm	cu in.	16.39
gal U.S.	cu in.	4.329×10^{-3}
liters	cu in.	1.639×10^{-2}
quarts	cu in.	1.732×10^{-2}
cu ft.	cu meters	35.31
cu in.	cu meters	61023
gas U.S.	cu meters	264.17
liters	cu meters	1.0×10^{3}
meters	ft	3.048×10^{-1}
cm/sec	ft/min	5.080×10^{-1}
km/hr	ft/min	1.829×10^{-2}
mph	ft/min	1.136×10^{-2}
cm/sec	ft/sec	30.48
gravity	ft/sec	32.1816
km/hr	ft/sec	1.097
knots	ft/sec	5.925×10^{-1}
mph	ft/sec	6.818×10^{-1}
Btu	gram-calories	3.969×10^{-3}
joules	gram-calories	4.184
kg/m	gram/cm	1×10^{-1}
lb/ft	gram/cm	6.721×10^{-2}
lb/in.	gram/cm	5.601×10^{-3}
kg/cu m	grams/cu cm	1000
lb/cu ft	grams/cu cm	62.43
slugs	grams/cu cm	1.940
Btu/hr	horsepower	2546
Btu/sec	horsepower	7.073×10^{-1}
ft-lb/min	horsepower	33,000
ft-lb/sec	horsepower	550

kilowatts	horsepower	7.457×10^{-1}
m-kg/sec	horsepower	76.04
metric hp	horsepower	1.014
Btu-sec	horsepower, metric	6.975×10^{-1}
horsepower	horsepower, metric	9.863×10^{-1}
kilowatts	horsepower, metric	7.355×10^{-1}
m-kg/sec	horsepower, metric	75.0
Btu	horsepower-hours	2.545×10^{3}
ft-lb	horsepower-hours	1.98×10^{6}
kilowatt-hours	horsepower-hours	7.457×10^{-1}
m-kg	horsepower-hours	2.737×10^{5}
cm Hg @ 32°F (0°C)	in H_2O @ 40°F (4°C)	1.868×10^{-1}
in Hg @ 32°F (0°C)	in H_2O @ 40°F (4°C)	7.355×10^{-2}
kg/sq m	in H_2O @ 40°F (4°C)	25.40
lb/sq ft.	in H_2O @ 40°F (4°C)	5.202
lb/sq in.	in H_2O @ 40°F (4°C)	3.613×10^{-2}
atmospheres	in H_2O @ 60°F (15°C)	2.455×10^{-3}
in Hg @ 32°F (0°C)	in H_2O @ 60°F (15°C)	7.349×10^{-2}
atmospheres	in Hg @ 32°F (0°C)	3.342×10^{-2}
ft H_2O 40°F (4°C)	in Hg @ 32°F (0°C)	1.133
in. H_2O @ 40°F (4°C)	in Hg @ 32°F (0°C)	13.60
kg/sq m	in Hg @ 32°F (0°C)	3.453×10^{2}
lb/sq ft.	in Hg @ 32°F (0°C)	70.73
lb/sq in.	in Hg @ 32°F (0°C)	4.912×10^{-1}
cm	in	2.54
Btu	Joules	9.480×10^{-4}
ft-lb	Joules	7.376×10^{-1}
hp-hr	Joules	3.725×10^{-7}
kg calories	Joules	2.389×10^{-4}
kg m	Joules	1.020×10^{-1}
watt-hr	Joules	2.778×10^{-4}
grams/cu cm	kg/cu m	1×10^{-3}
lb/cu ft	kg/cu m	62.43×10^{-3}
in H_2O @ 40°F (4°C)	kg/sq cm	3.28×10^{-7}
in Hg @ 32°F (0°C)	kg/sq cm	28.96
lb/sq ft.	kg/sq cm	2.048×10^{3}
lb/sq in.	kg/sq cm	14.22
Btu	kilogram-calories	3.9685
ft-lb	kilogram-calories	3087

m-kg	kilogram-calories	4.269×10^2
grams	kilograms	1×10^3
lb	kilograms	2.205
oz	kilograms	35.27
cm	kilometers	1×10^5
ft	kilometers	3.281×10^3
miles	kilometers	6.214×10^{-1}
nautical miles	kilometers	5.400×10^{-1}
Btu/sec	kilowatts	9.480×10^{-1}
ft-lb/sec	kilowatts	7.376×10^2
horsepower	kilowatts	1.341
kg calories/sec	kilowatts	2.389×10^{-1}
ft/sec, fps	km/hr	9.113×10^{-1}
knots	km/hr	5.396×10^{-1}
m/sec	km/hr	2.778×10^{-1}
mph	km/hr	6.214×10^{-1}
ft/sec, fps	knots	1.68
km/hr	knots	1.853
m/sec	knots	5.148×10^{-1}
mph	knots	1.151
grains	lb, avdp	7000
grams	lb, avdp	453.6
ounces	lb, avdp	16.0
poundals	lb. avdp	32.174
slugs	lb, avdp	3.108×10^{-2}
kg/cu m	lb/cu ft	16.02
grams/cu cm	lb/cu in.	27.68
lb/cu ft	lb/cu in.	1728
atmospheres	lb/sq in.	6.805×10^{-2}
ft H_2O @ 40°F (4°C)	lb/sq in.	2.307
in Hg @ 32°F (0°C)	lb/sq in.	2.036
kg/sq m	lb/sq in.	7.031×10^2
cu cm	liters	1×10^3
cu ft	liters	3.532×10^{-2}
cu in	liters	61.03
gal Imperial	liters	2.200×10^{-1}
gal U.S.	liters	2.642×10^{-1}
quarts	liters	1.057
ft-lb	meter-kilogram	7.233

joules	meter-kilogram	9.807
ft/sec, fps	meter/sec	3.281
km/hr	meter/sec	3.600
miles/hr	meter/sec	2.237
ft	meters	3.281
in.	meters	39.37
miles	meters	6.214×10^{-4}
in.	microns	3.937×10^{-5}
ft	miles	5280
kilometers	miles	1.609
nautical mile	miles	8.690×10^{-1}
in Hg @ 32°F (0°C)	milibars	2.953×10^{-2}
ft/sec, fps	mph	1.467
km/hr	mph	1.609
knots	mph	8.690×10^{-1}
m/sec	mph	4.470×10^{-1}
ft	nautical mile	6076.1
m	nautical mile	1852
miles	nautical mile	1.151
grains	ounces, avdp	4.375×10^{2}
grams	ounces, avdp	28.35
lb, avdp	ounces, avdp	6.250×10^{-2}
cu cm	ounces. fluid	29.57
cu in.	ounces. fluid	1.805
degree (arc)	radians	57.30
degrees/sec	radians/sec	57.30
rev/min, rpm	radians/sec	9.549
rev/sec	radians/sec	15.92×10^{-2}
radians/sec	rev/min. rpm	1.047×10^{-1}
radians	revolutions	6.283
lb	slug	32.174
sq ft	sq cm	1.076×10^{-3}
sq in.	sq cm	1.550×10^{-1}
sq cm	sq ft	929.0
sq in.	sq ft	144.0
sq cm	sq in.	6.452
Btu/sec	Watts	9.485×10^{-4}

Appendix C-11

Guide for Estimating Inlet Air Cooling Affects of Water Fogging

Consider a gas turbine with a mass flow of 500 pounds/second (pps), compressor efficiency of 80%, a compression ratio of 14:1, and a compressor discharge pressure of 200 psia. Also consider ambient air at 95°F and 100% Relative Humidity (RH). Calculate the expected compressor discharge temperature.

From the Pyschrometric Chart the Humidity Ratio is 0.0366

$$\frac{\text{\# moisture}}{\text{\# dry air}}$$

First solve for CDT in dry air at 95°F:

$$\text{Using } \eta_c = \frac{R_c^{\sigma} - 1}{\left(\frac{CCT}{CIT}\right) - 1}..$$

Where

Rc = Compression Ratio

CIT = Compressor Inlet Temperature (°R)

CDT = Compressor Discharge Temperature (°R)

$$\sigma = \frac{K-1}{K}$$

$$K = \frac{C_p}{C_v} = 1.406$$

C_p = Ratio of Specific Heat at constant pressure

C_v = Ratio of specific Heat at constant volume

Solving for CDT yields 886°F.

At 95% and 100% RH the air flow into the gas turbine includes 16.6 pps of water. At a latent heat of vaporization of 970 Btu/# this water will give up 16,050 Btu/sec (cp = 0.24672).

Using $Q = W_A \bullet c_p \bullet \Delta T$

Solving for ΔT yields 130°F.

Therefore, the dry bulb temperature of 886°F will be reduced to 756°F.

This addresses the maximum amount of water that can be expected to be evaporated in the inlet system upstream of the gas turbine-compressor inlet. Now consider adding water for wet compression.

Consider first adding 90 gpm of water to this inlet condition. This is approximately 2.5% of the inlet flow.

Within the compressor this water is evaporated through latent heat of evaporation.

$$\text{At 90 gmp} = 12.5 \text{ pps; then } 12.5_{pps} \square 970 \frac{Btu}{sec}$$
$$= 12{,}125 \text{ Btu/sec } \textit{Heat Absorbed}$$

Again solving for ΔT yields 99°F. Therefore, the injection of 90 gpm into the air stream at the compressor inlet will result in a compressor discharge temperature of 656°F.

Increasing amounts of water can be added until either the compressor surges or the discharge from the compressor contains liquid water droplets.

Consider adding 10% water at the original inlet conditions.

10% water = 50 pps = 360 gpm.

Repeating the above calculations yields a compressor discharge temperature of 383°F. At 200 psia & 380°F the water will be at the Liquid-Vapor Line (this can be confirmed on the Mollier Chart).

Too much liquid water injected into the compressor will result in changing the airfoil configuration as water accumulates leading to compressor surge. Too much liquid water in the combustor will result in flame out. The amount of water resulting in either surge or flame out must be defined for each compressor model.

Appendix C-12

Overall A-weighted Sound Level Calculation

An example calculation of the overall A-weighted level is presented. It is based on having a set of octave band levels that are un-weighted; that is, there is no filter weighting applied to the measurement. If the levels are already A-weighted then skip to the summation equation (C.12-2) that follows Table C.12-2.

Table C.12-1 Octave Band Sound Pressure Levels (SPL), dB

Octave Band Center Frequency, Hz	Sound Pressure Level, dB (Measured or Calculated)
31.5	74.0
63	66.0
125	71.0
250	61.0
500	60.0
1000	75.0
2000	82.0
4000	80.0
8000	87.0

Next, convert the sound pressure levels to octave band A-weighted levels by:

$$L_A = Lp + A - wt\ dB \qquad \text{(C.12-1)}$$

Table C.12-2 Conversion of SPL to A-wtd Levels, dB

Octave Band Center Frequency, Hz	Lp, dB	A-Wtg.	LA dB
31.5	74	-39.4	34.6
63	66	-26.2	39.8
125	71	-16.1	54.9
250	61	-8.6	52.4
500	60	-3.2	56.8
1000	75	+0.0	75.0
2000	82	+1.2	83.2
4000	80	+1.0	81.0
8000	87	-1.1	85.9

The last column is now summed logarithmically to arrive at the overall A-weighted level:

$$L_A = 10\ Log\ \{\Sigma\ 10^{(Li/10)}\}\ dB \qquad \text{(C.12-2)}$$

$L_A = 10\ \text{Log}\ \{10^{(34.6/10)} + 10^{(39.8/10)} + 10^{(54.9/10)} + 10^{(52.4/10)} + 10^{(56.8/10)} + 10^{(75.0/10)} + 10^{(83.2/10)} + 10^{(81.0/10)} + 10^{(85.9/10)}\}\ \text{dB}$

which results in...

$$L_A = 88.9\ \text{dB}$$

Appendix C-13

Occupational Noise Exposure

Occupational Safety and Health Act of 1970 (OSHA) regulates employee exposure to noise as described in the U.S. code of federal regulations. The code states, "The employer shall administer a continuing, effective hearing conservation program, as described whenever employee noise exposures equal or exceed an 8-hour time-weighted average sound level (TWA) of 85 decibels...or equivalently, a dose of fifty percent." This is termed the *action level* when an employer must institute a hearing conservation program and provide hearing protection to employees. The following table lists selected maximum allowable noise level exposure times (see the CFR for a full listing). When an employee is exposed to levels and/or times that exceed those listed then the employer is mandated to reduce employee noise exposure.

Table C.13-1 Permissible Noise Exposure Limits (without hearing protection)

Decibel Level, n	80	85	90	92	95	97	100	102	105	110	115
Exposure Time, Tn	32	16	8	6	4	3	2	1.5	1	0.5	0.25

Tn is in hours. For other sound levels, the maximum allowable exposure time (T) based on a particular (measured) sound level (L) is calculated by:

$$T = 8/2^{(L-90)/5} \text{ hours} \qquad \text{(C.13-1)}$$

When employees are exposed to variable levels of noise then the total noise dose is calculated by,

$$D = 100\Sigma(Cn/Tn) \qquad \text{(C.13-2)}$$

This summation is done to measure the employee's total noise exposure where Cn is the actual exposure time, in hours, at a maximum sound level interval n, and Tn is the maximum allowable exposure time at that sound interval level; and when the dose (D) computation exceeds 50%, then an effective hearing conservation program is required.

As an example, an employee will spend his day exposed to the following sound levels: 4 hours at 80 dB, 1.5 hours at 95 dB and 2.5 hours at 92 dB. The dose computation is as follows:

$$D = 100 \times (4/32 + 1.5/4 + 2.5/6) = 91.6\%$$

The values in the calculation are all units of time. The first term. 4/32, is based on 4 hours at 80 dB where the maximum allowable exposure time is 32 hours.

The eight-hour time weighted average (TWA) sound level exposure is then computed by:

$$TWA = 16.61 \log (D/100) + 90 \text{ dB} \qquad \text{(C.13-3)}$$

For D= 91.6, TWA = 89.4 dB which is greater than 85 dB; therefore, a hearing conservation program is required to be implemented.

Appendix C-14

Resume: Elden F. Ray Jr., P.E.

Mr. Ray is a Professional Engineer having over 20 years experience in acoustical design, noise control and analyses, product design and development, testing and analyses of equipment including industrial sites and power plants. In addition, Mr. Ray has over ten years experience in the construction industry. Mr. Ray is Board Certified by INCE, a national and international professional organization that is made up of noise control and acoustical professionals across industry.

Mr. Ray has over eight years direct experience working with combustion turbine noise control and inlet and exhaust system aerodynamics including the design of inlet and exhaust system silencers. He computer models industrial and power plant sites and performs acoustical analysis of such sites using Cadna A®, and other in-house developed computer models to determine the necessary noise controls to bring a site into compliance with regulatory requirements or to be compatible with the existing acoustical environment. He performs certification surveys for compliance, environmental sound surveys, and assesses community—environmental noise impacts.

Mr. Ray's experience includes over ten years in machinery noise control and analyses on military ships and systems including steam turbines, generators, propulsion systems, electronic control systems, and other mechanical and fluid control systems, and has several years experience with jet engine nacelle acoustical design and testing. His degrees in engineering and architecture, and construction and mechanical systems experiences give him unique qualifications in designing for noise control or mitigation applications and works to achieve cost effective solutions.

Publications:

"Measurement Uncertainty in Conducting Environmental Sound Level Measurements," *Noise Control Eng. J.* Vol. 48, No. 1 (Jan-Feb 2000)

"Remediation of Excessive Power Plant Noise Emissions," Inter-noise 99, 6-8 Dec. 1999, Ft. Lauderdale, FL

"Active Control of Low Frequency Turbine Exhaust Noise," Walker, Hersh, Celano, and Ray, Noise-Con 2000, 3-5 December 2000, Newport Beach, CA

Professional:
Professional Engineer, Mechanical
Registered, Delaware, No. 7793
February, 1991

Education:
B.S. Ocean Engineering, Aug. 1980
Concentrating in Noise Control & Acoustics
Florida Atlantic Univ. Boca Raton, FL

A.S. in Architecture, May 1976
St. Petersburg Junior College, FL

Continuing Education:
Flow Generated Aero-Dynamic Noise; Penn State, Jun. '97
Aerodynamic Noise of Turbomachines; Penn State, Jul. '94
Acoustic Boundary Element Modeling; Univ. of Kentucky, Jan. '91

Affiliations:
American Society of Mechanical Engineers (ASME) Member
Institute of Noise Control Engineers (INCE) Member, Board Certified
ANSI B133.8 Subcommittee Member, Gas Turbine Noise Emissions

Appendix C-15

Microturbine Manufacturers

Manufacturer	Product	Internet URL
Bowman Power Systems USA distributor: Kohler Power Systems Americas 444 Highland Drive Kohler, WI 53044 Tel: (920) 457 4441	Turbogen^{TM} family of microturbines 25 kWe –80 kWe	http://www.bowmanpower.com
DTE Energy Technologies 37849 Interchange Drive Farmington Hills, MI 48335	ENT 400 kW microturbine	http://www.dtetech.com/
Capstone Turbine Corp. 21211 Nordhoff Street Chatsworth, CA 91311 Tel: 818-734-5300 Fax: 818-734-5320	C30 microturbines C60 microturbines	http://www.microturbine.com/
Elliott Energy Systems 2901 SE Monroe St. Stuart, FL 34997 Tel: 772-219-9449 Fax: 772-219-9448	TA 80, TA 100 microturbines	http://www.tapower,com/
GE Power Systems Corporate Headquarters 1 River Road/Building 37-6 Schenectady, NY 12345 Tel: (914) 278-2200	Developing 175 kW microturbine	http://www.gepower.com/
Ingersoll-Rand Energy Systems 800-A Beaty Street Davidson, NC 28036 Tel: (704) 896-5373	PowerWorks 70 & 250 microturbines	http://www.irenergysystems.com/
Turbec Headquarters Regnvattengatan 1 200 21 Malmö, Sweden Tel: +46 40 680 00 00	Developing 50kW microturbine	http://www.turbec.com/

Appendix C-16

Estimate WHR Boiler Size and Cost

The following details the steps necessary to arrive at a preliminary estimate of a combined cycle/cogeneration system comprising a gas turbine, a WHR boiler, steam turbine, and condenser. This example itemizes the information that is required (given conditions or assumptions and the decisions to be made by the designers) and how that information is used in the various calculations.

Given

- Standard Day Conditions
- Gas Turbine Simple Cycle Horsepower = 16,330
- Gas Turbine Exhaust Flow = W_A = 126.2 lbs/sec = 454,320 #/hr
- Gas Turbine Simple Cycle Heat Rate = 7618 Btu/HPxHr
- Gas Turbine Exhaust Temperature = 808°F

Determine

- Total Power
- Overall Heat Rate (Efficiency)

Also to be determined

- Pinchpoint
- Steam Rate
- Required WHR Boiler Surface Area
- WHR Boiler Size
- Estimated WHR Boiler Cost
- Stack Exhaust Temperature
- Parasitic Losses

Assumptions To Be Made

- T_{FW} = Feedwater Temperature = 125°F

- P_{FW} = Feedwater Pressure = 2 psia
- T_{COND} = Condenser Temperature = 125°F
- P_{COND} = Condenser Pressure = 2 psia
- Steam Pressure = 200 psia
- Steam Temperature = 600°F
- T_{PP} = Pinch point Temperature = 20°F

Use Steam Tables

From Superheated Steam Table:

- h_1 = 1322 Btu/lb @ 200 psia & 600°F
- h_2 = 1199 Btu/lb @ Stm Sat Line & 200 psia
- h_3 = 355 Btu/lb @ Sat Water & 200 psia

From Saturated Steam: Pressure Table:

- T_3 = 382°F @ Stm Sat Line & 200 psia

From Saturated Steam: Temperature Table:

- h_4 = 93 Btu/lb @ 125°F & 2 psia

These values can also be found on the Mollier Chart.

Calculate Gas Temperature At Pinchpoint

See Mollier, Figure C-1.

T_{G3} = Gas Temperature At Pinchpoint = $T_3 + T_{PP}$

- = 382°F + 20°F
- = 402°F

Specific Heat vs. Temperature

See Figure C-2.

Heat Recovery To Pinchpoint

Heat Recovery To Pinchpoint = Q_1

$Q_1 = W_A \bullet C_P \bullet \Delta T$

Q_1 = 454,320#/hr • 0.253 Btu/#°F•(808°F – 402°F)

Q_1 = 46,666,842 Btu/Hr

Steam Rate

Steam Rate = $W_S = Q_1/\Delta h = Q_1/(h_1 - h_3)$

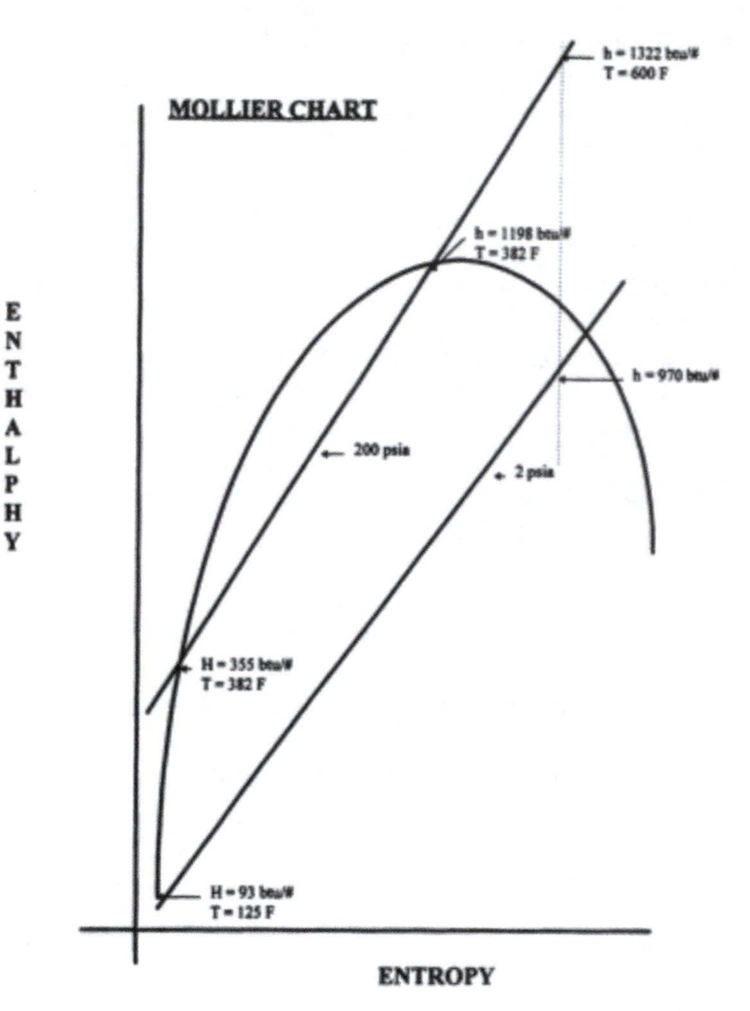

Figure C-1

W_S = 46,666,842 Btu/Hr/(1322 Btu/lb – 355 Btu/lb)
W_S = 48,279 #/hr

Superheater Heat Rate

Superheater Heat Rate = $Q_2 = W_s(h_1 - h_2)$
Q_2 = 48,279 #/hr • (1322 Btu/lb – 1198 Btu/lb)
Q_2 = 5,938,317 Btu/Hr

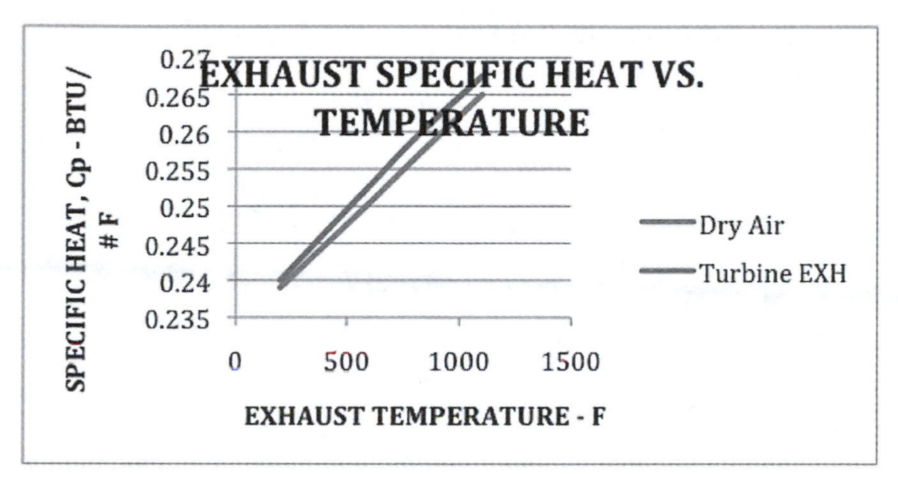

Figure C-2

Superheater ΔT

$\Delta T = Q_2/(W_A \bullet C_P)$

AND C_P = 0.258 Btu/#°F @ T = 808°F

ΔT = 5,938,317 Btu/Hr/(454,320 #/hr • 0.258 Btu/#°F)

ΔT = 51°F

Gas Temperature Entering The Evaporator

Gas Temperature Entering The Evaporator =

- = 808°F – 51°F
- = 757°F

Economizer Heat Rate

Economizer Heat Rate = Q3 = $W_S \bullet (h_3 - h_{FW})$

Q_3 = 48,279#/hr • (355°F – 93°F)

Q_3 = 12,668,410 Btu/Hr

Economizer Gas Δt

$\Delta T = Q3/(W_A \bullet CP)$

C_P = 0.245 Btu/#°F @ T = 350°F

ΔT = 12,668,410Btu/Hr/(454,320#/Hr • 0.245 Btu/#°F)

ΔT = 114°F

Stack Exhaust Temperature

Stack Exhaust Temperature

$= T_{G3}$ – Economizer Gas ΔT

= 402°F – 114°F

= 288°F

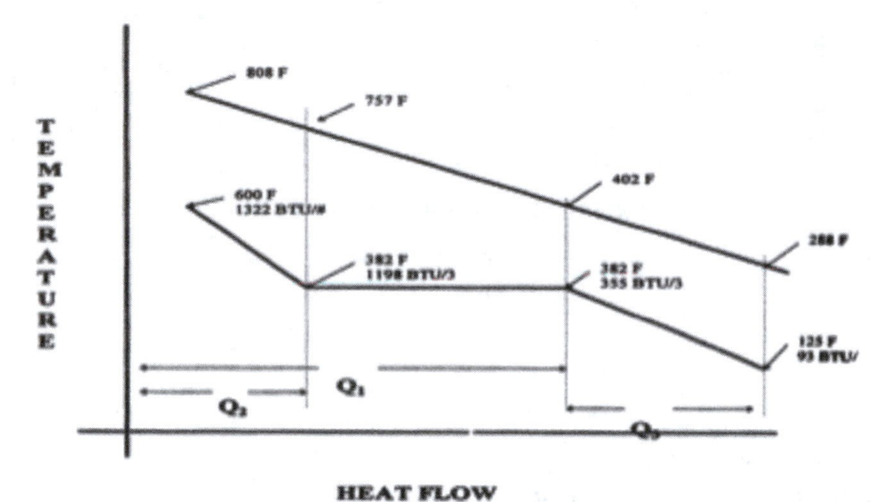

Figure C-3

Calculate Steam Power

Previously Obtained Information:

Steam Rate = 48,279 #/Hr

Steam Pressure = 200 psia

Steam Temperature = 600°F

Theoretical Steam Rate = TSR

TSR = 2545/Δh

Where $\Delta h = h_1 - h_{EXH}$

TSR = 2545/(1322 Btu/lb – 970 Btu/lb)

TSR = 7.23 #/HPxHr

Steam Turbine Efficiency = η_{STM}

η_{STM} = 75.7% (from C-4)

Actual Steam Rate = ASR =TSR/η_{STM}

ASR = 7.23 #/HP•Hr/0.757

ASR = 9.55 #/HP•Hr

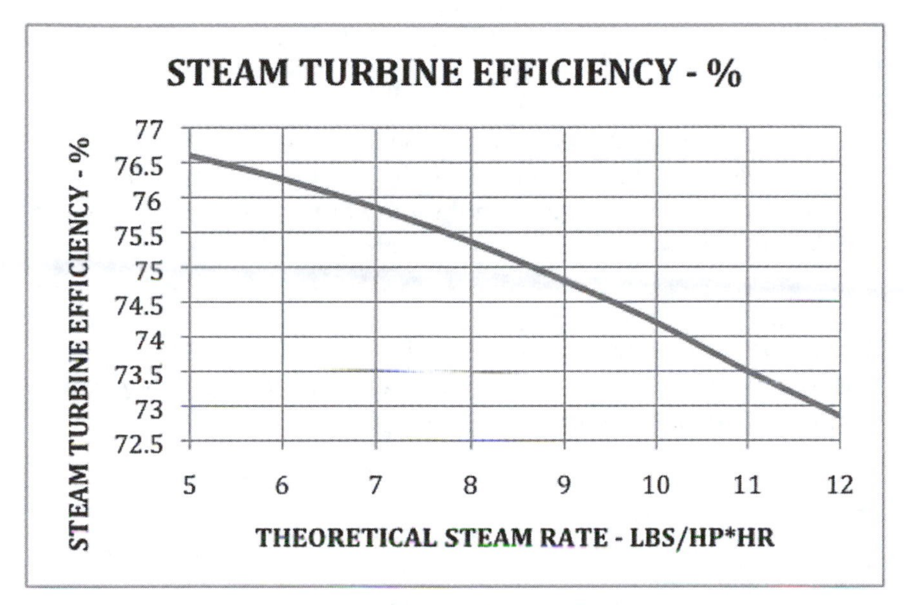

Figure C-4

Steam Power = Steam Rate/ASR
Steam Power = 48,279 #/Hr/9.55#/HP•Hr
Steam Power = 5,055 HP

Total Power
Total Power = Simple Cycle Power + Steam Power
Total Power = 16,330 HP + 5,055 HP
Total Power = 21,385 HP

Overall Heat Rate = HR_{TOTAL}
HR_{TOTAL} = (Simple Cycle Heat Rate • Simple Cycle HP)/Total Power
HR_{TOTAL} = (7618 Btu/HPxHr • 16,330 HP)/21,385 HP
HR_{TOTAL} = 5817 Btu/HP•Hr

Overall Efficiency = η_{TOTAL}
η_{TOTAL} = 2545/Heat Rate
η_{TOTAL} = 2545/5817 Btu/HP•Hr
η_{TOTAL} = 43%*
*Up From 33% Simple Cycle Efficiency

Parasitic Losses

New Design

New Design – calculate losses based on:

OEM information (duct losses, filter losses, etc.); pump horsepower for chemical injection pumps; make-up water; forced circulation/re-circulation pumps.

Existing Plant

Existing Plant – calculate losses based on:

Differential pressure across inlet & exhaust duct; Volt-Amps of pumps (injection, make-up water, etc.).

Duct Losses vs. Power

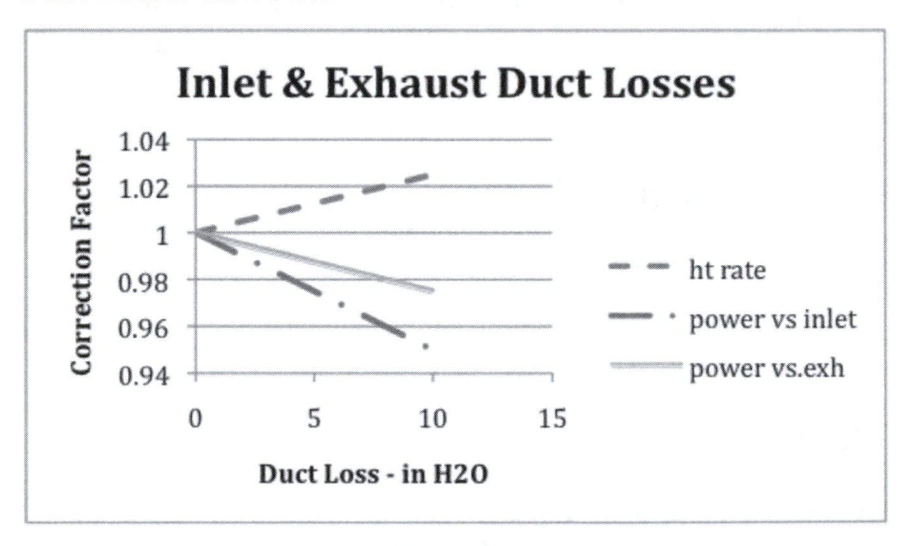

Figure C-5

Pump Horsepower

Where HP = pump horsepower

G = flow rate, gal/min

H = head

Sg = specific gravity

η = pump efficiency

Steam Turbine Efficiency
Based on Actual Data

Existing Steam Turbine Efficiency

- Where Δh actual = $h_1 - h_{EXH}$
- Δh theoretical = $h_1 - h_{EXH}$

Determine Required Surface Area = SA

- SA = Q/LMTD • μ
- where LMTD = Log Mean Temperature difference, and
- μ = Overall Heat Transfer Coefficient
- (Btu/Ft^2 •Hr•°F)

Calculating LMTD

- LMTD = $(\Delta T_{OUT} - \Delta T_{IN})/(2.3 \text{ Log10}(\Delta T_{OUT}/\Delta T_{IN}))$
- $LMTD_{SH}$ = 283.7°F
- $LMTD_{EV}$ = 121.3°F
- $LMTD_{EC}$ = 68.2°F

Assume the following overall heat transfer coefficients:

- μ_{SH} = 8.7 (Btu/Ft^2 •Hr•°F)
- μ_{EV} = 12.8 (Btu/Ft^2 •Hr•°F)
- μ_{EC} = 8.5 (Btu/Ft^2 •Hr•°F)

The overall heat transfer coefficients are the reciprocals of the sum of the reciprocals of the internal film coefficients, external film coefficients, internal fouling resistances and external fouling resistances. The values listed here are adequate for preliminary estimates.

Required Surface Areas

- SA_{SH} = 5,938,317 Btu/Hr/(283.7°F • 8.7 Btu/Ft^2 •Hr•°F)
- SA_{SH} = 2,406 Ft^2
- SA_{EV} = (46,666,842 - 5,938,317 Btu/Hr)/(121.8°F • 12.8 Btu/Ft^2 •Hr•°F)
- SA_{EV} = 26,124 Ft^2
- SA_{EC} = 12,668,410 Btu/Hr/(68.2°F • 8.5 Btu/Ft^2 •Hr•°F)
- SA_{EC} = 21,853 Ft^2
- Total Surface Area = 50,383 Ft^2

Determine WHR Boiler Size

- From fin tube selection chart C-6 for
- 2.0 inch O.D. tube with fin height @ 0.25 inch

$A_o/L = 3.5\ Ft^2/Ft$

Fin Tube Selection

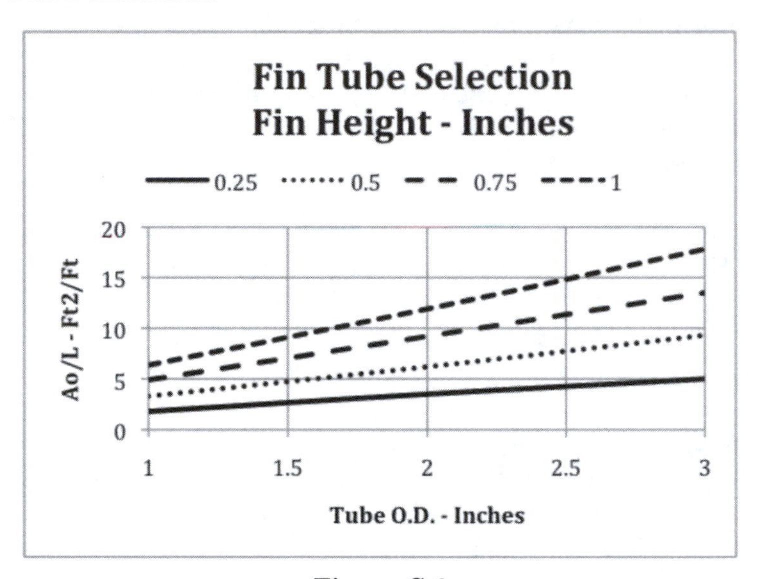

Figure C-6

- Superheater Size = 2406 $Ft^2/3.5\ Ft^2/Ft$
- Evaporator Size = 26,124 $Ft^2/3.5\ Ft^2/Ft$
- Economizer Size = 21,853 $Ft^2/3.5\ Ft^2/Ft$
- Assume HRSG width of 10 Ft and 2.5 inches between the 2.0 inch O.D. tube centerline and 0.5 inches to wall.
- Therefore, 10 Ft/2.5" x Ft/12" = 48 tubes

Superheater Size

- 687 Ft/48 tubes = 14.3' high
- ~ 18' high, one row too high for most roads
- 687 Ft/48 tubes = 7.15' high
- ~ 10' high, two rows, and
- 6"/row > 2' length

10' wide • 10' high • 2' long

Evaporator Size

- 7464 Ft/22 rows x 48 tubes = 7.07’ high
- ~ 10’ high, 22 rows, and
- 6”/row > 12’ length

10’ wide • 10’ high • 12’ long

Economizer Size

- 6,244 Ft/18 rows x 48 tubes = 7.22’ high
- ~ 10’ high, 18 rows, and
- 6”/row > 9’ length

10’ wide • 10’ high • 9’ long

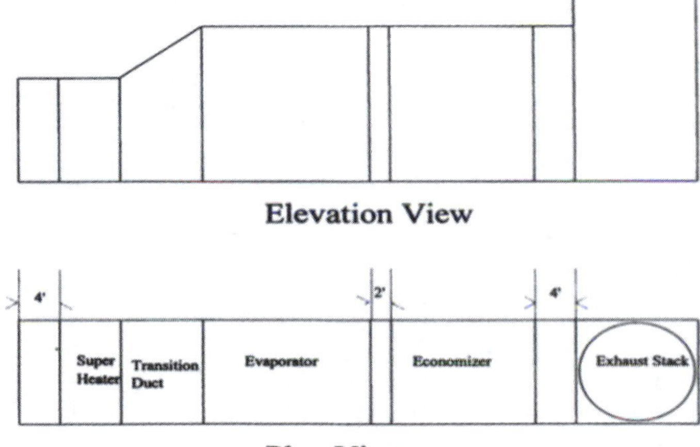

HRSG General Layout

Figure C-7

Estimated Cost

The estimated cost of the heat recovery steam generator based on $10.00/Ft^2 (manufacturing cost not purchase price) and a total surface area of 50,383 Ft^2 is $503,830.00.

Appendix D-1

Technical Organizations

he following is a list organizations that address gas turbines or their environment. The organizations highlighted in bold type are involved with gas turbines on a regular basis.

1. Air & Waste Management Association, P.O. Box 2861, Pittsburgh, PA 15232
2. American Association for the Advancement of Science (AAAS), 1200 New York Avenue, N.W., Washington, D.C. 20005, Phone (202)326-6400
3. American Chemical Society (ACS), 1155 16th Street, N.W., Washington, D.C., Phone (202)872-4600, Fax (202)872-4615
4. American Gas Association, 400 N. Capitol Street, N.W., Washington, DC 20001, Phone (202) 824-7000, Fac (202) 824-7115
5. **American Gear Manufacturers Association, 1500 King Street, Suite 201, Alexandria, Virginia, 22314**
6. American Institute of Aeronautics and Astronautics (AIAA), 1801 Alexander Bell Drive, #500, Reston, Virginia 20191, Phone (703)264-7500, Fax (703)264-7551
7. American Institute of Chemical Engineers (AIChE), 345 East 47th Street, New York, NY 10017-2395, Phone (212)705-7338, Fax (212)752-3294
8. American Institute of Steel Construction, One East Wacker Drive, Suite 3100, Chicago, IL 60601-2001
9. American Iron and Steel Institute, 1000 16th Street, N.W., Washington, D.C. 20036
10. American National Standards Institute (ANSI), 11 West 42nd Street, New York, NY 10036
11. **American Petroleum Institute (API), 1220 L Street, N.W., Washington, D.C. 20005**

12. American Society for Testing and Materials (ASTM), 1916 Race Street, Philadelphia, PA 19103
13. American Society of Heating, Refrigerating And Air-Conditioning Engineers, Inc. (ASHRAE), 1791 Tullie Circle, NE, Atlanta, GA 30329, Phone (404) 636-8400
14. **American Society of Mechanical Engineers (ASME), 345 East 47th Street, New York, NY 10017**
15. American Society of Metals International, 9639 Kinsman Road, Materials Park, Ohio 44073-0002, Phone USA & Canada (800)336-5152, Phone Europe (800)368-9800, Fax (216) 338-4634
16. American Society of Naval Engineers (ASNE), 1452 Duke Street, Alexandria, Virginia 22314-3458, Phone (703)836-6727, Fax (703)836-7491
17. American Welding Society, 550 N. W. Le Jeune Road, Miami, FL 33135
18. Anti-Friction Bearing Manufacturers Association, 1101 Connecticut Avenue, N.W. Suite 700, Washington, D.C. 20036
19. Association of Computing Machinery (ACS), 1515 Broadway, 17th Floor, New York, NY 10036-5701, Phone (212)626-0500, Fax (212)944-1318
20. Association of Computing Machinery (ACS), European Service Center, 108 Cowley Road, Oxford, OX4 IJF, United Kingdom
21. Association of Energy Engineers (AEE), 4025 Pleasantdale Road #420, Atlanta, GA 30340
22. Canadian Council of Technicians & Technologists (CCTT), 285 McLeod Street, Ottawa, Canada K2P 1A1, Phone (613)238-8123, Fax (613)238-8822
23. Canadian Standards Association (CSA), 178 Rexdale Boulevard, Etobicoke (Toronto), Ontario, Canada, M9W 1R3, Phone (416)747-4044, Fax (416)747-2475
24. Danish Hydraulic Institute (DHI), 5, Agren Alle, DK-2970 Horsholm, Denmark. Phone +45 45 76 95 55, Fax +45 45 76 25 67
25. **Electrical Power Research Institute (EPRI), 3412 Hillview Avenue, Palo Alto, CA 94304, Phone (415) 855-2419, FAX (415) 855-8572**
26. Factory Mutual Research Corporation (FM), 1151 Boston-Providence Turnpike, P.O. Box 9102, Norwood, MA 02062. Phone

(627)762-4300, Fax (627)762-9375

27. Gas Processors Suppliers Association, 1812 First Place, Tulsa, Oklahoma 74103, Phone (918)582-5112
28. Gas Technology Institute (formerly the Gas Research Institute), 1700 South Mount Prospect Road, Des Plaines, IL 60018-1804, Phone (847) 768-0500
29. Gas Turbine Association, P.O. Box 1408, Great Falls, VA 22066, Phone (703) 623-0698, Fax (703) 757-8274
30. Independent Power Producers Association, 1112 I Street #380, Sacremento, CA 95814
31. **International Gas Turbine Institute (IGTI), 6085 Barfield Road, Suite 207, Atlanta, GA 30328, Phone (404)847-0072, Fax (404)843-2517**
32. Institute of Electrical and Electronic Engineers (IEEE), 345 East 47th Street, New York, NY 10017
33. Institute of Environmental Sciences, 940 East Northwest Hwy, Mount Prospect, Illinois 60056. Phone (847)225-1561, Fax (847)225-1699
34. Institute of Industrial Engineers, 25 Technology Park/Atlanta, Norcross, GA 30092. Phone (404)449-0460, Fax (404)263-8532
35. Institution of Electrical Engineers, Savoy Place, London, WC2R OBL, United Kingdom, Phone +44(0)171 240 1871, Fax +44(0)171 240 7735
36. Instrument Society of America (ISA), P.O. Box 12277, Research Triangle Park, North Carolina, 27709
37. Insulated Power Cable Engineers Association, 283 Valley Road, Montclair, NY 07042
38. International Society of Standards (ISO)
39. International Union of Materials Research Societies (IUMRS)
40. Manufacturers Standardization Society of the Valve and Fittings Industry, 127 Park Street, Vienna, Virginia 22180
41. Materials Research Society (MRS), 9800 McKnight Road, Pittsburgh, PA 15237-6006, Ph. (412) 367-3004, Fax (412) 367-4373
42. Minerals, Metals, & Materials Society (TMS), 420 Commonwealth Drive, Warrendale, PA 15086, Phone (412)776-9000, Fax (412)776-3770
43. National Academy of Engineering (NAE), 2101 Constitution Avenue, N.W., Washington, D.C. 20418

44. National Association of Corrosion Engineers (NACE), P.O. Box 218340, Houston, TX 77218
45. National Electrical Manufacturers Association (NEMA), 2101 L Street, N.W., Washington, D.C. 20037
46. National Fire Protection Association (NFPA), 1 Batterymarch Park, Quincy, MA 02269-9101
47. National Rural Electric Cooperative Association, 4301 Wilson Blvd., Arlington, VA 22203-1860, Phone (703) 907-5500
48. National Society of Professional Engineers (NSPE), 1420 King Street, Alexandria, VA 22314-2794, Phone (703)684-2800, Fax (703)836-4875
49. Occupational Safety and Health Administration (OSHA), U.S. Department of Labor, 200 Constitution Avenue, N.W. Washington, D.C. 20210
50. Society of Automotive Engineers, 400 Commonwealth Drive, Warrendale, PA 15096-2683
51. Society of Environmental Management & Technology (SEMT), 4301 Connecticut Avenue, N.W., #300, Washington, D.C. 20008
52. Society of Environmental Engineering
53. Society of Environmental Engineers (SEE), Owles Hall, Buntingford, Herts SG9 9PL, Phone +44-1763-271209, Fax +44-1763-273255
54. Society of Plastics Engineering
55. Society of Plastics Engineers (SPE), 14 Fairfield Drive, Brookfield, CT 06804, Phone (203)775-0471, Fax (203)775-8490
56. Society of Reliability Engineers, 15 Winslow Road, Fredericksburg, VA 22406-4222
57. Society of Woman Engineers (SWE), 120 Wall Street, 11th Floor, New York, NY 10005-3902 Phone (212)509-9577, Fax (212)509-0224
58. Steel Structures Painting Council, 4400 Fifth Street, Pittsburgh, PA 15213-2683
59. Tubular Exchanger Manufacturers Association, 25 North Broadway, Tarrytown, New York 10591
60. Underwriters Laboratory, Inc. (UL), 1207 W. Imperial Hwy, Brea, CA Phone (714)447-3176

Appendix D-2

Technical Articles

The following lists are provided to give the reader additional insight into specific gas turbine topics addressed in this book. This lists are separated into the following topics:

- Material
- Combustion
- Fuels
- Controls
- Diagnostics
- Rotors, Bearings, And Lubrication
- Filtration And Cleaning
- Design Techniques
- Inlet And Turbine Cooling
- Emissions
- Performance
- Repair Techniques
- Instrumentation
- Fire Protection

MATERIALS

"Nickel Base Superalloys Single Crystal Growth Technology For Large Size Buckets In Heavy Duty Gas Turbines," H. Kodama; A. Yoshinari; K. Tijima; K. Kano; H. Matsursaki. ASME 91-GT-022.

"Thermal Strain Tolerant Abradable Thermal Barrier Coatings," T.E. Strangman. ASME 91-GT-039.

"Thermal Barrier Coating Life Prediction Model Development," S.M. Meier; D.M. Nissley; K.D. Sheffler; T.A. Cruse. ASME 91-GT-040.

"Nickel Base Alloy GTD-222, A New Gas Turbine Nozzle Alloy," D.W. Seaver; A.M. Beltran. ASME 91-GT-073.

"Life Time Prediction For Ceramic Gas Turbine Components," S. Wittig; G. Sturner; A. Schulz. ASME 91-GT-096.

"Fabrication And Testing Of Ceramic Turbine Wheels," K. Takatori; T. Honma; N. Kamiya; H. Masaki; S. Sasaki; S. Wada. ASME 91-GT-142.

"Gas Generator With High-Temperature Path Ceramic Components," A.V. Sudarev; V.H. Dubershtein; V.P. Kovalevsky; A.E. Ginsburg. ASME 91-GT-152.

"Optimization Of The Fatigue Properties Of INCONEL Alloy 617," G.D. Smith; D.H. Yates. 91-GT-161.

"Development Of Ceramic Components For Power Generating Gas Turbine," Y. Hara; T. Tsuchiya; F. Maeda; I. Tsuji; K. Wada. ASME 91-GT-319.

"High-Temperature Gas Generator With Ceramic Components For Stationary Equipment Gas Turbine Unit," G.A. Shishov; A.V. Sudarev; V.N. Dubershtein; A.N. Tsurikov. ASME 91-GT-362.

"Advanced Ceramic Engine Technology For Gas Turbines," W.D. Carruthers; J.R. Smyth. ASME 91-GT-368.

"Study Of The Ceramic Application For Heavy Duty Gas-Turbine Blades," T. Ikeda; H. Okuma; N. Okabe. ASME 91-GT-372.

"Design Of Ceramic Gas Turbine Components," G. Stuerner; M. Fundus; A. Schulz; S. Wittig. ASME 90-GT-048.

"Development Of Materials And Coating Processes For High Temperature Heavy Duty Gas Turbines," S. Nakamura; T. Fukui; M. Siga; T. Kojima; S. Yamaguch. ASME 90-GT-079.

"The LCF Behavior Of Several Solid Solution Strengthened Alloys Used In Gas Turbine Engines," S.K. Srivastava; D.L. Klarastron. ASME 90-GT-080.

"Testing Of Ceramic Components In The Daimler-Benz Research Gas Turbine PWT 10," M. Stute; H. Burger; N. Griguscheit; E. Holder; K.D. Moergenthaler; F. Neubrand; M. Radloff. ASME 90-GT-097.

"A Comparison Of Forming Technologies For Ceramic Gas `Turbine Engine Components, R.R. Hhengst; D.N. Heichel; J.E. Holowczak; A.P. Taglialavore; B.J. McEntire. ASME90-GT-184.

"Fabrication Of ATTAP Ceramic Turbine Components," R.W. Ohnsorg; J.E. Funk; H.A. Lawler. ASME 90-GT-185.

"Development Of Corrosion Resistant Coatings For Marine Gas Turbine Application," B.A. Nagaraj; D.J. Wortman; A.F. Maricocchi; J.S. Patton; R.L. Clarke. ASME 90-GT-200.

"Materials Keep Pace With Industrial Gas-Turbine Needs," Harry B. Gayley. ASME 61-GTP-15.

"GASTEAM—A Combination Of Gas And Steam Turbines," L. Eelmenius. ASME 66-GT/CMC-70

"Evaluation Of Heat-Resistant Alloys For Marine Gas Turbine Applications," L.J. Fiedler; R.M.N Pelloux. ASME 66-GT-81.

"Metallurgical Considerations Involved In The Development Of Nickel-Base High-Temperature Sheet Alloys," E.G. Richards. ASME 66-GT-104.

"The Effects Of Environmental Contaminants On Industrial Gas Turbine Thermal Barrier Coatings," H.E. Eaton; N.S. Bornstein; J.T. De-Masi-Marcin. ASME 96-GT-283.

"Development Of Ceramic Gas Turbine Components For The CGT301 Engine," Mitsuru Hattori; Tsutomu Yamamoto; Keiichiro Watanabe; Masaaki Masuda. ASME 96-GT-449.

"Research And Development Of Ceramic Gas Turbine (CGT 302)," Isashi Takehara; Isao Inobe; Tetsuo Tatsumi; Yoshihiro Ichikawa; Hirotake Kobayashi. ASME 96-GT-477.

"Evaluation of Commercial Coatings on Mar M-002, IN-939 and CM-247 Substrates," J.G. Gordjen, G.P. Wagner. ASME 96-GT-458

Role of Environmental Deposits in Spallation of Thermal Barrier Coatings on Aeroengine and Land-based Gas Turbines Hardware," Marcus P. Borom, Curtis A. Johnson, Louis A. Peluso. ASME 96-GT-285

"Oxidation Resistance and Critical Sulfur Content of Single Crystal Superalloys," James L. Smialek. ASME 96-GT-519

"Development of Cobalt Base Superalloy for Heavy Duty Gas Turbine Nozzles," M. Sato, Y. Kobayasbi, M. Matsuzaki, K. Shimomura. ASME 96-GT-390

"GTD111 Alloy Material Study," Joseph A. Daleo, James R. Wilson. ASME 96-GT-520

"Inconel Alloy 783: An Oxidation Resistant, Low Expansion Superalloy for Gas Turbine Applications," K.A. Heck, J.S. Smith. ASME 96-GT-380

"Applications of Continuous Fiber Reinforced Ceramic Composites in Military Turbojet Engines," Patrick Spriet, George Habarou. ASME 96-GT-284

"HP/HVOF as a Low Cost Substitute for LPPS Turbine MCrAlY Coatings," Ronald J. Honick, Jr., Richard Thorpe. ASME 96-GT-525

"European Standardization Efforts On Fibre-Reinforced Ceramic Matrix Composites," Marc Steen. ASME 96-GT-269.

"ASTM Standards For Monolithic And Composite Advanced Ceramics: Industrial, Governmental And Academic Cooperation," Michael G. Jenkins; George D. Quinn. ASME 96-GT-270.

"International Standardization Activities for Fine Ceramics—Status Of ISO/TC 206 On Fine Ceramics," Takashi Kanno. ASME 96-GT-321.

"Standards For Advanced Ceramics And Pre-Standardization Research—A Review," Roger Morrell. ASME 96-GT-320.

COMBUSTION

"Development Of An Innovative High-Temperature Gas Turbine Fuel Nozzel," G.D. Myers; J.P. Armstrong; C.D. White; S. Clouser; R.J. Harvey. ASME 91-GT-036.

"Evaluation Of A Catalytic Combustor In A Gas Turbine-Generator Unit," S. Kajita; T. Tanaka; J. Kitajima. ASME 90-GT-089.

"Second Generation Low-Emission Combustors For ABB Gas Turbines: Burner Development And Tests At Atmospheric Pressure," Th. Sattelmayer; M.P. Felchlin; J. Hausmann; J. Hellat; D. Styner. ASME 90-GT-162.

"A Simple And Reliable Combustion System," B. Becker; F. Bonsen; G. Simon. ASME 90-GT-173.

"Catalytically Supported Combustion For Gas Turbines," Viktor Scherer; Timothy Griffin. ASME 94-JPGC-GT-1.

"Reliability Analysis Of A Structural Ceramic Combustion Chamber," J.A. Salem; J.M. Manderscheid; M.R. Freedman; J.P. Gyekenyesi. ASME 91-GT-155.

"Low NO_x Burner Design Achieves Near SCR Levels," Jerry Lang, Industrial Power Conference 1994.

"Development of a Small Low NOx Combustor with a Staging System For an Aircraft Engine (Part 1)," Masayoshi Kobayashi, Masahi Arai, Toshiyuki Kuyama. ASME 96-GT-043

"Development of A Dry Low NOx Combustor for 2MW Class Gas Turbine," J. Hosoi, T. Watanabe, H. Tob. ASME 96-GT-053

"Development of A Low-Emission Combustor for a 100-kW Automotive Ceramic Gas Turbine," Masafumi Sasaki, Hirotaka Kumakura, Daishi Suzuki, Hiroyuki Ichikawa, Youichiro Obkubo, Yuusaku Yoshida. ASME 96-GT-119

"Development and Fabrication of Ceramic Gas Turbine Components," Koichi Tanaka, Sazo Tsuruzono, Toshifumi Kubo, Makoto Yoshida. ASME 96-GT-446

"Catalytic Combustion of Natural Gas Over Supported Platinum: Flow Reactor Experiments and Detailed Numerical Modeling," Tami C. Bond, Ryan A. Noguchi, Chen-Pang Chou, Rijiv K. Mongia, Jyb-Yuan, Robert W. Dibble. ASME 96-GT-130

"Development of a Catalytic Combustor for a Heavy-Duty Utility Gas Turbine," Ralph A. Della Betta, James C. Schlatter, Sarento G. Nickolas, Martin B. Cutrone, Kenneth W. Beebe, Yutaka Puruse, Toshiaki Tsuchiya. ASME 96-GT-485

"A Cold-Water Slurry Combustor For A 5MW Industrial Gas Turbine," C. Wilkes; C.B. Santanam. ASME 91-GT-206.

"Two Stage Slagging Combustor Design For A Coal-Fueled Industrial Gas Turbine," L.H. Cowell; R.T. LeCren; C.E. Tenbrook. ASME 91-GT-212.

"Low NO_x Combustor For Automotive Ceramic Gas Turbine Part 1-Conceptual Design," M. Sasaki; H. Kumakura; D. Suzuki. ASME 91-GT-369.

"Recent Test Results In The Direct Coal-Fired 80 MW Combustion Turbine Program," R.L. Bannister; P.W. Pillsbury; R.C. Diehl; P.J. Loftus. ASME 90-GT-058.

"Coal-Fueled Two-Stage Slagging Combustion Island And Cleanup System For Gas Turbine Application," L.H. Cowell; A.M. Hasen; R.T. LeCren; M.D. Stephenson. ASME 90-GT-059.

"Effect Of Burner Geometry And Operating Conditions On Mixing And Heat Transfer Along A Combustion Chamber," A.H. Hafez; F.M. El-Mahallawy; A.M. Shehata. ASME Cogen Turbo-Power '94.

FUELS

"Fuel Oil Storage Effect And Impact On Utility's Gas Turbine Operation," O. Backus. ASME 91-GT-020

"The Use Of Gaseous Fuels On Aero-Derivative Gas Turbine Engines," P. Mathieu; P. Pilidis. ASME 91-GT-044.

"Temperature Effects On Fuel Thermal Stability," J.S. Chin; A.H. Lefebvre; F.T.Y Sun. ASME 91-GT-097.

"Combustion Characteristics Of Multicomponent Fuels Under Cold Starting Conditions In A Gas Turbine," M.F. Bardon; V.K. Rao; J.E.D. Gauthier. ASME 91-GT-109.

"Separating Hydrogen From Coal Gasification Gases With Alumina Membranes," B.Z. Egan; D.E. Fain; G.E. Roettger; D.E. White. ASME 91-GT-132.

"Gas Turbines Above 150 MW For Integrated Coal Gasification Combined Cycles (IGCC)," Dr. B. Becker; Dr. B. Schetter. ASME 91-GT-256.

"Coal Mineral Matter Transformation During Combustion And Its Implications For Gas Turbine Blade Erosion," S. Rajan; J.K. Raghavan. ASME 90-GT-169.

"The Behavior Of Gas Turbine Combustion Chambers While Burning Different Fuels," C. Kind. ASME 60-GTP-10.

"Biomass-Gasifier/Aeroderivative Gas Turbine Combined Cycles: Part A—Technologies And Performance Modeling," Stefano Consonni; Eric D. Larson. ASME Cogen Turbo Power 94.

"Biomass-Gasifier/Aeroderivative Gas Turbine Combined Cycles: Part B—Performance Calculations And Economic Assessments," Stefano Consonni; Eric D. Larson. ASME Cogen Turbine Power '94.

"Gas Turbine Power Generation From Biomass Gasification," Mark A. Paisley; Robert D. Litt; Ralph P. Overend; Richard L. Bain. ASME Cogen Turbine Power '94.

"Performance Evaluation Of Biomass Externally Fired Evaporative Gas Turbine System," Jinyue Yan; Lars Eidensten; Gunnar Svedberg. ASME Cogen Turbine Power '94.

"Alternate Fuel Cofiring In Fluidized Bed Boilers," Charles R. McGowin; William C. Howe Industrial Power Conference 1994.

"Combined Cycle Gas And Steam Turbine Processes For Solid Fuels With Pressurized Fluidized Bed Combustion," Franz Thelen. ASME 94-JPGC-FACT-1.

"Gas Turbine Firing Medium BTU Gas From Gasification Plant," Piero Zanello; Andrea Tasseli; Fiat Avio. ASME 96-GT-008.

"Operation Of Combustion Turbines On Alternate Fuel Oils," K.W. Johnson;. ASME 96-GT-007.

"Development Of A Dual-Fuel Injection System For Lean Pre-Mixed Industrial Gas Turbines," Luke H. Cowell; Amjad Rajput; Douglas C. Rawlins. ASME 96-GT-195.

"Experimental Investigation Of The Liquid Fuel Evaporation In A Premix Duct For Lean Premixed And Prevaporized Combustion," Micharl Brandt; Kay O. Gugel; Christoph Hassa. ASME 96-GT-383.

"Biomass Gasification Hot Gas Filter Testing Results," Benjamin C. Wiant; Dennis M. Bachovchin. ASME 96-GT-336.

"Westinghouse Combustion Turbine Performance In Coal Gasification Combined Cycles," Richard A Newby. ASME 96-GT-231.

"Coal/Biomass Fuels And The Gas Turbine: Utilization Of Solid Fuels And Their Derivatives," Mario DeCorso; Don Anson; Richard Newby; Richard Wenglarz. ASME 96-GT-076.

"Design And Performance Of Low Heating Value Fuel Gas Turbine Combustors," Robert A Battista; Alan S. Feltberg; Michael A. Lacey. ASME 96-GT-531.

"Deposit Formation From No. 2 Distillate At Gas Turbine Conditions," Anthony J. Dean; Jean E. Bradt; John Ackerman. ASME 96-GT-046.

"Jet Fuel Oxidation And Deposition," Colette C. Knight; James C. Carnaban; Kevin Janora; John F. Ackerman. ASME 96-GT-183.

"Techniques To Characterize The Thermal-Oxidation Stability Of Jet Fuels And The Effects Of Additives," V. Vilimpoc; B. Sarka; W.L. Weaver; J.R. Gord; S. Anderson. ASME 96-GT-044.

CONTROLS

"The Design And Development Of An Electrically Operated Fuel Control Valve For Industrial Gas Turbines,," A.G. Salsi; F.S. Bhinder. ASME 91-GT-064.

"Active Suppression Of Rotating Stall And Surge In Axial Compressors," I.J. Day. ASME 91-GT-087.

"Highly Efficient Automated Control For An MGR Gas Turbine Power Plant," X.L. Yan; L.M. Lidsky. ASME 91-GT-296.

"A Combined Gas Turbine Control and Condition Monitor," S. DeMoss. ASME 91-GT-299.

"Design Of Robust Controllers For Gas Turbine Engines," D.E. Moellenhoff; S.V. Rao; C.A. Skarvan. ASME 90-GT-113.

"Control System Retrofit Completes 10 Years Of Gas-Turbine Reliability Enhancements," A.A. Kohlmiller; W.B. Piercy. ASME 90-GT-275.

"The Present Status And Prospects For Digital Power Plant Control Technology," R. Wetzl; Th. Kinn. ASME 94-JPGC-PWR-63.

"An Industrial Sensor For Reliable Ice Detection In Gas Turbines," J.W. Freestone; M. Weber. ASME Cogen Turbo Power '94.

"Fuel Trim Control For Multiburner Combustors Based On Emissions Sampling," Thomas F. Fric; Sanjay M. Correa; Elwin C. Bigelow. ASME Cogen Turbo Power '94.

"Adaptive Control Of A Gas Turbine Engine for Axial Compressor Faults," G. Lombardo. ASME 96-GT-445

"Multivariable Control Of Industrial Gas Turbines: Field Test Results," Ravi Rajamani; Bruce G. Norman. ASME 96-GT-289.

DIAGNOSTICS

"Monitoring Gas Turbines," A.J. Giampaolo, Diesel & Gas Turbine Worldwide, August 1982

"Analysis Of Cracked Gas Turbine Blades," H.L. Bernstein; J.M. Allen. ASME 91-GT-016.

"Detection Of Axial Compressor Fouling In High Ambient Temperature Conditions," I. Haq; H.I.H. Saravanamuttoo. ASME 91-GT-067.

"Implanted Component Faults And Their Effects On Gas Turbine Engine Performance," J.D. MacLeod; V. Taylor; J.C.G. Laflamme. ASME 91-GT-041.

"Remote Condition Monitoring Of A Peaking Gas Turbine," R.E. Harris; H.R. Simmons; A.J. Smalley; R.M. Baldwin; G. Quentin. ASME 91-GT-185.

"Optimization Of LM2500 Gas Generator And Power Turbine Trim Balance Techniques," B.D. Thompson. ASME 91-GT-240.

"Vibration Analysis Of Cracked Turbine And Compressor Blades," W. Ostachowicz; M. Krawczuk. ASME 90-GT-005.

"Service Temperature Estimation Of Turbine Blades Based On Microstructural Observations," J.M. Aurrecoechea; W.D. Brentnall; J.R. Gast. ASME 90-GT-023.

"Navier-Stokes Analysis Of Turbine Blade Heat Transfer," R.J. Boyle. ASME 90-GT-042.

"Detection Of A Rotating Stall Precursor In Isolated Axial Flow Compressor Rotors," M. Inoue; M. Kuroumaru; T. Iwamoto; Y. Ando. ASME 90-GT-157.

"Combustor Exit Temperature Distortion Effects On Heat Transfer And Aerodynamics Within A Rotating Turbine Blade Passage," S.P. Harasgama. ASME 90-GT-174.

"An Analytical Approach To Reliability Failure Forecasting And Product Quality," J.L. Byers. ASME 90-GT-190.

"Inlet Flow Distortion Effects On Rotating Stall," J. Fortis; W.C. Moffatt. ASME 90-GT-215.

"Enhanced On-Line Machinery Condition Monitoring Through Automated Start-up/Shutdown Vibration Data Acquisition," R.G. Harker; G.W. Handelin. ASME 90-GT-272.

"Fault Detection And Diagnosis In Gas Turbines," G. Herrington; O.K. Kwon; G. Goodwin; B. Carlsson. ASME 90-GT-339.

"The Role For Expert Systems In Commercial Gas Turbine Engine Monitoring," D.L. Dundaas. ASME 90-GT-374.

"Gas Turbine Component Fault Identification By Means Of Adaptive Performance Modeling," A. Stamatis; K. Mathioudakis; M. Smith; K. Papailiou. ASME 90-GT-376.

"Development Of Advanced Diagnostics For Turbine Disks," J.R. Dunphy; W.H. Atkinson. ASME 90-GT-390.

"Vibration Baseline, A Useful Concept For Condition Monitoring Of Rotating Equipment," J. Gorter; A.J. Klijn. ASME 85-GT-106.

"A Preliminary Assessment Of Compressor Fouling," H.I.H. Saravanamuttoo; A.N. Lakshminarasimha. ASME 85-GT-153.

LM1600 Gas Turbine Combustor Induced Vibration In Natural Gas Transmission Pipeline Applications," Barney Ong Industrial Applications Of Gas Turbines 1993 Symposium October 13-15, 1993.

"Small Gas Turbine Generator Set Starting," Colin Rogers. ASME Cogen Turbo Power '94.

"Analyzing Gas Turbine Failures To Identify Training Shortfalls," Lane Hunter Coward; James R. Hardin. ASME 96-GT-329.

"The Effects Of Environmental Contaminants On Industrial Gas Turbine Thermal Barrier Coatings," H.E. Eaton; N.S. Bornstein; J.T. DeMarcin. ASME 96-GT-283.

"Capacitive Measurement Of Compressor And Turbine Blade Tip To Casing Running Clearance," D. Muller; S. Mozumdar; E. Johann, A.G. Sheard. ASME 96-GT-349.

"Effects Of Inlet Distortion On The Flow Field In A Transonic Compressor Rotor," Chunill Hah; Douglas C. Rabe; Thomas J. Sullivan. ASME 96-GT-547.

"Microwave Blade Tip Clearance Measurement System," Richard Grzybowski; George Foyt; William Atkinson; Hartwig Knoell. ASME 96-GT-002.

"Thermal Stresses In Gas Turbine Exhaust Duct Expansion Joints," Michael D. Ninacs; Rodney P. Bell. ASME 96-GT-398.

"Failure Analysis Of A Burner Component," Kenneth L. Saunders; Brian P. Copley. ASME 96-GT-399.

"Analytical Solution Of Whirl Speed And Mode Shape For Rotating Shafts," S.M. Young; G.J. Sheu; C.D. Yang. ASME 96-GT-264.

"Indicators Of Incipient Surge For Three Turbofan Engines Using Standard Equipment And Instrumentation," Adam J. Baran; Michael G. Dunn. ASME 96-GT-056.

"Laser-Doppler Velocimetry Measurements In A Cascade Of Compressor Blades At Stall," Garth V. Hobson; Andrew J.H. Williams; Humberto J. Ganaim. ASME 96-GT-484.

"A Distortion Problem In A Single Shaft Military Turbojet," Petros Kotsiopoulos; Andreas Kottarakos; Pericles Pilidis. ASME 96-GT-432.

"Integrated Predictive Diagnostics: An Expanded View," Robert J. Hansen; David L. Hall; G. William Nickerson; Shashi Phoba. ASME 96-GT-034.

"The Negative Information Problem In Mechanical Diagnostics," David L. Hall; Robert J. Hansen; Derek C. Lang. ASME 96-GT-035.

"The Effects Of Engine Degradation On Creep—Using A Transient Engine Simulation," P. Darrin Little; Riti Singh. ASME 96-GT-033.

"Development Of A Diagnostic Tool To Trouble Shoot LM2500 Performance & Controls Problems," Bruce D. Thompson; Richard Raczkowski. ASME 96-GT-213.

"Test Technics And Evaluation Requirements For Marine Gas Turbine Air Inlet Systems," Howard Harris, Ivan Pineiro; Alan Oswald. ASME 96-GT-501.

"Power Turbine Remaining Life Analysis: Experience And Practical Considerations," Robert J. Klova;. Hermes Galatis. ASME 96-GT-211.

"Sulfur And Sea Salt Attack Of Turbine Blades," C.A. Dalton. ASME 65-GTP-7.

"Pinpointing Vibration In Rotating Machinery," Hugh W. Ness. ASME 65-GTP-24.

ROTORS, BEARINGS, AND LUBRICATION

"Generalized Polynomial Expansion Method For The Dynamic Analysis Of Rotor-Bearing Systems," T.N. Shiau: J.L. Hwang. ASME 91-GT-006.

"Multiobjective Optimization Of Rotor-Bearing System With Critical Speeds Constraints," T.N. Shiau; J.R. Chang. ASME 91-GT-117.

"Experimental And Computed Performance Characteristics Of High Speed Silicon Nitride Hybrid Ball Bearings," G.W. Hosang. ASME 90-GT-110.

"Rolling Wear Of Silicon Nitride Bearing Materials," J.W. Lucek. ASME 90-GT-165.

"Lubrication Oil Reservior Mist Elimination," R.D. Belden. ASME 90-GT-183.

"The Application Of Active Magnetic Bearings To A Power Turbine," G.E. Brailean; W.M. Grasdal; V. Kulle; C.P. Oleksuk; R.A. Peterson. ASME 90-GT-199.

"Status Of The Industrial Applications Of The Active Magnetic Bearings Technology," M. Dussaux. ASME 90-GT-268.

"Triaryl Phosphates—The Next Generation Of Lubricants For Steam And Gas Turbines," W.D. Phillips. ASME 94-JPGC-PWR-64.

"Tilting Pad Journal Bearing Pivot Design For High Load Applications," John C. Nichols; Karl D. Wygant. 24th Turbomachinery Symposium, September 26-28, 1995.

"Thermal Effects In Hydrodynamic Journal Bearings Of Speed Increasing And Reduction Gearboxes," Christophe Bouchoule, Michel Fillon, Daniel Nicholas and Fouad Y. Zeidan. 24th Turbomachinery Symposium September 26-28, 1995.

"Troubleshooting Bearing And Lube Oil System Problems," Thomas H. McCloskey. 24th Turbomachinery Symposium, September 26-28, 1995.

"Experimental Analysis Of Journal Bearings," A.H. Elkholy; A. Elshakweer. ASME Cogen Turbo Power '94.

"Stability Analysis Of Symmetrical Rotor-Bearing Systems With Internal Damping Using Finite Element Method," L. Forrai. ASME 96-GT-407.

FILTRATION AND CLEANING

"Compressor Cleaning Effectiveness For Marine Gas Turbines," A. Abdelrazik; P. Cheney. ASME 91-GT-011.

"US Navy On-Line Compressor Washing Of Marine Gas Turbine Engines," R. Margolis; S. Costa. ASME 91-GT-309.

"The Mechanism Of Fouling And The Cleaning Technique In Application To Flow Parts Of The Power Generation Plant Compressors," A.D. Mezheritsky; A.V. Sudarev. ASME 90-GT-103.

"Effect Of Particle Size Distribution On Particle Dynamics And Blade Erosion In Axial Flow Turbines," W. Tabakoff; A. Hamed; M. Metwally. ASME 90-GT-114.

"Technical And Other Considerations For The Selection Of Inlet Air Filtration Systems For High-Efficiency Industrial Combustion Turbines," C.H. Goulding; M.G. Rasmussen; F.M. Fritz, Jr. ASME 90-GT-176.

"A New Air Filtation System Concept For Space Limited Applications And Retrofits," L. Cuvelier; M.D. Belcher. ASME 90-GT-177.

"Interim Report On Dust Louvre Inertial Separator Application On 8500-hp U.P.R.R. Gas-Turbine Locomotive," F.D. Bruner; J.K. Sparrow; W.B. Moyer. ASME 63-AHGT-39.

"The Gas Turbine-Air-Cleaner Dilema," Mark G. Mund; Thomas E. Murphy. ASME 63-AHGT-63.

"Application Of Particle-Size Analysis Data For The Determination Of Air/Cleaner Performance," Lawrence J. Czerwonka; Jack M. Carey. ASME 65-GTP-6.

"Evaluation Of Air Filter Test Methods," James W. May. ASME 65-GTP-10.

"Axial Compressor Washing Of Aero Derivative Gas Turbines. A Review Of Existing Practices," Gordon F. Aker, Canadian Gas Association Industrial Application Of Gas Turbines Symposium, Banff, Alberta, October 13-15, 1993.

"Newly Developed Filter Products For Gas Turbine Intake Air Filtration," Thomas Schroth; Antje Rudolph; Carl Freudenberg. ASME 96-GT-517.

"High Contaminant Air Filtration System For US Navy Hovercraft Gas Turbines," Jerome Ehrhardt; Ivan Pineiro. ASME 96-GT-502.

"An Analysis Of Axial Compressors Fouling And A Cleaning Method Of Their Blading," A.P. Tarabrin; V.A. Schurovsky; A.I. Bodrov; J.P. Stalder. ASME 96-GT-363.

"On-Line Gas Turbine Inlet Air Filter Tests," Tony Giampaolo, Canadian Gas Association Industrial Application Of Gas Turbines Symposium, Banff, Alberta, November, 1992

"Increasing Capacity Sales With Inlet Air Cooling," Todd A. Sundbom; Jerry A. Ebeling. ASME Cogen Turbo Power '94.

DESIGN TECHNIQUES

"Acoustic Design Of Lightweight Gas Turbine Enclosures," R.D. Rawlinson. ASME 90-GT-054.

"Three-Dimensional Flow In An Axial Turbine. Part 1—Aerodyanamic Mechanisms," H.D. Joslyn; R.P. Dring. ASME 90-GT-056.

"Three-Dimensional Flow In An Axial Turbine. Part 2—Profile Attenuation," H.D. Joslyn; R.P. Dring. ASME 90-GT-057.

"Thermogasdynamic Effects Of The Engine Turbines With The Contra-Rotating Turbines," Y.V. Sotsenko. ASME 90-GT-063.

"Basic Analysis Of Counter-Rotating Turbines," R. Cai; G. Fang; W. Wu. ASME 90-GT-108.

"Garrett's Turboshaft Engine And Technologies For 1990's," T.D. Pyle; D.R. Aldrich. ASME 90-GT-204.

"Advanced Technology Programs for Small Turboshaft Engines: Past, Present, Future,"E.T. Johnson; H. Lindsay. ASME 90-GT-267.

"Considerations For The Use Of Variable Geometry In Gas Turbines," J.E.A. Roy-Aikins. ASME 90-GT-271.

"Salt Water Aerosol Separator Development," R.E. Kaufman. ASME 65-GTP-9.

"Sea Experience With Pratt And Whitney Aircraft FT12 Gas Turbine in LCM-8," G.H. Nolte. ASME 65-GTP-26.

"Experimental Evaluation Of A Low Emissions, Variable Geometry, Small Gas Turbine Combustor," K.O. Smith; M.H. Samii; H.K. Mak;. ASME 90-GT-085.

"Procedures For Shape Optimization Of Gas Turbine Disks," T.C. Cheu. ASME 90-GT-281.

"Acoustical Design And Analysis Techniques For Combustion Turbine Products," Lisa A. Beson; George A. Schott. ASME Cogen Turbo Power '94.

"Optimization Scheme For Gas Turbine Nozzle Design," Francisco J. Cunha; David A. DeAngelis. ASME Cogen Turbo Power '94.

"Designing For High Reliability And Availbility In New Combustion Turbines," G. McQuiggan. ASME 96-GT-14.

"Development Of The Next Generation 1500C Class Advanced Gas Turbine For 50HZ Utilities," S. Aoki; Y. Tsukuda; E. Akita; Y. Iwasaki; R. Tomat; M. Rosso; C. Schips. ASME 96-GT-314.

"An International Initiative To Catalyze An Intercooled Aeroderivative (ICAD) Gas Turbine Launch Order," George A. Hay; Barry Davidson; Dan Whitney; Clark Dohner, Collaborative Advanced Gas Turbine (CAGT) Program Status Presented At Power-Gen '95. Anaheim, Ca., December 7, 1995.

"The Jet Plane Is Born," T.A. Heppenheimer, Invention & Technology, Fall 1993.

"Technology Development Programs For The Advanced Turbine Systems Engines," Ibor S. Diakunchak; Ronald L. Bannister; David J. Huber; D. Fran Roan. ASME 96-GT-005.

"A Review Of Recent Research On Contra-Rotating Axial Flow Compressor Stage," P.B. Sharma; A. Adekoya. ASME 96-GT-254.

"Development Of Ceramic Turbine Rotors And Nozzles For The 100kW Automotive Ceramic Gas Turbine Program," Satoru Yamada; Keiichiro Watanabe; Massaki Masuda. ASME 96-GT-295.

"Progress On The European Gas Turbine Program-AGATA," Rolf Gabielsson; Goran Holmquist. ASME 96-GT-362.

"Status Of The Automotive Ceramic Gas Turbine Development Program—Year Five Progress," Tsubura Nisiyama; Norio Nakazawz; Masafumi Sasaki; Masumi Iwai; Haruo Katagiri; Noritoshi Handa. ASME 96-GT-036.

"Radial Turbine Development For The 100kW Automotive Ceramic Gas Turbine," Norio Nakazawa; Hiroshi Ogita; Masayuki Takahashi; Takaaki Yoshizawa; Yasumasa Mori. ASME 96-GT-366.

"An Assessment Of Humid Air Turbines In Coal-Fired High Performance Power Systems," Julianne M. Klara; Robert M. Enick; Scott M. Klara; Lawrence E. Van Bibber. ASME 96-GT-091.

"Externally Fired Evaporative Gas Turbine With A Condensing Heat Exchanger," Jinyue Yan; Lars Eidensten; Gunnar Svedberg. ASME 96-GT-077.

"Externally Fired Combined Cycles (EFCC)—Part A Thermodynamics And Technological Issues," Stefano Consonni; Ennio Macchi; Francesco Farina; Ansaldo Rierche. ASME 96-GT-092.

"Externally Fired Combined Cycles (EFCC). Part B Alternative Configurations And Cost Projections," Stefano Consonni; Ennio Macchi; ASME 96-GT-093.

"Performance Modeling Of Aeroderivative Steam-Injected Gas Turbines And Combined Cycles Fueled From Fixed Or Fluid-Bed Biomass Gasifiers," Eric D. Larson; Wendy E.M. Hughes. ASME 96-GT-089.

"Next-Generation Integration Concepts For Air Separation Units And Gas Turbines," Arthur R. Smith; Joseph Klosek; Donald W. Woodward. ASME 96-GT-144.

"A New Approach For A Low-Cooled Ceramic Nozzle Vane," C. Gutmann; A. Schulz; S. Wittig. ASME 96-GT-232.

"Development Of Ceramic Stator Vane For 1500 C Class Industrial Gas Turbine," Takashi Machida; Masato Nakayama; Katsuo Wada; Tooru Hisamatsu; Isao Yuri; Kazunori Watanabe. ASME 96-GT-459.

"Structural Design And High Pressure Test Of A Ceramic Combustor For 1500 C Class Industrial Gas Turbine," I. Yuri; T. Hisamatsu; K. Watanabe; Y. Etori. ASME 96-GT-346.

"Gas Turbine Silencers," Oliver C. Eckel. ASME 63-AHGT-17.

INLET AND TURBINE COOLING

"Field Evaluation Of On-Line Compressor Cleaning In Heavy Duty Industrial Gas Turbines," G.L. Haub; W.E. Hauhe, Jr. ASME 90-GT-107.

"Options In Gas Turbine Power Augmentation Using Inlet Air Chilling," I.S. Ondryas; D.A. Wilson; M. Kawamoto; G.L. Haub. ASME 90-GT-250.

"Gas Turbine Performance Improvement. Direct Mixing Evaporative Cooling System." American Atlas Cogeneration Facility, Rifle, Colorado J.P. Nolan; V.J. Twombly. ASME 90-GT-368.

"A Qualitative Analysis Of Combustion Turbine Inlet Air Cooling Alternatives with Case Histories," Jerry A. Ebeling. ASME 94-JPGC-GT-4.

"Effects Of Closed-Circuit Gas Turbine Cooling Systems On Combined Cycle Performance," Takashi Ikeguchi; Kazuhiko Kawaike. ASME 94-JPGC-GT-8.

"The Advanced Cooling Technology For The 1500 C Class Gas Turbines—The Steam Cooled Vanes And The Air Cooled Blades," H. Nomoto; A. Koga; S. Shibuya; S. Ito; Y. Fukuyama; F. Otomo; M. Sato; Y. Kobayashi; H. Matsuzaki. ASME 96-GT-016.

"Economics Of Gas Turbine Inlet Air-Cooling System For Power Enhancement," Motoaki Utamura; Yoshio Nishimura; Akira Ishikawa; Nobuo Ando. ASME 96-GT-516.

"Inlet Conditioning Enhances Performance Of Modern Combined Cycle Plants For Cost-Effectve Power Generation," Septimus Van Der Linden; David E. Searles. ASME 96-GT-298.

"The Application Of Convective Cooling Enhancement Of Span-Wise Cooling Holes In A Typical First Stage Turbine Blade," D.J. Stankiewicz; T.R. Kirkham. ASME 91-GT-012.

"Minimization Of The Number Of Cooling Holes In Internally Cooled Turbine Blades," G.S. Dulikravich; B. Kosovic. ASME 91-GT-052.

"Film Cooling On A Gas Turbine Rotor Blade," K. Takeishi; S. Aoki; T. Sato; K. Taukagoshi. ASME 91-GT-279.

"Full Coverage Discrete Hold Film Cooling: The Influence Of The Number Of Holes And Pressure Loss," G.E. Andrews; A.A. Asere; M.L. Gupta; M.C. Mkpadi; A. Tirmahi. ASME 90-GT-061.

"Film Cooling Effectiveness Downstream Of Single Row Of Holes With Variable Density Ratio," A.K. Sinha; D.J. Bogard; M.E. Crawford. ASME 90-GT-043.

"Gas Turbine Film Cooling: Flowfield Due To A Second Row Of Holes," A.K. Sinha; D.G. Bogard; M.E. Crawford. ASME 90-GT-044.

"Prediction And Measurement Of Film Cooling Effectiveness For A First-Stage Turbine Vane Shroud," D. Granser; T. Schulenberg. ASME 90-GT-095.

"Discharge Coefficient Of Turbine Cooling Holes: A Review," N. Hay; D. Lampard. ASME 96-GT-492.

"Adiabatic Effectiveness And Heat Transfer Coefficient On A Film-Cooled Rotating Blade,"Vijay K. Garg . ASME 96-GT-221.

"Effects Of Surface Roughness On Film Cooling," Donald L. Schmidt; Basav Sen; David G. Bogard. ASME 96-GT-299.

"Effects Of Geometry On Slot-Jet Film Cooling Performance," Daniel G. Hyams; Kevin T. McGovern; James H. Leylek. ASME 96-GT-187.

"An Experimental Study Of Turbine Vane Heat Transfer With Water-Air Cooling," Nirm V. Nirmalan; John A. Weaver; Larry D. Hylton. ASME 96-GT-381.

"Analysis And Validation Of Turbine Disc Cooling," John M. Hannis; John D. Maltson; Robert J. Corry, Nigel Johnson. ASME 96-GT-097.

"Aerodynamic and Thermodynamic Effects of Coolant Injection on Axial Compressors," P.G. Hill, November 1963.

"Water Spray Injection of an Axial Flow Compressor," I.T. Wetzel and B.H. Jennings, Mid-West Power Conference, April 18 - 20, 1949.

"Technical Performance of an Axial-Flow Compressor in a Gas-Turbine Engine Operating with Inlet Water Injection," Reece V. Hensley, Technical Note 2673, March 1962, National Advisory Committee For Aeronautics.

"Effect of Water Spraying on Operation of the Compressor of a Gas turbine Engine," L.I. Slobodyanyuk, Energeticka, No. 1; 1973.

"Gas Turbines with Heat Exchanger and Water Injection in the Compressed Air," Gasparovic and Hellemans; Combustion, Dec 1972.

"Axial-Compressor Flow Distortion with Water Injection"; Tsuchiya, et al.; AIAA paper No AIAA-83-0004; 1983.

"Direct and System Effects of Water Injection into Jet Engine Compressors"; Murthy, et al.; Presented at AIAA/ASME 4th Joint Fluid Mechanics, Plasma Dynamics and Laser Conference, Mat 12-14, 1896.

"Water Injection into Axial Flow Compressors," Technical ReportA-FAPL-TR-7677; Aug 1976.

"Value of Wet Compression in Gas-Turbine Cycles"; R. Kleinschmidt; Mechanical Engineer, vol. 69, No. 2; pp.115-116, ASME 1946.

"Benefits of Compressor Inlet Air Cooling for Gas Turbine Cogeneration Plants, Delucia et al., ASME Paper No. 95-GT-311, 1995.

"The Theory and Operation of Evaporative Coolers for Industrial Gas Turbine Installations; R. Johnson, Sr., pp. 327-334, Journal for Engineering for Gas Turbines and Power, vol. 111; April 1989.

"Field Evaluation Of On-Line Compressor Cleaning In Heavy Duty Industrial Gas Turbines," G.L. Haub; W.E. Hauhe, Jr. ASME 90-GT-107.

"Options In Gas Turbine Power Augmentation Using Inlet Air Chilling," I.S. Ondryas; D.A. Wilson; M. Kawamoto; G.L. Haub ASME 90-GT-250.

"Gas Turbine Performance Improvement. Direct Mixing Evaporative Cooling System. American Atlas Cogeneration Facility, Rifle, Colorado J.P. Nolan; V.J. Twombly ASME 90-GT-368.

"A Qualitative Analysis Of Combustion Turbine Inlet Air Cooling Alternatives with Case Histories," Jerry A. Ebeling ASME 94-JPGC-GT-4.

"Effects Of Closed-Circuit Gas Turbine Cooling Systems On Combined Cycle Performance," Takashi Ikeguchi; Kazuhiko Kawaike ASME 94-JPGC-GT-8.

"The Advanced Cooling Technology For The 1500C Class Gas Turbines - The Steam Cooled Vanes And The Air Cooled Blades," H. Nomoto; A. Koga; S. Shibuya; S. Ito; Y. Fukuyama; F. Otomo; M. Sato; Y. Kobayashi; H. Matsuzaki ASME 96-GT-016.

"Economics Of Gas Turbine Inlet Air-Cooling System For Power Enhancement," Motoaki Utamura; Yoshio Nishimura; Akira Ishikawa; Nobuo Ando ASME 96-GT-516.

"Inlet Conditioning Enhances Performance Of Modern Combined Cycle Plants For Cost-Effectve Power Generation," Septimus Van Der Linden; David E. Searles ASME 96-GT-298.

"The Application Of Convective Cooling Enhancement Of Span-Wise Cooling Holes In A Typical First Stage Turbine Blade," D.J. Stankiewicz; T.R. Kirkham ASME 91-GT-012.

"Minimization Of The Number Of Cooling Holes In Internally Cooled Turbine Blades," G.S. Dulikravich; B. Kosovic ASME 91-GT-052.

"Film Cooling On A Gas Turbine Rotor Blade," K. Takeishi; S. Aoki; T. Sato; K. Taukagoshi ASME 91-GT-279.

"Full Coverage Discrete Hold Film Cooling: The Influence Of The Number Of Holes And Pressure Loss," G.E. Andrews; A.A. Asere; M.L. Gupta; M.C. Mkpadi; A. Tirmahi ASME 90-GT-061.

"Film Cooling Effectiveness Downstream Of Single Row Of Holes With Variable Density Ratio," A.K. Sinha; D.J. Bogard; M.E. Crawford ASME 90-GT-043.

"Gas Turbine Film Cooling: Flowfield Due To A Second Row Of Holes, A.K. Sinha; D.G. Bogard; M.E. Crawford ASME 90-GT-044.

"Prediction And Measurement Of Film Cooling Effectiveness For A First-Stage Turbine Vane Shroud," D. Granser; T. Schulenberg ASME 90-GT-095.

"Discharge Coefficient Of Turbine Cooling Holes: A Review," N. Hay; D. Lampard ASME 96-GT-492.

"Adiabatic Effectiveness And Heat Transfer Coefficient On A Film-Cooled Rotating Blade,"Vijay K. Garg AASME 96-GT-221.

"Effects Of Surface Roughness On Film Cooling," Donald L Schmidt; Basav Sen; David G. Bogard ASME 96-GT-299.

"Effects Of Geometry On Slot-Jet Film Cooling Performance," Daniel G. Hyams; Kevin T. McGovern; James H. Leylek ASME 96-Gt-187.

"An Experimental Study Of Turbine Vane Heat Transfer With Water-Air Cooling," Nirm V. Nirmalan; John A. Weaver; Larry D. Hylton ASME 96-GT-381.

"Analysis And Validation Of Turbine Disc Cooling," John M. Hannis; John D. Maltson; Robert J. Corry Nigel Johnson AASME 96-GT-097.

EMISSIONS

"Design And Operating Experience Of Selective Catalytic Reduction Systems For NO_x Control In Gas Turbine Systems," S.M. Cho; A.H. Seltzer; Z. Tsutsui. ASME 91-GT-026.

"Evaluation Of NO_x Mechanisms For Lean, Premixed Combustion," R.A. Corr; P.C. Malte; N.M. Marinov. ASME 91-GT-257.

"A Predictive NO_x Monitoring System For Gas Turbines," W.S.Y. Hung. ASME 91-GT-306.

"Combustion Performance Of A Water-Injected MS7001E Gas Turbine Operting At A NO_x Emission Level Of 25 ppmvd," D.O. Fitts; R.A. Symonds; E.R. Western. ASME 90-GT-071.

"An Ultra-Low NO_x Combustion System For A 3.5 MW Industrial Gas Turbine," C. Wilkes; H.C. Mongia; C.B. Santaman. ASME 90-GT-083.

"Assessment Of Hot Gas Cleanup Technologies In Coal-Fired Gas Turbines," E.L. Parsons, Jr; H.A. Webb; C.M. Zeh. ASME 90-GT-111.

"Ultra-Low NOx Rich-Lean Combustion," N.K. Rizk; H.C. Mongia. ASME 90-GT-087.

"Calculating NH_3 Slip For SCR Equipped Cogeneration Units," C.M. Anderson; J.A. Billings. ASME 90-GT-105.

"Unburned Hydrocarbon, Volatile Organic Compound, And Aldehyde Emissions From General Electric Heavy-Duty Gas Turbines," R.E. Pavri; R.A. Symonds. ASME 90-GT-279.

"Experimental Evaluation Of A Low NO_x LBG Combustor Using Bypass Air," T. Nakata; M. Sato; T. Ninomiya; T. Abe; S. Mandai; N. Sato. ASME 90-GT-380.

"NO_x Exhaust Emissions For Gas-Fired Turbine Engines," P.C. Malte; S. Bernstein; F. Bahlmann; J. Doelman. ASME 90-GT-392.

"Minimizing Emissions In Supplementary Fired HRSG's And Auxillary Package Boilers In Gas Turbine Based Cogeneraton Systems," Jon C. Backlund And James H. White, Industrial Power Conference 1994.

"Predictive Emission Monitoring System (PEMS): An Effective Alternative To In-Stack Continuous NO_x Monitoring," Wilfred Y. Hung, Industrial Power Conference 1994.

Reducing Gas Turbine Emissions Through Hydrogen-Enhanced, Steam Injected Combustion," James R. Maughan, John H. Bowen, David H. Cooke, And John J. Tuzson. ASME Cogen Turbo Power '94.

"The Control Of NO_x And CO Emissions From 7-MW Gas Turbines With Water Injection As Influenced By Ambient Conditions," W.S.Y. Hung And D.D. Agan. ASME 85-GT-85.

"Development Of A Dry Low NO_x Combustion System," V. Mezzedimi; L. Bonciani; G. Ceccherini; R. Modi. Canadian Gas Association Symposium On Industrial Application Of Gas Turbines, Banff, Alberta, October 1993."

"A Model For The Prediction Of Thermal, Prompt, And Fuel NO_x Emissions From Combustion Turbines," J.L. Toof. ASME 85-GT-29

"Stochastic Modelling Of NO_x And Smoke Production In Gas Turbine Combustors," I.M. Aksit; J.B. Moss. ASME 96-GT-186.

"Dry Low NO_x Combustion Systems For GE Heavy-Duty Gas Turbines," L. Berkeley Davis. ASME 96-GT-027.

"An Advanced Development Of A Second-Generation Dry, Low-NO_x Combustor For 1.5MW Gas Turbine," Shin-ichi Kajita; Shin-ichi Ohga; Masahiro Ogata; Satoru Iraka; Jun-ichi Kitajima; Takeshi Kimura; Atsushi Okuto. ASME 96-GT-049.

"Sensor Based Analyzer For Continuous Emissions Monitoring In Gas Pipeline Applications," Paul F. Schubert; David R. Sheridan; Michael D. Cooper; Andrew J. Banchieri. ASME 96-GT-481.

"Dry Low Nox Combustion Systems Development And Operating Experience," Torsten Strand. ASME 96-GT-274.

PERFORMANCE

"Simulation Of An Advanced Twin-Spool Industrial Gas Turbine," P. Zhu; H.I.H. Saravanamuttoo. ASME 91-GT-034.

"Repowering Application Considerations," J.A. Brander; D.L. Chase. ASME 91-GT-229.

"Humidity Effects On Gas Turbine Performance," W. Grabe; J. Bird. ASME 91-GT-329.

"FT8A, A New High Performance 25 MW Mechanical Drive Aero Derivative Gas Turbine," A. Prario; H. Voss. ASME 90-GT-287.

"A New Method Of Predicting The Performance Of Gas Turbine Engines," Y. Wang. ASME 90-GT-337.

"Performance Deterioration In Industrial Gas Turbines," I.S. Diakunchak. ASME 91-GT-228.

"Transient Performance And Behaviour Of Gas Turbine Engines," G. Torella. ASME 90-GT-188.

"The Advanced Cycle System Gas Turbines GT24/GT26 The Highly Efficient Gas Turbines For Power Generation," Beat Imwinkelried; Rainer Hauenschild. ASME 94-JPGC-GT-7

"An Investigation Of The Dynamic Characteristic Of Axial Flow Compressors," G.C. Tang; B. Hu; H.M. Zhang. ASME 90-GT-210.

"Performance Prediction Of Centrifugal Impellers," Abraham Engeda. ASME Cogen Turbo Power '94.

"Second Law Efficiency Analysis Of Gas-Turbine Engine For Cogeneration," S.C. Lee; R.M. Wagner. ASME Cogen Turbo Power '94.

"A Parametric Study Of Steam Injected Gas Turbine With Steam Injector," Niklas D. Agren; H.U. Frutschi; Gunnar Svedberg. ASME Cogen Turbo Power '94.

"Humid Air Gas Turbine Cycle (HAT Cycle): Irreversibility Accounting And Exergetic Analysis." ASME Cogen Turbo Power '94.

"The Hydraulic Air Compressor Combustion Turbine," Jerry L. Shapiro. ASME Cogen Turbo Power '94.

"A Combined Cycle Designed To Achieve Greater Than 60 Percent Efficiency,"Michael S. Bbriesch; Ronald L. Bannister; Ihor S. Diakunchak; David J. Huber. ASME Cogen Turbo Power '94.

"RAM—Performance Of Modern Gas Turbines," Axel W. Von Rappard; Heinz G. Neuhoff; Salvatore A. Della Villa. ASME 96-GT-416.

"Performance Analysis Of Gas Turbines Operating At Different Atmospheric Conditions," Umberto Desideri. ASME Cogen Turbo Power '94.

"Typical Performance Characteristics Of Gas Turbine Radial Compressors," Colin Rogers. ASME 63-AHGT-14.

REPAIR TECHNIQUES

"The Utilization Of Computer Aided Drafting (CAD) In Quality Control Of Gas Turbine Component Repairs," D.J. Baldwin. ASME 91-GT-138.

"Fracture Mechanics Approach To Creep Crack Growth In Welded IN738LC Gas Turbine Blades," W.P. Foo; R. Castillo. ASME 91-GT-119.

"Innovations In Refurbishing Gas Turbine Components," C.F. Walker; J.S. Cosart. ASME 90-GT-202.

"User Experience In Upgrading Early Models Of Aero Derived Gas Turbine Pipeline Compressor Units To Current Standards," L.J. Williams; D.J. Pethrick. ASME 90-GT-291.

""Recycling" Of Gas Turbines From Obsolete Aircraft," J.M. Morquillas; P. Pilidis. ASME 90-GT-309.

"An Evaluation Of Waspaloy Expander Blade Tip Weld Repairs." D.W. Cameron. ASME 90-GT-317.

"Chemical Stripping Of Honeycomb Airseals. Overview & Update," Z. Galel; F. Brindisi; D. Nordstrom. ASME 90-GT-318.

"Blade Attachment And Seal Area Replacement For First Stage Turbine Discs Of Westinghouse Model 501AA Combustion Turbines," William F. Cline; Richard League, Combustion Turbine Operations Task Force Meeting September 25-28, 1995 Portland, Oregon.

"Procedure Development For The Repair Of GTD-111 Gas Turbine Bucket Material," Joseph M. Hale. ASME Cogen Turbo Power '94

"Flash Weldability Studies Of Incoloy Alloy 908," Jerry E. Gould; Timothy Stotler; Michael Steeves. ASME Cogen Turbo Power '94.

"Development Of A Weld Repair For Airfoils Of MAR-M 247 Material," Daniel P. Rose, Jr.; Joe A. Saenz. ASME Cogen Turbo Power '94.

"Nondestructive Evaluation Of Locally Repaired Coating For Gas Turbine Blades," Y. Shen; N. Pietranera; U. Guerreschi; S. Corcoruto; E. Gandini. ASME 96-GT-220.

"High Strength Diffusion Braze Repair For Gas Turbine Components," Roger D. Wustman; Jeffrey S. Smith; Leonard M. Hampson; Marc E. Suneson. ASME 96-GT-427.

"LM2500 High Pressure Turbine Blade Refurbishment," Matthew J. Driscoll; Peter P. Descar Jr.; Gerald B. Katz; Walter E. Coward. ASME 96-GT-214.

INSTRUMENTATION

"Measuring Rotor And Blade Dynamics Using An Optical Blade Tip Sensor," H.R. Simmons; K.E. Brewer; D.L. Michalsky; A.J. Smalley. ASME 90-GT-091

"Unsteady Velocity And Turbulence Measurements With A Fast Response Pressure Probe," G. Ruck; H. Stetter. ASME 90-GT-232.

"Aero/Aeroderivative Engines: Internal Transducers Offer Potential For Enhanced Condition Monitoring And Vibration Diagnostics," M.J. Werner. ASME 90-GT-273.

"A Fast-Response Total Temperature Probe For Unsteady Compressible Flows," D.R. Buttsworth; T.V. Jones. ASME 96-GT-350.

"Hot Wire Measurements During Rotating Stall In A Variable Pitch Axial Flow Fan," Eduardo Blanco-Marigorta; Rafael Ballesteros-Tajadura; Carlos Santolaria. ASME 96-GT-441.

"A High-Frequency-Response Pressure Probe For The Measurement Of Unsteady Flow Between Two Rotors In A Hydrodynamic Turbomachine," C. Achtelik; J. Eikelmann. ASME 96-GT-412.

"Laser Vibrometry Measurements Of Rotating Blade Vibrations," Andrew K. Reinhardt; J.R. Kadambi; Roger D. Quinn. ASME Cogen Turbo Power '94.

"Improved Method For Flame Detection In Combustion Turbines," Richard J. Roby; Andrew J. Hamer; Erik L. Johnsson; Shawn A. Tilstra. ASME Cogen Turbo Power '94.

"Application Of Analytical Redundancy To The Detection Of Sensor Faults On A Turbofan Engine," Ronald E. Kelly. ASME 96-GT-003.

"Simple Instrumentation Rake Designs For Gas Turbine Engine Testing," Peter D. Smout; Steven C. Cook. ASME 96-GT-032.

"5,000-Hour Stability Tests Of Metal Sheathed Thermocouples At Respective Temperatures Of 538 C and 675 C," T.P. Wang; A. Wells; D. Bediones. ASME 91-GT-182.

FIRE PROTECTION

"Experience With External Fires In Gas Turbine Installations And Implications For Fire Protection," R.E. Dundas. ASME 90-GT-375.

ADDITIONAL ARTICLES

"Micro, Industrial, and Advanced Gas Turbines Employing Recuperators," # GT2003-38938, by Steven I. Freedman, Jim Kesseli, Jim Nash, Tom Wolf

"Regenerative Heat Exchangers for Microturbines, and an Improved Type," # GT2003-38871, by Dave Wilson

"Ceramic Reliability for Microturbines Hot-Section Components," http://www.eere.energy.gov/de/pdfs/microturbine_ceramic_reliability.pdf

"Behavior of Two Capstone 30 kW Microturbines Operating in Parallel with Impedance Between Them," # LBNL-55907, Prepared by Robert J. Yinger, Southern California Edison, July 2004

"Clean Distributed Generation Performance and Cost Analysis" Prepared for Oak Ridge National Laboratory And U.S. Department of Energy, Prepared by DE Solutions Inc, April 2004

"Microturbines Power Conversion Technology Review," by R.H. Staunton, B. Ozpineci, April 8, 2003

"Blackouts, Lighting and Microturbines" by Robin Mackay, September 2003

"Behavior of Capstone and Honeywell Microturbines Generators during Load Changes," Prepared by Robert J. Yinger, Southern California Edison, July 2001

"U.S. Installation, Operation, and Performance Standards for Microturbines Generator Sets," # PNNL-13277, by A.M. Borbely-Bartis, J.G. DeSteese, S. Somasundaram, Prepared for the U.S. Department of Energy, Contract # DE-AC06-76RL01830

"Distributed Energy Technology Simulator—Microturbines Demonstration," by Mindi Farber De Anda, Christina TerMaath, Nde K. Fall, October 2001

Glossary

ADIABATIC

A process in which no heat loss crosses the system boundary in either direction.

APPROACH POINT

The difference in temperature between the exhaust gas temperatures and the superheater outlet temperature and the economizer inlet temperature at that point.

BOTTOMING CYCLE

The lower temperature cycle is called the bottoming cycle. All bottoming cycles are Rankine Cycles.

CRITICAL POINT

Above which the substance exists only as a vapor regardless of the pressure & temperature.

ENTHALPY

Enthalpy (H) is the sum of the internal energy (U) and the product of pressure and volume (PV) given by the equation: $H = U + PV$. When a process occurs at constant pressure, the heat evolved (either released or absorbed) is equal to the change in enthalpy.

ENTROPY

A quantitative measure of the amount of thermal energy not available to do work in a closed thermodynamic system.

HEAT CAPACITY

The amount of heat required to raise the temperature of a unit mass of a substance one-degree.

HEAT OF FUSION

Heat related to melting or freezing a solid (for water = 143 Btu/lb.).

HEAT OF VAPORIZATION

Heat related to vaporizing or condensing a liquid (Water = 970 Btu/lb).

ISENTROPIC

Without change in entropy; at constant entropy. Constant entropy—reversible adiabatic.

LATENT HEAT

The quantity of heat absorbed or released by a substance undergoing a change of state, such as ice changing to water or water to steam, at constant temperature and pressure.

PINCHPOINT

The difference in temperature between the saturated steam temperature and the gas temperature at that point.

SENSIBLE HEAT

Heat that changes the temperature of a substance (Btu/lb).

SPECIFIC HEAT

The ratio of the heat capacity of a substance to the heat capacity of water Btu/lb-°F (cal/gmC).

TOPPING CYCLE

The high temperature cycle is commonly called the topping cycle and is either Otto, Brayton or Rankine cycles.

Index

A

absolute or dynamic viscosity 110
accelerate 94
acceleration 98, 115, 116
  problems 251
  schedule 88
  schedule limit 89
  scheduler 87
accelerometers 115, 116
accessories 93
accessory equipment 113
accessory gearbox 95, 96
acoustic 204
acoustical materials 168, 170, 171, 172
adiabatic 493
  processes 47
aero-derivative 15, 21, 22, 26, 27, 32, 39, 41, 42, 43
  turbine 23
aerodynamic coupling 19
aero engines 13, 15
afterburners 223
age-hardening 34
aging effect 153
airflow 235, 237, 243, 244
airflow parameter corrected 233
airflow resistance 172
  of acoustical materials 177
air inlet cooling system 93
air leaks 130
air/oil coolers 107
air/oil seals 249
air pollution 141
alarm level 89
American National Standards Institute ANSI 164, 165, 176
ammonia 149
  injection 152
  injection system 93
  slip 153
  sulfates 153
amplitude 114, 116
analog signals 81
angle of attack 61
angle of incidence 61
anhydrous 149
  ammonia 151
annular 37
  combustor 262
  design 262
  gap 101
anti-friction bearings 24, 39, 102
anti-icing 88
  valve 81
approach points 213, 493
APUs 182
aqueous 149
  ammonia 151
area classifications 96
ash content 111
aspect ratio 61
atmospheric contaminants 123, 142
automobiles 22
A weighting 161
axial 27
axial-flow turbine 73

B
back pressure turbines 228
baffle, parallel 167
ball and roller bearing 40
bare tubing 228
barrier or pad type filters 125
barrier type filter 126, 127, 129
baseline data 232
baseplate resonance 115
basket 40
battery 96
bearings 39, 188
  air 188, 189
  anti-friction 188
  hydrodynamic 188, 189
  metal temperature 120
  worn 231
Bernoulli's Law 72
BHP Corrected 233
blade damage 241
blade passing frequency (bpf) 157
blades 76
blades or vanes
  partially or wholly missing 231
blade tip clearances, excessive 231
bleed valve 81
  malfunction 250
boiler 203, 217, 218, 225
boiler hardware 224
boiling point 215, 216
boroscope 41, 232, 256
  flexible fiber-optic 263
  rigid-type, 90-degree 264
boroscope inspection 240, 253, 276
  ports 257, 264
  probe 255
bottoming cycle 194, 195, 208, 493
Brayton 210
Brayton Cycle 45-46, 52, 206-208
burner 72, 73
  efficiency 247
burning 262
bypass stack 229

C
calcium carbonate 125, 135
can-annular 37
  combustors 37, 41
  design combustor 261
cannister filter 126, 127, 129
carbo-blast 245
carbon monoxide 143, 144
carbon residue 111
Carnot 207
Carnot Cycle 206
Carnot, Nicolas 206
cascade contro 82
catalytic combustor 329
catalytic reduction 205
central processing unit 81
centrifugal 27
  compressor 19, 61, 180, 182, 187
  pump 21
ceramics 329
changeable pitch fan blades 107
chiller 132
chopper circuit 99
circulation pump 219
cleaning agents 245
climatic conditions 198
climatic influences 197
cloud point 109
coal 328, 330
coalescer 135
coating 259
CO catalyst 229
cogeneration 21, 193, 207, 213, 297, 309

barriers to 203
cycle 196, 204
cold end drive 16, 17, 18, 284
combined cycle 16, 193, 213
mode 229
plant 204, 224
process 196
combustion efficiency 73
combustion process 142
combustion system 288
combustor 37, 41, 72, 73, 98, 179, 185, 186, 290
boroscope ports 261, 262
distress and plugged fuel nozzles 247
efficiency 38, 72, 248
cracked and warped 231
component efficiencies 49
compressibility factor 59
compressor 17, 20
aerodynamics 132
air extraction 292
blades 259
blades and stators 258
bleed-air flow-control 82
cleaning 128
diaphragms 42
discharge pressure 80, 235, 243, 244, 251
discharge pressure corrected 233
discharge temperature 237, 243, 245
efficiency 49, 129, 243, 244, 256, 258
fouling 134, 243
inlet pressure 80
inlet temperatures 80
pressure ratio 49, 237
stators 41
surge 250, 259
surge line 147
total discharge 49
total discharge pressure 49
total inlet 49
total inlet pressure 49
turbine 79, 80, 87, 92
washing 244
computer controls 81
or electronic controls 251
concave 54
condensate pump 213, 219, 226
condenser 213, 223, 225, 226, 229
condensing steam turbines 213
condensing turbine 196, 228
constrained-gap 101
contaminants 128, 130, 135, 141
contamination 109, 110, 111, 258
control(s) 180, 186, 187, 231
functions 83, 86, 117
load 83
map 89
speed 83
start 83
stop 83
system 292
temperature 83
unload 83
converters 180, 188
convex 54
cooler 103
design 107
cooler effectiveness 133
cooling fans 107
corrosion 32, 123, 128, 134, 231, 235, 239, 240, 241, 256, 262, 264
cracking 256
cracks 263

cranking speed 98
creep 34, 77, 117
critical point 493
critical press 216
critical speed 80
critical temperature 215, 216
cross-over tubes 98
CTOL 327
Curtis Turbine 74
CV 327
C weighting 161, 163
cyclone separator 221

D

deaerator 104
deceleration limit 90
decibels 155, 158, 159, 160, 164, 166, 169
deionized 147
demineralized 147
demineralized/deionized (DI) water 123
density 58
dependent variables 231
 parameters 232
depth of field 256
detectable faults 231
diaphragms 42
diesel motors 94, 97
diffuser efficiency 72
digital signals 81
direct drive 93
directionally solidified or DS 35, 36
directivity correction 160
disc creep 90
discs 29
distorted temperature pattern 247
domestic object damage 247
downcomer fluid 217
drag torque 94
driven unit output 86
dual pressure boiler 226
dual spool-split output shaft 16
 gas turbine 20
dynamic insertion loss 169

E

economizer 196, 201, 216, 217, 218, 219, 220, 221, 222, 228
effects of operating parameters 152
efficiency 49
electrical type controls 81, 83
electric generation 16, 17, 18, 96
electric motor 94, 96, 105
 alternating current 94, 95
 direct current 94
electric power 234
elements of an airfoil 59, 60
emergency lube system 43
emissions 145
emulsion 111
engine parameters 231
enthalpy 45, 209, 235, 493
entropy 209, 211, 225, 493
environment 130
environmental restraints 204
EPA 164, 176, 177
equiaxed 36
 casting 35
erosion 123, 128, 231, 235, 239, 240, 256, 262, 264
evaporative cooler 123, 132, 134-135
 effectiveness 132
evaporator 201, 216, 217, 218, 219, 220, 221, 222, 228
exciters 98, 99
exhaust gas temperature 117, 235, 237, 243, 245, 247, 263

corrected 234
exhaust gas treatment 205
exhaust products 142
exhaust smoke 119
exhaust temperatures 80
profile 262
expansion turbine 223
external leaks 119
extraction & induction turbines 228

F

feedwater heaters 228
feedwater pump 229
field of view 255
filter 103
performance 130
filters 106
barrier or pad type 125, 127, 129
cannister type 126, 127, 129
high efficiency barrier 126, 128
inertial separator type 130
intermediate 126
self-clean cannister 126
finned tubing 228
fins 107
fire point 109
first stage turbine nozzles 41
fir-tree 76
First Law of Thermodynamics 205
flame 40
flame-out 88, 90, 147
flame tubes 98
flammability 215, 216
flanking noises 173
flash point 109, 216
flexible boroscope 255
flexible fiber-optic boroscope 263
flexible fiber optic scopes 254
flexible shaft 113
flock point 109
fluid boiler 196
fluidized bed technology 229
FOD 250
fogger nozzles 135
forced circulation 218
boiler 219, 223
once-through cycle 220
forced recirculation boiler 222, 223
foreign object 135
damage 231, 247, 256, 263
fouling 123, 128, 231, 246, 249
FOV 255
free power turbine 20, 80, 234
freezing point 215, 216
frequencies
broad-band 161
critical 175
discrete 161, 163
octave 161, 164
frequency bandwidths 161
frequency coincidence 174
frequency forcing 174
fretting 262
friability 171
fuel bound nitrogen (FBN) 143
fuel flow 80, 87, 88, 235, 237, 243, 245, 251, 263
fuel flow control 82
fuel flow corrected 233
fuel nozzle 37, 38, 41, 72, 73, 257, 262
port 262
burnt, plugged or "coked-up" 231, 262
fuel system 286
fuel valve 81, 187
full-load 88

G
gas generator 11, 19, 79
  discharge total pressure 51
  discharge total temperature 51
  turbine 80
  turbine adiabatic efficiency 51
gas horsepower (GHP) 79, 80
gasifier 283, 287, 290
gas path 253
  analysis 232, 240, 241, 263
  vibration analysis 256
gas temperature 152
gas turbine 11, 12, 284
  compressor 223
  performance 129
  power plants 21
  small 94, 97
gear 105
gearbox 19, 93, 104, 105
gear noise 115
generator 180, 183, 187, 229, 284
  induction 180, 187
  synchronous 180, 187
generator drives 19, 20
governor lines 89
governor malfunctions 251
gravitational constant 48
GT compressor surge control 86

H
head 58, 59
heat 45
heat capacity 493
heat exchanger 228, 229
heat of fusion 493
heat of vaporization 147, 494
heat recovery boilers 216
heat recovery steam generator (HRSG) 280, 283, 298, 301, 303, 310, 312
heat transfer tubes 203
heavy frame industrial 30
heavy industrial 27
  gas turbine 15, 16, 41, 42, 43
heavy industrial units 26
heavy turbines 39
helicopters 20
hertz 161
high efficiency barrier filter 126
high efficiency filters 125
high frequency capacitor 99
high frequency/high tension transformer 99
high pressure compressor 237
high tension capacitor 100
high tension transformer 100
hollow-air cooled 76
horsepower 48, 130
hot end drive 16, 18, 20
hot starts 250
hung start 250
hybrid 30
hybrid gas turbines 15
hybrid industrial gas turbine 16, 23, 39
hydraulic 105
  controls 251
  impulse turbine 97
  motors 94, 97, 105
  pump 105
hydrodynamic bearings 24, 39, 40, 102
hydrogen 330, 331
hydromechanical controls 81, 82

I
ice 88
ideal head 62

Idle-No-Load 88
IGCC 277, 281
igniter 94, 98, 99
  plug 100, 101
  ignition 98, 216
  system 93
imbalance 115
immersed oil temperature 120
impact damage 235, 239, 246
impingement starting 98
impulse 73
impulse-reaction 74
  turbine 30, 74
independent variables 231, 232
Indianapolis 500 Race 13
indirect drive 93
induction motor 95
inductive AC and DC systems 99
inductive ignition 100
industrial gas turbine 23
inertial separators 125
infra-sound 156
injection 147
  of steam or water 123
inlet air temperature 132
inlet cooling 134
integrated gasification-combined-cycle (IGCC) 277
intermediate filters 125, 126
intermediate turbine temperature 50
internal leaks 119
internal turbine airfoil cooling 132
International Organization of Standardization (ISO) 130, 233
investment casting 34
isentropic 494
isochronous 90
isolated generation 199, 200
ISO Standard 233, 234

J

jet engines 11
jet propulsion 12
joule 99

K

kilojoules 48
kilowatts 48
kinematic viscosity 110
kinetic energy 45

L

lacing wire 76
latent heat 215, 216, 494
Laws of Thermodynamics
  First Law of Thermodynamics 205
  Second Law of Thermo-dynamics 206
  Zeroth Law 205
liner 40
load control 86
lobe 105
looseness 115
louvers 108
low frequency noise 156, 164, 176
low pressure compressor 237
low tension systems 99
lube oil coolers 107
lubrication 231
  oil 102, 104
  system 93

M

magnetic flux 100
marine propulsion 16
mass flow rate 45

material loss 240
mechanical and electric controls 251
mechanical bearing losses 234
mechanical drive 11, 16
  gas turbine 15
mechanical loss of bearing 235
megawatts 130
methane 330, 331
microprocessors 83
microturbine 179, 180
  applications 180, 181
  cost 180, 181
mineral oil 102
mini-turbine 179
misalignment 115
Mk. V Speedtronic® turbine control systems 292, 294
Mollier diagram 208, 209, 211, 212
Mollier, Richard 209
monotube boiler 220
monotube cycle 222
motors
  diesel 97
  direct current (DC) 96
  hydraulic 97
  pneumatic
  impulse-turbine 96
  vane pump 96
multi-function control 83

N

National Electrical Manufacturers Association (NEMA) 130
natural circulation 217
  boiler 217, 218
natural frequency 114, 241
natural or resonant frequency 113
natural recirculation system 218
net positive suction head 105
Neutralization Number 111
nitrogen 143
nitrogen oxide emissions 144
nitrogen oxides 143
noise
  casing 166
  exhaust 157, 166, 176
  inlet 158
  reduction 166, 167, 168, 170, 173, 175
non-condensing steam turbine 196, 213, 228
NO_x 147, 149, 151, 153
  formation of 143
NO_x catalyst 229
NO_x control 93, 205
NO_x emissions 143, 312
nozzle 34
  area
  changes in 235
  guide vanes 30
  orifice size 135

O

oil contamination 108
oil frothing 249
oil pressure 119
oil pump 104
oil throw-off temperature 120
oil viscosity 102
oil whirl 115
once through 217
once-through monotube cycle 221
once-through spillover cycle 221
open area ratio, of perforated sheets 168
operating line 89
operating parameters

effects of 152
operating point 89
organic bottoming cycle 195, 197, 203
disadvantages of 203
organic fluid 196, 203
organic NO 143, 147
organic system 198
OSHA (Occupational Health and Safety Act) 165
overriding clutch 96
overspeed 79, 89, 251
oxidation 32, 111, 231, 235, 239, 240, 256, 262, 264

P

pad and fogger 135
parallel baffle 167
Pelton Wheel 97
pinchpoint 213, 214, 494
plugged fuel nozzle 247, 249, 256
pneumatic controls 251
pneumatic motors 94, 96
pollution 142
positive replacement pumps 105
potable (drinking) water 123
potential energy 47
pour point 109
power augmentation 93, 146, 149
power extraction turbine 79, 80, 87, 92
power input 234, 235
power output 129, 130
power turbine 19, 20
efficiency 51
horsepower 51
precipitation strengthening 34
prefilters 125, 129
pressure 81
pressure compounded impulse turbine 74
pressure correction factor 233
pressure limit controller 87, 89
pressure ratios 27
prime mover 223
process control 83
programmable logic controllers 83
protection control 86
protection controller 89
proximity probes 115
psychometric chart 133
pump 17, 19, 20, 103, 105
backup (redundant) 104
direct-drive 104
indirect-drive 105
indirect-drive shaft driven 104
pre- and post-lube 105
pump cavitation 104
pumps
positive replacement 105
rotary 105
purge 93, 94
time 98

R

radial turbine 180
radiation and convection 235
heat loss 234
radio interference filter 99
Rankine 210
Rankine cycle 206, 207, 208, 213, 226
rapid acceleration 73
Rateau Turbine 74
ratio of specific heats 49
rayls/meter 172
reaction 73, 74
turbine 74

recirculation rate 223
rectifier 99, 187
recuperator 180, 185
regenerators 28
regulator 103, 106
reheat 226
reheat-condensing turbines 228
reheaters 228
reservoir 103, 104, 105
residence time 135
resistance temperature detectors 120
rigid boroscope 253, 255
riser fluid 217
risers 218
rotary 105
rotational mechanical equipment 113
rotor blades 29
rotor disc or blad
    cracked 231
rotor speed 117, 132, 235, 237, 243, 263
routine operation control 86, 87
rupture 34

S

SCR 151
    catalysts 150
    efficiency 152
screw 105
Second Law of Thermodynamics 206
seals, worn 231
seismic probes 115
selective catalytic reactor 145
selective catalytic reduction (SCR) 149, 151, 229
self-clean cannister filter design 126
self-generation 199, 200
self-noise 169
self-sustaining speed 87, 93, 94, 96, 97, 98
sensible heat 88, 494
sequence controller 86, 87
sequencer 86
sequencing control 86
    load 86
    start 86
    stop 86
    unload 86
shaft bowing 43
shaft horsepower (SHP) 80, 87, 88
shaft power 11
shafts 29
shaft speed 80
ship's propulsion power plant 22
shutdown level 89
simple cycle efficiency 48
simple cycle mode 229
simple cycle plant 204
single combustor 38
single crystal 36
single pressure boiler 224
single spool-integral output shaft 16
    gas turbines 18
single spool-split output shaft 16, 20
    gas turbine 19, 21
sinks 144
smoke 249
sound power level 155, 158, 159, 164, 166, 169, 175
sound pressure level 158, 159, 163, 164
South Coast Air Quality Management District—

SCAQMD 205
spark plug 99, 100, 101
specific fuel consumption
corrected 233
specific gravity 108
specific heat 215, 494
at constant pressure 49, 51
at constant volume 49
speed 81, 251
control 86
control governor 87
speed corrected 233
speed-droop governor 90
speed increaser/decreaser 17
speed increaser gearbox 97
speed limit 89
speeds 244
spillover 217
cycle 222
split-shaft 76
squealer cut 76
stall 53
margin 235, 237
standard conditions 233
standard day 233
starter motor torque 250
starters 93
starting system 93
static pressure 72
station control 83
stators 42
stator vanes 27, 29
steam 105
steam/air condenser 223
steam boiler 223
steam bottoming cycle 195, 197, 202, 203
steam cycle 202
steam injection 145, 148, 149, 150, 310, 312
nozzles 290
system 93
steam turbine 15, 213, 223, 225, 228, 284
steam turbine-compressor 223
stiff shaft 113
stone wall 53
storage capacitor 99, 100
STOVL 327
sulfidation attack 123
sulfidation corrosion 22, 32, 128, 246
superheated 203, 225
superheater 196, 201, 217, 218, 219, 220, 221, 222, 223, 228
supplementary tired boiler 223
surge 53, 79, 88
synchronize 87
synchronous generation 199
SYNGAS 285, 287
synthetic heat transfer fluids 214
synthetic lubricants 103
synthetic oil 40, 41, 103

T

temperature 81, 251
control 86
control valve 108
corrected 233
correction factor 233
temperature limit controller 87, 89
temperature profile 117
thermal cycling 32
thermal efficiency 48
thermal NOx 143, 145
thermal stability 215, 216
thermal stresses 88
thermocouple 120, 250

thermodynamics 205
  of waste heat recovery 207
thermodynamic cycles 206
thermodynamic gas path 231
  elements 113
thrust 11
tip shrouds 76
topping cycle 194, 195, 197, 208, 494
torque 94, 96, 98
total base number (TBN) 111
toxicity 215
transformer 99
transition duct 72, 73
transmission loss 170, 173, 175, 177
tube 40
tube and shell coolers 108
turbine 73
turbine blade 30, 32, 34, 128
  and nozzle cooling 237
  and nozzles 262
turbine control 83
turbine efficiency 50, 235, 237, 249
turbine health 113
turbine horsepower 50
turbine inlet gas flow 51
turbine inlet temperature 50, 80, 88, 117, 132, 146, 237
turbine inlet total pressure 51
turbine inlet total temperature 51
turbine light oil 41
turbine nozzle 30, 128
  area 256
turbine or compressor efficiencies 256
turbine temperature profile 256
turbine vanes or blades
  burned or warped 231
turbocharger 180, 182
turbo-expander 196
turbofans 11, 12, 20
turbojets 11, 12
turboprop 11, 12, 13, 19, 182
turbulators 107

U

ultraviolet (UV) flame detectors 250

V

vane 105
vapor pressure 216
variable geometry actuator 81
variable geometry malfunction 250
velocity 115, 116, 130
  compounded impulse turbine 74
  pressure 72
vibration 32, 81, 113, 115, 231, 242, 249, 257, 262
  and trending analysis techniques 253
  measurements 114
viscosity 110
volts 99

W

waste heat boiler 201, 213, 226
waste heat recovery 193, 194, 199
  boiler or HRSG 197
  generator 223
  systems 201
water and steam injection 143
water content 111
water flow 135
  nozzles 135
water hardness 135
water injection 145, 148, 149

system 93
water (or steam) injection 205
water vapor concentration 152
water-wash 245
watt 99
weight flow 235
work 45
work per unit mass 45

Y
Young's modulus 175

Z
Zeroth Law 205